Baboon Perspectives on Early Human Ancestors

Glenn E. King

Baboon Perspectives on Early Human Ancestors

One Approach to Reconstructing Early Hominin Ecology and Behavior

Springer

Glenn E. King
Department of History and Anthropology
Monmouth University
West Long Branch, NJ, USA

ISBN 978-3-031-36768-7 ISBN 978-3-031-36769-4 (eBook)
https://doi.org/10.1007/978-3-031-36769-4

This Springer imprint is published by the registered company Springer Nature Switzerland AG
The registered company address is: Gewerbestrasse 11, 6330 Cham, Switzerland

To the memory of Sherwood Washburn and the other pioneers in the scientific study of baboon ecology and behavior.

To the many colleagues who have continued that work and brilliantly expanded upon it.

And to my Special Friend, Barbara Barrett King.

Preface

Yu wapi nyani?

I had three goals in writing this book. The first was the quixotic intention of gathering in one place as much of the current knowledge about baboon ecology and behavior as I could. Baboons are defined here as monkeys of the genus *Papio*. The second goal was to illustrate the potential of baboon studies for reconstructing many different aspects of the life of our early ancestors and their closest relatives, classified as hominins. Much of that discussion revolves around the genus *Australopithecus*, but it also includes predecessors of *Australopithecus* as well as the earliest members of the genus *Homo*. The comparative purpose dictated an emphasis on free-living baboons under relatively natural conditions, but some of the most important social and psychological experiments are also included. The third goal was to make information and ideas about the subject available to a relatively wide audience, including behavioral and social scientists in various fields as well as sophisticated non-specialists.

From an early age, I was fascinated by animals and "cave men." I learned in college that I could combine the two in a fascinating career, studying nonhuman primates to better understand our ancestors and their closest relatives. All primates are of interest for that purpose, but baboons stand out along with chimpanzees and bonobos. The genus *Pan* (chimpanzees and bonobos) and the genus *Papio* (baboons) contribute to the subject in different ways. The *Pan* species are our closest living relatives, descendants of the last common ancestor that we share with any other living species. As such, some of the similarities between the genera probably arise from genetic systems that they both derive from their common origin. Baboons are much more distant relatives of humans. The main reason for comparison with early human ancestors is that baboons have expanded throughout much the same wide range of environments in Africa. They face problems that our ancestors must have faced and suggest solutions that our ancestors might have arrived at.

I chose to specialize in baboons and went to Berkeley to study with Sherwood Washburn (and to be reminded of the significance of other primates by Phyllis Dolhinow). When I told Sherry that I planned to study Swahili in preparation for research in East Africa, he said that I only needed one phrase: "Yu wapi nyani?" (Where are the baboons?). Fortunately, along with the occasional quip, he prepared me for the field with deeper knowledge about baboons, early hominins, and how they were connected with each other.

Baboons in Africa

When I got to Africa, I found out that I didn't have to ask where the baboons were. They were pretty much everywhere outside the cities. In fact, one male baboon tried to become my roommate as I moved into temporary lodging at the Serengeti Inn. Other animals quickly introduced themselves, such as a hyrax strolling by and hyenas cackling outside my window on the first night.

I watched baboons in several places in Tanzania, but I especially enjoyed beautiful Lake Manyara National Park. I found pleasure in mundane things like trying to count the number of baboons in several troops. There were a lot of them in the wooded northern part of the park, more than a hundred in at least one troop (I was told that there were even larger troops further south). Some incidents were ridiculously amusing. Parked in the midst of a troop feeding on hard fruits or nuts, I listened to the mass crunching sound building to a crescendo until I found myself laughing helplessly (okay, maybe there was a little culture shock involved).

Other single events fed my interest in baboon behavior. My favorite memory in that regard is that of a male baboon sitting next to a female and her infant and suddenly picking up the infant with both hands. He raised it above his head, very much like a human father playing with his baby. The infant uttered no more than one surprised "geck!"—apparently unfrightened by this handling. The male was presumably a "special friend" of the mother—a relationship among baboons that may have something to tell us about the origins of the human family.

Sometimes, the male relationship to mother and infant was less obvious. On one occasion, a mother and infant were frightened by a dustcloth flapped outside the vehicle window. She tucked the infant onto her abdomen and disappeared over a low hill. Moments later a large male came galloping over the hill toward us. He sat down outside the door of the vehicle and gave us what is sometimes known as the "stink eye" (Fig. 1). This lasted for several minutes until a tourist truck came along the narrow road and forced him to move. He only went as far as the other side of the vehicle, where he resumed his pose. It wasn't scientific data by itself, but it taught me something about baboon attention and persistence.

Fig. 1 An adult male olive baboon directing a threat-stare at the author and driver. He reacted to the flight of a mother and infant that we inadvertently frightened. (Photo by Glenn King. Manyara, Tanzania)

Who Wants to Know?

I tried to write this book for anyone interested in reconstructing the behavior and ecology of early human ancestors, including implications for the behavior of modern humans. Some of the disciplines I had in mind were anthropology (biological and cultural), sociology, psychology (standard, comparative, and evolutionary), political science, and linguistics. To reach such an audience, I have tried to minimize highly specialized terminology and to explain any such terms that were necessary. Many of my sources from various disciplines compared findings about baboons to our knowledge of modern humans and/or early hominins.

Organization of the Book

This book follows a rather standard progression from ecology through social life to mental processes. The first three chapters discuss the general approach that connects early hominins with baboons. Chapter 1 introduces baboons and early hominins; Chap. 2 explains the basis for comparing them; and Chap. 3 discusses conceptual frameworks for organizing comparisons, as well as objections to such reasoning.

Ecological comparisons are expanded in the next three chapters. Chaps. 4 and 5 are mainly about exploiting a great variety of food sources, including plant products and animal prey. Chapter 4 also discusses environmental stresses, including food supply, heat, and predators. Chapter 6 goes into more detail about danger from predators and ways to counter it.

Discussion of social organization (Chap. 7) revolves largely around the transition from relatively simple to multi-level societies, which includes complex and changing relationships of adult males to one another. This is followed by an examination of individualized relationships between particular males and females (Chap. 8). Chapter 9 focuses primarily on females, especially their relationships with each other. The latter two chapters encompass relationships of adults to their young.

The last part of the book considers the intertwined subjects of communication and cognition, beginning with non-linguistic communication, particularly vocalizations and gestures (Chap. 10). The importance of grunting in humans may come as a surprise to some readers. This chapter is followed by theories of language origins in which diverse components, including major cognitive functions, become coordinated with one another (Chap. 11). Consideration of cognition continues in Chap. 12, including domain-general functions and more specific mechanisms devoted to coping with environment. Finally, Chap. 13 surveys a variety of social cognition mechanisms.

The book concludes with an Epilogue that indulges the author in a more personal look at the history and prospects of baboon-hominin comparisons. An appendix provides connections with current field research projects on baboon ecology and behavior.

Acknowledgments

I am very grateful to many people and institutions who have made this book possible.

Colleagues

First and foremost, there are my colleagues in various disciplines who have studied baboons and early hominins, especially those researchers who have generously shared the results of their work.

Second, there are those who participated in what might be called the dress rehearsal for this book: my recent paper for the National Academy of Sciences (King 2022). Meredith Bastian played a vital role in preparation for submission, providing a great deal of patient advice. Larissa Swedell guided the development of the paper with extensive critique and commentary. She and several anonymous reviewers gave me important suggestions that have carried over to this book.

Dietmar Zinner read several chapters of the manuscript and responded with numerous comments that were the epitome of constructive criticism. I hope that he will be one of the anonymous reviewers of every paper that I write from here on. Julia Fischer made an important comment on the issue of distinguishing social entities and provided permission for an important illustration.

Of course, none of these colleagues are responsible for any errors that I failed to eliminate.

At Monmouth University, the late Ken Stunkel, history professor and dean, and Bill Mitchell, anthropology professor and frequent chair of my department, were enthusiastic supporters. They made it possible for me to teach human evolution and primatology within a diverse program. Both have been esteemed and cherished friends.

Kenneth Teng, Senior Editor in Life Sciences, Springer Nature Group, was patient, prompt, and personable in responding to numerous queries. Kate Lazaro and Vishnu Prakash were also very helpful in response to similar bombardments.

Benefactors

The National Institutes of Health provided a fellowship, administered by Sherwood Washburn, that made it possible for me to obtain my doctorate from the University of California, Berkeley.

Grants and sabbaticals from Monmouth University facilitated research and writing. I am indebted to the various faculty committees and administrators who were responsible. A Judith H. Stanley Travel Fellowship, administered by Monmouth University, made research in East Africa possible.

Rob Blumenschine and John Cavallo facilitated my work in Tanzania by providing a vehicle and (along with several of their Rutgers colleagues) valuable advice. Daniel drove and repaired the vehicle and assisted with some observations. He also provided companionship and a lesson in overcoming language barriers (I am sad and embarrassed to have lost my notes with his last name).

Operation Wallacea arranged transportation to South Africa and accommodations there for a study of chacma baboons. When that project proved untenable, moral and culinary support from Jesse Felthuis became essential. Some data on baboons in close contact with humans in a different locality were obtained with the kind help of Xander and Susan Combrink.

The Monmouth University Library, ResearchGate, and Academia.edu have given me easy online access to valuable journal articles. I am grateful to them and to my colleagues who made their work so readily available.

Friends and Acquaintances

Many people in Kenya, Tanzania, and South Africa were hospitable, friendly, helpful, and showed me many kindnesses. I regret that I had to leave Nairobi too soon to accept a generous invitation to a dinner of roast goat. Hopeless with languages, I learned little Swahili beyond “Jambo” and nothing of IsuZulu, so I am grateful to the many people who spoke fluent English.

The Home Front

Barbara Barrett King made the arrangements for our wedding and reception while I was in South Africa. She has supported my work ever since, never begrudging any time or expense, including vital technical assistance. In this case, the main technical assistance was extensive preparation of the figures for publication. Were I a baboon or an early hominin, Barbara would have been the special friend who made me monogamous.

Photographic Credits

All photographs attributed to Curt Busse were taken in Okavango Delta National Park in Botswana. Dr. Busse has kindly made these and many others available to any interested parties at http://www.curtbusse.com/okavango/index.html.

All photographs attributed to Steven Gotz were taken in the Oakland Zoo in Oakland, California, USA. More Oakland Zoo photos and other work can be viewed at https://artofzootography.myportfolio.com/.

Photographs attributed to Glenn King were taken in Lake Manyara National Park in Tanzania; Tarangiri National Park in Tanzania; or Mkhuze Game Reserve, Kwa-Zulu Natal, Republic of South Africa. They are distinguished in the captions as "Manyara," Tarangiri, or "Mkhuze." More of my work can be found on the Facebook page "Primate Behavior and Human Origins" at: https://www.facebook.com/profile.php?id=100069554065139 and on ResearchGate: https://www.researchgate.net/profile/Glenn-King-2

The following figures are licensed by Wikimedia: 1.1, 1.2, 7.3, and 8.5: (https://en.wikipedia.org/wiki/Creative_Commons).

Dr. Julia Fischer and Dr. Dietmar Zinner made Figs. 1.9 and 7.4 available.

The beautiful cover photo was provided by Dr. Sandra Molesti. It was taken at the Station de Primatologie of the CNRS, Rousset, France. The dominant male of the group was grooming a female associate.

West Long Branch, NJ, USA — Glenn E. King

Reference

King GE. Baboon perspectives on the ecology and behavior of early human ancestors. Proc Natl Acad Sci. 2022;119(45):e2116182119. https://doi.org/10.1073/pnas.2116182119.

Contents

1 An Introduction to Early Hominins and Baboons ... 1
- 1.1 Introduction ... 1
- 1.2 An Overview of Early Hominins ... 1
- 1.3 Key Characteristics of Early Hominins ... 2
  - 1.3.1 Bipedalism and Terrestriality ... 2
  - 1.3.2 Body Size and Sexual Dimorphism ... 3
  - 1.3.3 Brain Size ... 4
  - 1.3.4 Canine Teeth ... 5
- 1.4 Hominin Evolution ... 5
  - 1.4.1 Hominin Origins ... 5
  - 1.4.2 *Ardipithecus ramidus* (Dated to About 4.8 to 4.3 mya in Ethiopia) ... 7
  - 1.4.3 *Australopithecus* (~4.2 to 2.0 mya) ... 7
  - 1.4.4 *Australopithecus anamensis* (~4.2 to 3.9 mya in Ethiopia and Kenya) ... 8
  - 1.4.5 *Australopithecus afarensis* (~4.2 to 2.9 mya; Ethiopia, Kenya, Tanzania) ... 9
  - 1.4.6 *Australopithecus in South Africa* (~3.7 to 2.0 mya) ... 9
  - 1.4.7 *Australopithecus (=Kenyanthropus platyops)* (~3.5 to 3.3 mya in Kenya) ... 10
  - 1.4.8 Other *Australopithecus* Species ... 10
  - 1.4.9 Early *Homo* (~2.8 to 1.7 mya; Ethiopia, Kenya, Tanzania, South Africa) ... 11
- 1.5 An Overview of Baboons ... 12
- 1.6 Key Characteristics of Baboons ... 13
  - 1.6.1 Positional Behavior ... 13
  - 1.6.2 Body Mass and Sexual Dimorphism ... 15

1.7 *Papio* Evolution and Species . . . 16
1.7.1 *Papio* Species . . . 18
1.7.2 Ecological Niches of *Papio* Species . . . 19
1.8 *Papio* Societies . . . 20
1.9 Summary and Discussion . . . 21
References . . . 23

2 Why Compare Early Hominins to Baboons? . . . 29
2.1 Introduction . . . 29
2.2 Environment . . . 30
2.2.1 Environmental Diversity . . . 30
2.2.2 Environmental Changes . . . 32
2.3 Positional Behavior . . . 34
2.3.1 Bipedalism and Terrestriality . . . 34
2.3.2 Arboreality . . . 37
2.4 Body Mass . . . 38
2.5 Sexual Dimorphism . . . 40
2.5.1 Sexual Dimorphism in Body Mass . . . 40
2.5.2 Sexual Dimorphism in Canine Teeth . . . 41
2.6 Summary and Discussion . . . 42
References . . . 43

3 Conceptual Frameworks and Criticisms . . . 47
3.1 Introduction . . . 47
3.2 Models . . . 47
3.2.1 Referential Models . . . 48
3.2.2 Conceptual Models . . . 50
3.3 Analogies . . . 50
3.3.1 Scenarios . . . 51
3.4 Criticism and Papiophobia . . . 52
3.4.1 Absolutist Arguments . . . 53
3.4.2 Relativist Arguments . . . 55
3.5 Summary and Discussion . . . 59
References . . . 60

4 Subsistence and Stress . . . 63
4.1 Introduction . . . 63
4.2 Dietary Variation and Trends . . . 64
4.2.1 Hominin Dietary Shifts . . . 64
4.2.2 Generalists and Specialists . . . 65
4.3 Food Sources . . . 66
4.3.1 Fruit . . . 67
4.3.2 High-Protein Plants . . . 69
4.3.3 Savanna Plants . . . 69
4.3.4 Underground Storage Organs . . . 70
4.3.5 Grasses and Sedges . . . 71

4.3.6 More About Plant Foods ... 73
4.3.7 Toxic Plants ... 74
4.4 Ingestion of Other Substances ... 75
4.4.1 Geophagy ... 75
4.4.2 Medicinal Plants ... 75
4.5 Foraging and Stress ... 75
4.5.1 Dietary Stress ... 76
4.5.2 Heat Stress ... 76
4.5.3 Dehydration ... 78
4.6 Summary and Conclusions ... 79
References ... 80

5 Animal Foods and the Origins of Meat Eating ... 85
5.1 Introduction ... 85
5.2 Invertebrates ... 85
5.2.1 Arthropods ... 86
5.2.2 Aquatic Invertebrates ... 87
5.3 Vertebrates: Fish ... 87
5.4 Terrestrial Vertebrates ... 88
5.4.1 Meat Eating ... 88
5.4.2 Foraging and Hunting ... 91
5.4.3 Hunting and Scavenging ... 92
5.4.4 Meat: Costs and Motivation ... 94
5.5 Summary and Discussion ... 95
References ... 97

6 Predation on Hominins ... 101
6.1 Introduction ... 101
6.2 Predators and Predation ... 101
6.2.1 Potential Predators of Hominins ... 102
6.2.2 Baboon Predators ... 105
6.2.3 Body Mass of Predators and Prey ... 107
6.3 Defenses Against Carnivorans ... 108
6.3.1 Avoidance and Flight ... 108
6.3.2 Active Defense ... 110
6.3.3 The "Male Rampart" ... 112
6.3.4 Use of Weapons ... 113
6.4 Other Predators ... 114
6.4.1 Raptors ... 114
6.4.2 Crocodiles ... 115
6.4.3 Snakes ... 115
6.5 Sleeping Sites ... 116
6.5.1 Sleeping Above the Ground ... 116
6.5.2 Sleeping on the Ground ... 118
6.6 Summary and Discussion ... 119
References ... 120

7 **Social Organization and Male–Male Relationships** 125
7.1 Introduction 125
7.2 Archeological Perspectives on Prehistoric Hominin Society 125
7.3 The Troop: A Baboon Model for Early Hominin Society 126
7.3.1 Relations Between Troops 127
7.3.2 Fission–Fusion 128
7.4 Multilevel Societies 129
7.4.1 Multilevel Society in Hamadryas Baboons 130
7.4.2 Multilevel Society in Guinea Baboons 131
7.4.3 Possible Societies of Early Hominins 132
7.5 Male Philopatry 133
7.5.1 Male Philopatry in Humans and Baboons 133
7.5.2 The Frontier Hypothesis 133
7.6 Male–Male Relationships in Troops 134
7.6.1 Aggression and Dominance Among Males 134
7.6.2 Coalitions 136
7.6.3 The Kinda Baboon Alternative (?) 137
7.7 Male–Male Affiliation in Multilevel Species 137
7.7.1 Male–Male Relationships in Hamadryas Baboons 138
7.7.2 Male–Male Relationships in Guinea Baboons 139
7.7.3 Men in Groups 141
7.8 Summary and Discussion 142
References 144

8 **Male–Female Relationships** 147
8.1 Introduction 147
8.2 Some Basic Features of Baboon Sexuality 147
8.3 Human Families and Primate Analogies 149
8.4 Male–Female Associations in Troops 150
8.4.1 Special Friends (Primary Associations) 150
8.4.2 Benefits of Special Friendships 151
8.4.3 From Special Friends to Families? 155
8.4.4 Conflicts and Reconciliations 155
8.5 Male–Female Interaction in Kinda Baboons 156
8.6 Male–Female Associations in Multilevel Societies 158
8.6.1 Male-Focal Groups in Guinea Baboons 158
8.6.2 One-Male Units in Hamadryas Baboons 159
8.7 Sexual Aggression 160
8.7.1 Forms of Sexual Aggression 161
8.7.2 Sexual Aggression in Troops 161
8.7.3 Sexual Aggression in Multilevel Societies 162
8.7.4 Concealment of Mating and Sexual Privacy 163
8.8 Summary and Discussion 164
References 165

9 Female–Female Relationships . . . 169
9.1 Introduction . . . 169
9.2 Female–Female Affiliation . . . 169
9.2.1 Female Affiliation in Humans . . . 170
9.2.2 Female–Female Dyads in Troops . . . 171
9.2.3 Female Groups and Networks in Humans and Baboons . . . 172
9.3 Female–Female Relationships in Multilevel Societies . . . 174
9.3.1 Female–Female Relationships in Guinea Baboons . . . 174
9.3.2 Female–Female Relationships in Hamadryas Baboons . . . 175
9.4 Further Evolution of Hominin Female Relationships: Two Scenarios . . . 176
9.4.1 The Swedell/Plummer Scenario . . . 176
9.4.2 The Newson/Richerson Scenario . . . 177
9.4.3 Female Bands with Males . . . 177
9.5 Female–Female Competition . . . 179
9.5.1 Competition for Food . . . 179
9.5.2 Competition for Males . . . 180
9.6 Summary and Discussion . . . 183
References . . . 184

10 Nonlinguistic Communication . . . 187
10.1 Introduction . . . 187
10.2 Vocalizations . . . 189
10.2.1 A Flexible Call System . . . 189
10.3 Loud Calls . . . 190
10.3.1 "Roars" in Humans . . . 190
10.3.2 "Wahoos" in Baboons . . . 191
10.3.3 Female Barks . . . 192
10.4 Grunts . . . 193
10.4.1 An Evolutionary Theory of Human Grunts . . . 193
10.4.2 Grunts and Social Interaction . . . 194
10.4.3 Grunts and Referential Communication . . . 195
10.5 Gestures . . . 196
10.6 Salutations . . . 197
10.6.1 Greetings . . . 197
10.6.2 Sequence Organization . . . 198
10.6.3 Leave-Taking . . . 199
10.7 Summary and Discussion . . . 199
References . . . 201

11 Language Origins . . . 203
11.1 Introduction . . . 203
11.2 Components of Language . . . 203
11.3 Language and Concepts . . . 204

11.4 Domain-General Functions in Language Origins 205
11.4.1 Memory 205
11.4.2 Categorization.................................. 206
11.4.3 Statistical Regularities 206
11.4.4 Analogical Reasoning............................ 207
11.4.5 Pragmatics and Semantics 207
11.5 Domain-Specific Components: Vocalization and Speech 208
11.5.1 Protophones and the Vocalization-First Theory 208
11.5.2 Vowel-Like Sounds............................... 209
11.5.3 Vowels Versus Consonants 210
11.6 The Gesture-first Theory 210
11.6.1 Human and Primate Evidence 210
11.6.2 Flexibility and Intent in Gestural Communication...... 211
11.6.3 Baboons in Gesture Experiments 212
11.7 Language and Laterality 213
11.7.1 Communicative Laterality Versus Manipulative Laterality 213
11.7.2 Language and Brain Laterality..................... 215
11.7.3 Wernicke's Area 215
11.7.4 Other Brain Structures 215
11.8 Language Learning 216
11.9 Summary and Discussion.................................. 217
References... 219

12 Cognition 223
12.1 Introduction 223
12.2 Quantification.. 223
12.3 Reasoning by Exclusion.................................. 225
12.4 Cognitive Flexibility 226
12.5 Exploratory Tendencies 227
12.6 Two More Experiments 228
12.7 Ecological Factors in Cognition.......................... 228
12.7.1 Foraging and Home Range......................... 229
12.8 Manipulation of Stones 230
12.8.1 Stone Tools in the Early Oldowan 230
12.8.2 Stone Manipulation by Baboons in Feeding 231
12.8.3 Use of Stones As Weapons by Baboons.............. 231
12.8.4 Possible Construction of Terrestrial Shelters by Early Hominins 232
12.9 Group Travel Decisions 232
12.9.1 Cognitive Maps.................................. 233
12.9.2 Initiation and Direction of Movement 233
12.9.3 Group Organization During Travel.................. 235
12.10 Summary and Discussion................................. 236
References... 238

13 Social Cognition 241
13.1 Introduction 241
13.2 Theories of Social Cognition 242
13.2.1 The Machiavellian Intelligence Hypothesis 242
13.2.2 The Social Brain Hypothesis 243
13.2.3 The Embodied Social Brain 244
13.2.4 Proximate Factors in Brain Size 245
13.2.5 Cultural Intelligence 245
13.3 Cognition in Social Interactions 246
13.3.1 Tactical Deception 246
13.3.2 Cooperation 247
13.3.3 Pointing 248
13.3.4 Social Facilitation 249
13.4 Self in Society 249
13.4.1 Possession and Ownership 250
13.4.2 Personality 251
13.4.3 Emotions 252
13.5 Social Information 252
13.5.1 Social Comparison 253
13.5.2 Information About Others 253
13.5.3 Information from Others 254
13.5.4 Culture? 254
13.6 Summary and Discussion 255
References 257

Epilogue: The Once and Future Baboon 261

Appendix: Ongoing Long-Term Baboon Research Projects 267

Index 273

About the Author

Glenn E. King was trained in anthropology at Cornell University (Ithaca) and UCLA, and in primatology with Sherwood Washburn at UC Berkeley. He taught anthropology and primatology at Monmouth University in New Jersey and is now Professor Emeritus. His research includes close observation of free-ranging olive baboons in Tanzania and pursuit of uncooperative chacma baboons in South Africa. Previously, he studied captive chimpanzees at Holloman AFB; and hamadryas baboons, geladas, and mandrills in the Los Angeles Zoo. Dr. King's primate behavior text (Routledge 2016) emphasizes the significance of primates in general for understanding the evolution of human behavior. His most recent publications appeared in the *Encyclopedia of Animal Cognition and Behavior* and the *Proceedings of the National Academy of Science*.

Chapter 1
An Introduction to Early Hominins and Baboons

1.1 Introduction

This chapter introduces the two subjects of the book, baboons and early hominins, primarily for those who are not familiar with them or with the extensive knowledge about them that has become available in recent years. Three qualifications must be stated immediately and emphatically so that there is no misunderstanding. First, there is no intention of setting forth baboons as the *only* or *most important* primates as models for early hominins. The ultimate conclusion will be that baboons are *one of many* important avenues toward understanding our ancestors. The second vital point is that baboon perspectives apply primarily to our *early* ancestors, that is, those that lived before about 1.7 *mya* (million years ago). Major environmental shifts after that time apparently led to significant anatomical and behavioral changes in the hominins. Third, the extent to which any of those early adaptations continued into later hominins or modern humans is a separate matter from the reconstruction of early hominins. This issue will be touched on at various points in the book, but it is not the main concern here.

1.2 An Overview of Early Hominins

Hominin is an informal term that refers to any and all species in a taxonomic group that contains *Homo*, technically the tribe hominini (Alméija et al. 2021). The hominin evolutionary lineage emerged from an ancestor shared with the genus *Pan* (chimpanzees and bonobos). The species ancestral to *Pan* and *Homo* is often called the *Last Common Ancestor* or *LCA* (i.e., the last ancestor shared by humans with any other living primates). This is a theoretical construct, since the particular ancestral species is unlikely to be discovered and/or clearly identified as such.

G. E. King, *Baboon Perspectives on Early Human Ancestors*,
https://doi.org/10.1007/978-3-031-36769-4_1

The separation of the *Pan* and *Homo* lineages probably took place between 9 and 5 mya, based on DNA comparisons of living species and fragmentary fossil evidence (Andrews 2020). The currently known fossil record of possible hominins extends back to about 6 or 7 mya (Bobe and Wood 2021). Projections from molecular comparisons of living species go as far back as 12 mya (Moorjani et al. 2016). A study of primate mutation rates placed the split between 10 and 7.5 mya (Chintalapati and Moorjani 2020).

The phrase *early hominins*, as used here, refers primarily to the human ancestors and their closest relatives that existed from the beginning of our evolutionary lineage through the earliest representatives of the genus *Homo*. Early hominins are represented by relatively well-studied and widely known fossil genera such as *Ardipithecus* (Mongle et al. 2019) and *Australopithecus* (Clarke et al. 2021; Du et al. 2020). The evidence for earlier forms is sparser and their hominin status is subject to more controversy (Boyle and Wood 2017; Macchiarelli et al. 2020). I place the end of this era at about 1.7 mya because of major changes in environment and hominin behavior around that time (Antón and Middleton 2023; Patterson et al. 2019).

The hominin lineage contains numerous branches (also called lineages) that include ourselves, our presumptive ancestors, and their closest (extinct) relatives. Scientists have divided the hominins into numerous genera and species, many of which are controversial. A brief survey follows, emphasizing taxa and features of particular importance for comparison to extant baboons.

1.3 Key Characteristics of Early Hominins

There is uncertainty as to which of the many known early hominins might have been our direct ancestors, if any of them (e.g., Vaneechoutte et al. 2023). Taken together, however, the evidence from paleontology and paleoecology suggests that most or all early hominins shared some basic traits that we can reasonably postulate to be characteristic of human ancestors. The descriptions below are drawn from taxa that are widely or unanimously considered to be hominins. They characterize an early hominin way of life shared by a number of species and genera. These are traits that must be taken into consideration when comparing early hominins with baboons.

1.3.1 Bipedalism and Terrestriality

Bipedalism (also called bipedality) in hominins can be described as the ability to stand and walk on two legs for substantial periods of time. It differs from brief episodes of two-legged locomotion that occur in many other primates. It differs still more from two-legged stances and locomotion in such diverse animals as birds (hopping) and kangaroos (tail balanced). Because many animals can stand briefly or take a few bipedal steps, the greatest significance attaches to *habitual* (usual) and *obligatory* (necessary) bipedalism. Some form of bipedalism was probably a feature of the hominin lineage from the beginning or nearly so (Stamos and Alemseged 2023; Casenave and Kivell 2023; Prang 2019). It may have evolved originally as an adaptation to arboreal life (Crompton et al. 2010; Johannsen et al. 2017).

Terrestriality (also called terrestrialism) refers to a behavioral pattern in which a large part of most days is spent on the ground. Although the origin of bipedalism might have been arboreal, numerous changes in the lower body and legs of early hominins were clearly adaptations to walking on the ground (Casenave and Kivell 2023; Stamos and Alemseged 2023). The terrestriality of hominins distinguishes them from the great majority of primate species, which spend most of their lives in trees.

1.3.2 Body Size and Sexual Dimorphism

Hominins can be considered medium sized land mammals (Jolly 2013), somewhere between shrews and elephants. The body mass of early hominins, represented by estimated weight, is comparable to that of large ape species living today (chimpanzees and orangutans, but not gorillas). Various methods have produced varied results (Jungers et al. 2016; Will et al. 2017), but much of the evidence (Table 1.1) seems to suggest a range of about 25–40 kg with a mean around 35 kg (precision is

Table 1.1 Body mass estimates for early hominin taxa

Taxon	Approx. dates	Body mass	In kilograms
		Mean	Range
Sahelanthropus tchadensis	~7 mya	No data	No data
Orrorin tugenensis	~6 mya	36	30–43 ($N = 2$)
Ardipithecus ramidus	~4.5 mya	32 ($N = 1$)	No data
Australopithecus anamensis	4.2 to 3.9 mya	46 ($N = 1$)	No data
A. afarensis	3.8 to 3.0 mya	39	26–63 ($N = 12$)
A. africanus	3.0 to 2.4 mya	31	23–43 ($N = 6$)
A. garhi	~2.5 mya	No data	No data
A. sediba	~2 mya	26	23–30 ($N = 3$)
Kenyanthropus platyops	3.5–3.3 mya	No data	In *Australopithecus* range?
Homo habilis	2.8 to 2.0 mya	33	27–38 ($N = 2$)
H. ergaster[a]		51	39–64 ($N = 3$)
H. erectus	1.8 mya to 140 kya	53	49–62 ($N = 8$)
H. sapiens (fossils)	300 kya to???	~67 ($N = 43$)	

The values are drawn from a survey of the fossil evidence by Boyle and Wood (2017), citing the results of numerous researchers. They are rounded off here for easier reference and comparison. Early hominins, as the term is used here, includes *Homo habilis*, *H. ergaster*, and *H. erectus*. *H. sapiens* are dated later than early hominins and are included for a comparative perspective

[a]*Homo ergaster* is a term some use for African *H. erectus*

not possible with such small samples). More recent work has pointed to some higher numbers, up to 75 kg for one species. However, the lower end of the range is still about 30 kg (Ruff et al. 2020; Simpson et al. 2019).

The degree of variation in the body mass calculated for early hominin species indicates that these taxa were sexually dimorphic for the trait (Plavcan 2018). That is, males on the average weighed significantly more than females. However, the full extent of the difference is subject to intense controversy (e.g., Reno and Lovejoy 2015; Ward et al. 2020). The degree of size dimorphism may have varied significantly among hominin taxa.

1.3.3 Brain Size

Brain size in fossils is approximated by calculating the *endocranial volume*, that is, the internal volume of the skull (Table 1.2). Hominin evolution probably began with brains of about 300–350 cc, well within the range of modern apes. Ultimately, after millions of years, early hominin brains exceeded apes by attaining about half the size of modern humans. Brain size is a crude indicator of mental capacity and behavior, but the magnitude of difference between modern humans and early

Table 1.2 Early hominin endocranial volumes

Taxon	Approx. dates	Endocranial volume		
		Mean	Range of Variation	*N*
Sahelanthropus tchadensis	~7 mya	365		1
Orrorin tugenensis	~6 mya	No data		
Ardipithecus ramidus	~4.5 mya	300–350		1
Australopithecus anamensis	4.2 to 3.9 mya	No data		
A. afarensis	3.8 to 3.0 mya	433	385–550	6
A. africanus	3.0 to 2.4 mya	454	391–568	9
A. garhi	~2.5 mya	450		1
A. sediba	~2 mya	420		1
Kenyanthropus platyops	3.5–3.3 mya	Suggested to be within *Australopithecus* range		
Homo habilis	2.8 to 2.0 mya	611	478–796	6
H. rudolfensis	2.0 to 1.8 mya	793	760–825	2
H. ergaster[a]		796	715–909	4
H. erectus	1.8 mya to 140 kya	981	656–1300	36
H. sapiens (fossils)	300 kya to???	1457	1090–1775	56

All values are in cubic centimeters. They are drawn from a survey of the fossil evidence by Boyle and Wood (2017), citing the results of numerous researchers. They are rounded off here for easier reference and comparison. Early hominins, as the term is used here, including *Homo habilis*, *H. ergaster*, *H. erectus*, and *H. sapiens*, are dated later than early hominins and are included for a comparative perspective

[a]*Homo ergaster* is a term some use for African *H. erectus*

hominins suggests significant behavioral limitations in the latter. This can be viewed as support for the use of nonhuman species as models for at least some early hominin behavior.

1.3.4 Canine Teeth

In contrast to most other primates and many other mammals, the *canine teeth* (eyeteeth to dentists) of hominins were relatively small, that is, similar in size to the incisors. Also, contrasting with many other primates, there was little or no difference between the sexes in this trait. Suwa et al. (2021) analyzed a set of more than 300 fossils spanning six million years with a probabilistic method that reduces bias in the estimation of moderate dimorphism. They concluded that reduction of male canine size occurred early in hominin evolution and there was little change after that.

1.4 Hominin Evolution

Broadly shared hominin characteristics cannot be fully understood without placing them in the context of hominin evolution, including all of the uncertainties and controversies. Putative hominin fossils from earlier than 4 mya are rare and fragmentary. For periods after 4 mya, fossils represent a lineage that proliferated in a geographic arc from Chad through East Africa to South Africa. Briefly stated, the subject here is an evolutionary lineage represented by fossils drawn from a time span of five million years or more, and assigned to at least three different genera. The evidence comes from three geological periods: the late Miocene (11.6 to 5.3 mya), Pliocene (5.3 to 2.6 mya), and early Pleistocene (2.6 to 0.8 mya).

The following review addresses three major issues pertinent to this book. First, which fossil taxa are considered hominids? Second, which of the probable hominids (if any) are likely to be on or close to the ancestry of *Homo sapiens*? Third, what are the distinctive characteristics of each taxon that might be pertinent to the concerns of this book?

1.4.1 Hominin Origins

There are widely divergent ideas about the nature of the Last Common Ancestor of *Pan* and *Homo*, due to the dearth of relevant fossils and the use of different research methods. One approach focuses on comparison with living African apes and leads to the conclusion that the LCA was much like a chimpanzee. Yaxley and Foley (2019) tried to improve on purely morphological comparisons with a genomic analysis that they applied to 14 continuous traits known to vary among great ape subspecies. They concluded that their work corroborates previous findings that the LCA was a "chimp-like" animal.

Evolutionary analysis by Hunt (2016) led to the conclusion that the LCA was more generalized than any living ape. His premise was that living apes have a distinctive suite of anatomical and behavioral characters that evolved relatively late and relatively independently of one another. Convergence occurred because all benefited from using varied postures to harvest ripe fruits from terminal tree branches. Hunt inferred from genetic evidence for rapid chimpanzee evolution that their specialized features evolved after the divergence of the hominin and *Pan* lineages diverged; consequently, the LCA could not have been like a chimpanzee. It might have been ape-like in some features.

Almécija et al. (2021) based their reconstruction on the fossil record. They saw living hominoids (apes and humans) as a narrow representation of a widely distributed evolutionary radiation (cf. Foley and Lahr 2023). None of those ancient taxa exhibit the entire suite of locomotor adaptations present in their extant relatives. Almécija and colleagues suggested that hominins originated from Miocene ape ancestors unlike any living species. Similarities among modern apes and humans might have evolved as parallel responses to similar selection pressures.

During the probable time of the LCA, most of Africa displayed a patchy distribution of forests, woodlands, and grasslands, together with altitudinal gradients of temperature and moisture. The LCA probably occupied seasonal woodland habitats, a conclusion from evidence for such conditions associated with their ape predecessors in the preceding Miocene and with early hominins in the Pliocene (Andrews 2020). "Woodland" is a broad category that can be thought of as an environment with tree cover that is intermediate between forest and grassland, that is, trees may be close together but do not form a continuous canopy of leaves.

The earliest taxa that can reasonably be attributed to the newly distinct hominin lineage existed in the period around 6–7 mya. They have been classified in the genera *Sahelanthropus* (in Chad), *Orrorin* (in Kenya), and *Ardipithecus kadabba* (in Ethiopia). Three lines of evidence, all highly debatable, have been used to assign them to the hominin lineage (Boyle and Wood 2017). First, the upper canine teeth are relatively small and do not hone themselves against the lower premolars as in living apes; however, similar changes took place during the Miocene in a number of taxa that were undoubtedly apes. Second, the location and orientation of the foramen magnum suggests bipedalism; however, comparison with other primates indicates that head carriage and brain size can also be factors in the characteristics of the foramen magnum. Third, some postcranial bones are also indicative of bipedalism, but these are fragmentary (e.g., part of a femur from *Orrorin* and a toe of *A. kadabba*).

1.4.2 Ardipithecus ramidus *(Dated to About 4.8 to 4.3 mya in Ethiopia)*

Two species have been assigned to *Ardipithecus* (Boyle and Wood 2017). Little is known about the earlier one (*A. kadabba*, see above). *A. ramidus* is the later species (earliest appearance at ~4.5 mya) and more likely to be a hominin (but see Alemseged 2023). Bipedalism is indicated by the foot, pelvis, and curve of the lower spine. Analysis of craniodental material strengthens support for *A. ramidus* as the *sister taxon* of all later hominins (i.e., in the lineage most closely related to later hominins) (Mongle et al. 2019). From here on, references to the genus pertain only to *A. ramidus*.

Ardipithecus was primarily a woodland animal, as shown by associated evidence for environmental conditions, even though it had access to some more open areas (Andrews 2020; Simpson et al. 2019). Certain features of the pelvis and legs in *Ardipithecus* indicate terrestrial bipedalism (Casenave and Kivell 2023), but this mode of locomotion may have been used only rarely (Stamos and Alemseged 2023). The hand provided evidence for *suspensory behavior*, hanging and moving beneath branches as all apes can do today. The foot, with an opposable *hallux* (big toe), could have facilitated vertical climbing (Prang et al. 2021). Arboreality is further indicated by relatively long arms and long curved phalanges on the hands and feet (Selby and Lovejoy 2017).

1.4.3 Australopithecus *(~4.2 to 2.0 mya)*

Members of this genus are considered to be hominins by almost all experts, but there is considerable controversy about the number of species and their relationships to each other. Distribution of the genus runs from Chad in the north down through eastern Africa to South Africa. The face of *Australopithecus* is usually described as "apelike" due to the protrusion of the lower face, fairly prominent brow ridges above the eyes, and a low-bridged nose (Fig. 1.1). Most *Australopithecus* species lived in environmentally mixed habitats like that of *Ardipithecus* but may have used them differently (Bobe et al. 2020; Manthi et al. 2020; Villasenor et al. 2020).

Vaneechoutte et al. (2023) argued for exclusion of the entire genus from human ancestry. Villmoare (2018) accepted some possible involvement in human ancestry but denied that *Australopithecus* is a valid taxon, based on the opinion that the species that have been included in the genus do not share a common ancestor unique to them. As noted earlier, no species of early hominin can be undeniably assigned to the direct line of evolution leading to *Homo sapiens*. To reconstruct human ancestry during this period, we can only (1) look for the common features of all known early hominins, (2) formulate tentative hypotheses from less common features, and (3)

Fig. 1.1 A reconstruction of *Australopithecus afarensis* from the "Lucy" fossil illustrates general features of the genus that are described in the text. By Shalom. Warsaw Museum of Evolution. Resized for publication. (Wikipedia license: https://creativecommons.org/licenses/by-sa/3.0/deed.en)

look for links to early *Homo*. Some important features of *Australopithecus* are consistent with what is known about early *Homo* (Kimbel and Villmoare 2016).

The following survey of proposed species uses conventional taxonomic terms for convenience of reference. No judgment is made or intentionally implied about taxonomic validity. The purpose is to organize material that may be pertinent to comparison with baboons.

1.4.4 Australopithecus anamensis *(~4.2 to 3.9 mya in Ethiopia and Kenya)*

This is the earliest known species of the genus. The relative numbers of *Australopithecus. anamensis* fossils at sites with differing paleoenvironments indicate that *A anamensis* thrived in mixed woodland/grassland habitats, as opposed to wetter and more closed conditions (Bobe et al. 2020). Chemical analysis of the fossils showed that their plant foods utilized a mode of photosynthesis found in trees, shrubs, and bushes. This indicates that woodland was their primary environment. *Paleosols* (ancient soil strata) show that the environment was highly seasonal. *A. anamensis* shares important features with its chronological successor *A. afarensis*. Some scientists have advanced evidence for an ancestor–descendant relationship (Du et al. 2020), but others view "*A. anamensis*" as an early form of *A. afarensis* rather than a separate species (Alemseged 2023).

1.4.5 Australopithecus afarensis *(~4.2 to 2.9 mya; Ethiopia, Kenya, Tanzania)*

This durable species is famous because of the specimen known as Lucy, a small female. The species seems to have occupied a wider geographic area and lived in a greater variety of environments than its predecessors. Because of this, and its generalized morphology, Alemseged (2023) considers *Au. afarensis* to be the probable ancestor of all later species of the genus, as well as of the genus *Homo*.

A comparative study showed that *A. afarensis*, like *A. anamensis*, were more abundant in mixed woodland/grassland environments than in more closed woodlands (Villasenor et al. 2020). At Woranso-Mille in Ethiopia *A. afarensis* lived in the familiar mixture of woodland and grassland (Su and Haile-Selassie 2022), but at about the same time another population existed in a drier environment at Laetoli in Tanzania (Su and Harrison 2015).

Foot fossils of the species demonstrate adaptations to terrestrial bipedalism, such as a stiff midfoot for leverage in striding and a calcaneus capable of dissipating force during heel strikes (DeSilva et al. 2018; DeSilva et al. 2020). Alignment of the hallux with other toes allowed for a strong push-off (McNutt et al. 2018). On the other hand, the humerus of one of the best-preserved individuals ("Lucy") displays a mixture of traits between the derived condition of humans and a more generalized arboreal pattern (Arias-Martorell et al. 2015).

1.4.6 Australopithecus in South Africa *(~3.7 to 2.0 mya)*

At one time, *A. africanus* seemed to be a chronological successor to *A. afarensis* and possibly a descendent. However, though capable of bipedal walking, this species had more arboreal features than its putative ancestor. A change in dating is said to resolve the apparent anomaly: new dates by a different method placed *A. africanus* before 3 mya and contemporaneous with *A. afarensis* (Granger et al. 2022). However, the proposed change is in dispute (Frost et al. 2022).

Clarke et al. (2021) argued for a distinct species called *Australopithecus prometheus* (with a proposed date of about 3.7 mya) and claimed that many of the South African fossils should be assigned to this species rather than to *A. africanus*. The proposed species is mainly represented by just one skeleton (StW 573), but that skeleton is more than 90% complete. Limb proportions and some skull features were reported to resemble *A. afarensis* (Clarke and Kuman 2019; Heaton et al. 2019), which seems to be in accord with the early date. Whatever the species assignment might be, the evidence suggests that this *Australopithecus* resembled others in combining arborealism with habitual bipedalism on the ground.

1.4.7 Australopithecus (=Kenyanthropus platyops) *(~3.5 to 3.3 mya in Kenya)*

This taxon is the strongest candidate for a distinct hominin species during the time of *A. afarensis*. The face is relatively flat and vertical and the molar teeth display several very distinctive traits. The discoverers placed it in a separate genus, *Kenyanthropus* (Leakey et al. 2001) and asserted that *Kenyanthropus* was a more likely ancestor for *Homo* than *A. afarensis*. On the other hand, White (2003) attributed the flat and vertical face to distortion of the fossil due to in situ pressures and assigned it to *A. afarensis*.

Spoor et al. (2016) used a quantitative method to compare the upper jaws of *Kenyanthropus* with those of two *Australopithecus* species. They concluded that *Kenyanthropus* is a distinct taxon, validating the species designation. They also considered the morphological distance between it and the *Australopithecus* species to be sufficient for the genus designation. Alemseged (2023) accepts the species status but contests the genus classification on the grounds that recent evidence has significantly narrowed the morphological gap.

Even if in a separate genus, *Kenyanthropus* seems to fit the general ecological pattern of early hominins. A study by Villasenor et al. (2020) suggested that *Kenyanthropus* and *Australopithecus afarensis* inhabited similar types of habitats across different river basins in the Rift Valley. It was viewed as part of a growing body of evidence suggesting that early hominins diverged from the great apes by "abandoning" woodland-dominated habitats. However, chemical evidence for diet indicated that *Kenyanthropus* foraged in woodland as well as more open areas (Martin et al. 2020). It seems more accurate to say that hominins expanded their *environmental range* to include more open areas, rather than leaving the woodlands.

1.4.8 Other Australopithecus *Species*

Australopithecus bahrelghazali from Chad may just be a regional variant of *A. afarensis* (Alemseged 2023; Boyle and Wood 2017). Even so, it is important in extending the geographic distribution of the genus to northern Africa. *Australopithecus garhi* is a possible descendent of *A. afarensis* (Alemseged 2023). Its most obvious distinctions are large canine crowns and large cheek teeth like those of the related hominin genus *Paranthropus*, which represents a distinct lineage evolving in a different direction. *Australopithecus deyiremeda* (~3.5 to 3.3 mya in Ethiopia) has been distinguished from *A. afarensis* on the basis of jaws and teeth that resemble later hominins (Haile-Selassie et al. 2015). However, the researchers attributed the similarities to *homoplasy*, that is, parallel or convergent evolution, rather than an ancestor–descendent relationship.

Australopithecus sediba, in South Africa at about 2 mya, seems to be too late to be ancestral to *Homo*. Rak et al. (2021) compared diagnostic features of two jaws (specifically, the ascending rami) and concluded that they represented two genera, *Australopithecus* and *Homo*. Associated postcrania indicate active selection for habitual, terrestrial bipedalism in at least one group. However, the gait was unique compared to other hominins and the short lower limb implies limited walking distance. Casenave and Kivell (2023) postulated continued stabilizing selection for arboreal capability, based on upper limb and hand morphology and mobile foot joints. Surviving to a relatively late date, this species seems to represent the durability of the basic *Australopithecus* pattern.

1.4.9 *Early* Homo *(~2.8 to 1.7 mya; Ethiopia, Kenya, Tanzania, South Africa)*

Fossils from Ethiopia indicate an origin for the genus *Homo* between 2.8 and 2.4 mya, contemporary with later *Australopithecus*. An exhaustive review by Antón and Middleton (2023) concluded on the basis of cranial remains that three species of the genus existed by about 2 mya: *Homo habilis*, *H. rudolfensis*, and *H. erectus* (concurring with Grine et al. 2019). Sparser remains of pelves seem to indicate two or three species, but at present they cannot be linked to the crania. New fossils and new analyses of older ones led Anton and Middleton to profound reassessments of early *Homo*. Compared to many previous interpretations, *H. erectus* is older (perhaps dating to 2.3 mya), smaller, and more variable in size and morphology. *H. habilis* and *H. rudolfensis*, sometimes lumped into a single species, are no different in size but differ significantly in cranial shape (Fig. 1.2).

Fig. 1.2 Reconstruction of *Homo rudolfensis*, one of three *Homo* species that existed near the beginning of the lineage (subsumed by the term "early Homo" in this book). Mauricio Anton. Cranium 23, 17–40. Resized for publication. (Wikipedia license: https://creativecommons.org/licenses/by/3.0/)

Despite the differences among the species, there are generalizations that can be made about the genus *Homo* at 2 mya (Antón and Middleton 2023). For instance, the average brain size of *Homo*, implied by the height of the cranial vault, is significantly greater than that of *Australopithecus*. The *Homo* palate is deep and broad and the front of the maxilla is round or square rather than triangular as in *Australopithecus*. Protrusion of the lower face in *Homo* is relatively mild and the premolar teeth are not *molarized* (i.e., reshaped and resized to resemble the molars).

Although there are anatomical details that distinguish *Homo* from *Australopithecus*, there is an "apparent absence of a major transitional shift at the boundary of the genera" (Antón and Middleton 2023:17). The ranges of brain size overlap and frontal lobe organization of *Homo* "remains apelike across species" (Antón and Middleton 2023:18). With the recognition that *H. erectus* was not as large as previously thought, the range of height (145–180 cm) overlaps *Australopithecus* at the lower end. There were small and large individuals in all *Homo* species. The more limited evidence from foot and pelvis does not indicate greater efficiency in locomotion than in earlier hominins.

These findings support the view of Kimbel and Villmoare (2016) that adaptations present at the emergence of the genus *Homo* were "simply amplifications or extensions of ancient hominin trends," such as dietary eclecticism and expansion of brain size. Fossils of earliest *Homo* do not show a clear shift toward human-like morphology in clavicles and scapulae, suggesting continuation of arboreal capability (Arias-Martorell 2018/9). Some early *Homo* lived in mixed environments like their ancestors (Magill et al. 2016), though others occupied habitats that were largely grassland (Plummer et al. 2009).

Schroeder and Ackermann (2017) cast doubt on adaptive explanations for early *Homo* characteristics. Using statistical tests based on quantitative genetics theory, they calculated that (depending on the particular tests) 51–95% of the variation among early *Homo* groups resulted from genetic drift. However, jaw shape repeatedly contradicted the null hypothesis of drift, indicating an important role for selection. The researchers suggested a causal role for dietary change, possibly due to environmental changes (Chap. 4 in this book).

1.5 An Overview of Baboons

The term "baboon" has been used to encompass a diverse array of relatively large and relatively terrestrial monkeys in the Old World. These include geladas (*Theropithecus*), mandrills, and drills (*Mandrillus*) and some species of macaques (*Macaca*). The evolutionary relationships among them were largely resolved after genetic data were added to comparative anatomy. The results set the genus *Papio* apart from similar monkeys. Restricting the term baboon to *Papio* ensures that important behavioral, morphological, and genetic differences are not obscured (Bergman et al. 2018). Though there has been some comparison of early hominins with the other genera, especially geladas, most such efforts have focused on *Papio* over the course of 70 years (King 2022; Fischer and Zinner 2020; Swedell and Plummer 2019; Washburn and DeVore 1961).

The genus *Papio* (and the other "baboons" along with many other primate species) are Old World monkeys. That is, they belong to the superfamily Cercopithecoidea, which is spread across Asia and Africa. Baboons and the closely related macaques (*Macaca*) are the most widely distributed genera. Old World monkeys are classified together with all apes and hominins as Catarrhines.

For many decades a certain image of baboons prevailed (Strum 2019), especially outside of primatology. Baboons belonged to a single widespread species that foraged in tropical grasslands called *savannas*. They lived in large troops that revolved around the dominance of adult males. Recent research has shown that the genus *Papio* is far more complex and varied than the old stereotype. It offers many more possibilities for hypothesizing about the behavior of early hominins.

1.6 Key Characteristics of Baboons

Despite the recently demonstrated variation among baboons, it is still possible to make generalizations that are useful in comparisons with early hominins. All baboon species share the characteristics reviewed below, though there is quantitative variation (Bergman et al. 2018; Fischer et al. 2019; Swedell 2011).

1.6.1 Positional Behavior

Positional behavior encompasses postures, movements, and the relationship of these patterns to environmental context. Baboons are quadrupeds, that is, they habitually stand and walk on four legs. The usual context for this behavior is the ground (Fig. 1.3), although it also appears in trees. Compared to most other primates, "Baboons are a highly terrestrial species" (Hammond et al. 2022). Rose (1976) studied positional behavior in one baboon species. He observed baboons in trees during only 1.8% of observation time and most of this activity was performed by infants and juveniles.

Though they are usually on the ground much of the day, baboons are also agile climbers (Fig. 1.4). They often feed in trees when desirable foods, such as fruits, are available. They also use trees as a refuge from predators and for safe places to sleep at night (Hammond et al. 2022).

Baboons could be called partly arboreal, except for the fact that many baboons sleep on cliffs at night and some seem to prefer the cliffs even where sufficient trees are available (Fig. 1.5). It might be appropriate to replace the commonly used misnomer arbo-terrestrial with *scando-terrestrial*, that is, climbing and ground-walking.

Fig. 1.3 Olive baboon foraging on the ground. (Photo by Glenn King. Manyara, Tanzania)

Fig. 1.4 Chacma baboons in a tree, demonstrating the arboreal capability of the genus. (Photo by Curt Busse. Okavango, Botswana)

Fig. 1.5 Hamadryas baboons on a wall in a zoo. In the wild, baboons often sleep on cliff sides rather than trees. (Photo by Steven Gotz. Oakland Zoo)

1.6.2 Body Mass and Sexual Dimorphism

Baboons, like hominins, can be considered medium-sized land mammals (Jolly 2013), placing somewhere between shrews and elephants (Fig. 1.6). They are the largest monkeys, except for the mandrills (*Mandrillus sphinx*). Considering all of the *Papio* species and both sexes, the low end of weight variation in adult baboons is about 10 kg but adult males can weigh up to 35 kg (Table 1.3). The largest baboons approximate the weight of a German shepherd dog. The smallest baboons are about the size of a beagle.

The largest baboons are males and the smallest are females. Baboons are *sexually dimorphic* in body mass (Fig. 1.7 and Table 1.3). Sexual dimorphisms are male-female differences that are characteristic of a species or larger grouping, such as the antlers of deer. From the female point of view, the female-to-male ratio in weight varies from about 0.5 to 0.7. Looked at from the other side, the male-to-female ratio varies from about 1.4 to almost 2.0.

Baboons are also sexually dimorphic in their canine teeth, which are significantly longer in males (Fig. 1.8). Across four baboon species, the upper canine length in males varied from about 2.7 cm to about 4.7 cm. The range for females of those species is 0.9–1.6 cm (Plavcan and Ruff 2008). Male canines are roughly three times longer than those of females.

Fig. 1.6 Size comparison of female olive baboon with human child. (Photo by Glenn King. Seronera Lodge, Serengeti National Park, Tanzania)

Table 1.3 Body mass estimates for baboon taxa [Revised 4-17-24]

Species	Male weight	Female weight	F/M Ratio	F/M Ratio	M/F RATIO
Chacma Baboon (*P. ursinus*)	25–35	12–20	52–60%	0.52–0.69	1.45–1.92
Olive Baboon (*P. anubis*)	22–30	14–18	52–55%	0.52–0.61	1.64–1.94
Yellow Baboon (*P. cynocephalus*)	20–28	8–13	~55%	0.46–0.60	1.68–2.17
Guinea Baboon (*P. papio*)	17–26	10–14	~59%	0.54[a]	1.86[a]
Hamadryas Baboon (*P. hamadryas*)	16–21	10–15	~58%	0.54–0.59	1.70–1.84
Kinda Baboon (*P. kindae*)	~16	~10	~63%	No data	No data

Figures for weight are in kilograms. Information in the first three categories was obtained from the compilation "Genus *Papio*" in Mittermeier et al. (eds) (2013). The figures in the last two columns were obtained from the survey of populations within species by Swedell (2011)
[a]One captive population

1.7 *Papio* Evolution and Species

According to fossil and molecular evidence, the *Papio* lineage became distinct from other monkeys, including its closest relatives, between 2.4 and 2 mya (Gilbert et al. 2018), Fossils and comparative evidence indicate that baboons first evolved in a tropical region of southern Africa (Elton and Dunn 2020; Kopp et al. 2023). They

Fig. 1.7 Male and female chacma baboons, illustrating sexual dimorphism in body size. (Photo by Glenn King. Mkhuzi, South Africa)

Fig. 1.8 A male hamadryas baboon displays his long canine teeth. (Photo by Steven Gotz. Oakland Zoo)

began to separate into distinct taxa between 2 and 1.5 mya (Fischer et al. 2019). Eventually the genus occupied a wide variety of habitats across the continent outside of the Sahara Desert and the Congo rainforest.

1.7.1 Papio *Species*

As of 2020, six species of *Papio* were generally recognized (Fig. 1.9). These taxonomic groups are widely accepted as valid, but their technical classification as *species* is in question (Fischer et al. 2019). According to an influential criterion, they are divisions of a single species, because they interbreed freely wherever they meet in the wild and they produce hybrid offspring that survive and are fertile. However, many contemporary experts agree that the species distinctions are more appropriate because the baboon groupings differ substantially from one another in physical traits, individual behavior, and characteristics of their societies (Fischer et al. 2019). The latter view is followed here.

Olive baboons (*Papio anubis*) are distributed across the widest geographic range, from Sierra Leone to Eritrea in the north and down into Tanzania. Guinea baboons (*P. papio*) occupy a small area west of the olive baboons, and hamadryas baboons (*P. hamadryas*) are northeast of the olive baboons in the Horn and in a southwestern

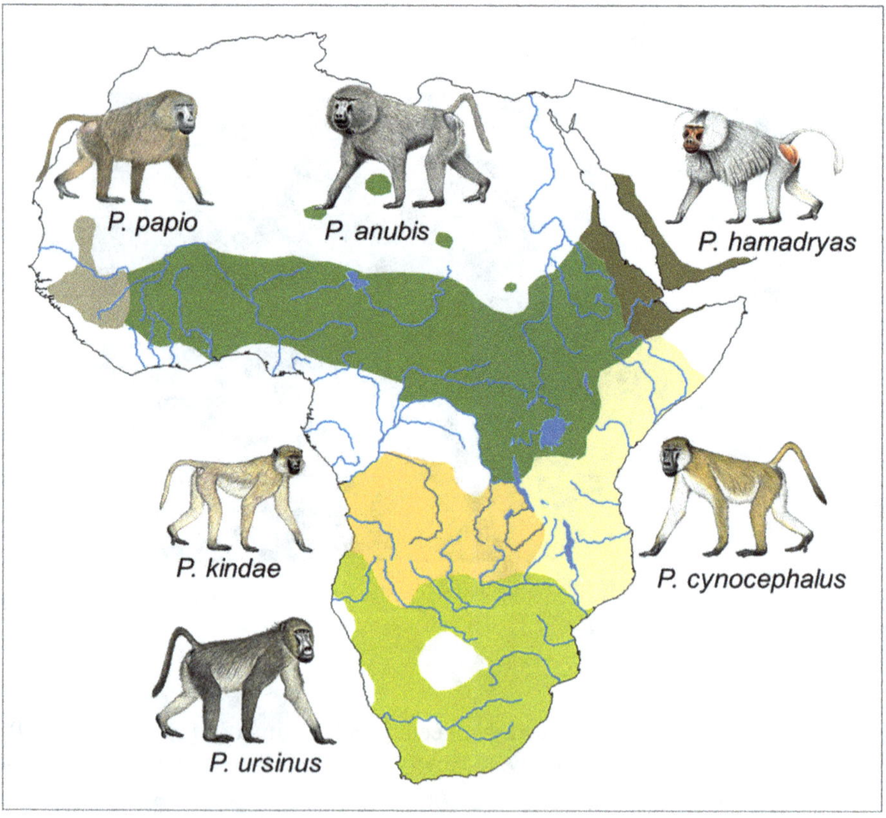

Fig. 1.9 Geographic distribution of the six baboon species identified in the text. African countries within those distributions are specified in the text. (By permission from Julia Fischer)

strip of the Arabian Peninsula. Yellow baboons (*P. cynocephalus*) inhabit a large part of eastern Africa and Kinda baboons (*P. kindae*) a small south-central area. Chacma baboons (*P. ursinus*) occupy the south, through Botswana and Mozambique down to Cape Horn of South Africa. The situation is complicated by extensive gene flow between the species in some areas, which may be analogous to some early hominin species (Martinez et al. 2019).

1.7.2 *Ecological Niches of* Papio *Species*

An important dimension of baboon species variation is represented by the application of ecological niche models. Fuchs et al. (2018) found that species distributions are correlated with climatic variables, especially seasonal variation in temperature and rainfall. Species that have been lumped together as "savanna baboons" in the past (thought to be especially significant for comparison with early hominins) actually occupy distinct niches. Based on niche breadth, some species may be more specialized than others. There is no significant correlation between the degree of niche overlap and the evolutionary divergence time for the same species. This indicates that the niche distinction is an ecological phenomenon rather than just resulting from separate evolution. The results of the Fuchs study corroborated and expanded earlier findings by Winder (2014) and Kamilar (2006). The following discussion is based on Fuchs et al. (2018) and Mittermeier et al. (2013) except where other sources are indicated.

Guinea baboons inhabited areas with the highest annual temperature and second highest annual rainfall out of the six species. They also occupied the smallest geographic range and scored low in niche breadth. DNA studies suggest that the unity of the population is connected with adaptation to unique and specific environmental conditions and relatively isolated evolution after divergence from the other baboon species. Although Guinea baboons have survived in the arid Sahel region, they are found mostly in forested habitats, varying from coastal mangrove and riverine evergreen to woodland savanna.

Kinda baboons ranked second in conformity to the niche model. Precipitation during the driest quarter of the year was the most important variable. That and low temperature could have socioecological effects such as elevated thermoregulatory requirements and decreased activity in heavy rainfall. Kinda baboons occur in light woodlands and gallery forests These results, combined with minimal overlap with other baboons, suggest a distinctive niche that supports recent recognition as a sixth species of the genus. Kinda baboons are also distinctive as the smallest species of *Papio* with the lowest degree of sexual dimorphism in body mass (Table 1.3).

Variables affecting chacma baboons indicated that they are better suited for cooler environments, potentially at higher altitudes than other baboons. Their habitats include cold mountain areas. Chacmas are the largest baboon species, slightly larger than olive baboons (Table 1.3). Size might function in heat retention if chacmas are in fact adapted to cooler temperatures.

The niche model for yellow baboons fits with the finding of Winder (2014) that the species inhabits lower latitudes with a cooler mean annual temperature compared to hamadryas, olive, and Guinea baboons. Nevertheless, yellow baboons occupy a variety of habitats, from steppes and dry bushland to coastal littoral.

The niche model for olive baboons did not apply as well as those described above and the value for niche breadth was higher than for the others. Both results suggest that the species is more ecologically flexible than any of the others. Olive baboons occupy the widest variety of habitats of any baboon species, although chacma baboons are close competitors. The optimal environment of olive baboons seems to be one of low temperature seasonality and moderate rainfall. They are flexible enough to live in areas of high rainfall, but apparently at the cost of longer interbirth intervals, higher infant mortality, decreased foraging time, and increased risk of disease (Higham et al. 2009). The relatively great adaptability of olive baboons might suggest that they are the best single-species model within the genus for reconstructing early hominin ecology and at least some hominin behavior. Of all the *Papio* species, olive baboons seem to come closest to matching early hominin success in expanding throughout diverse habitats.

Hamadryas baboons were associated with a moderately performing niche model, which may be due to a variety of factors. Niche breadth results suggested that hamadryas baboons are more ecologically flexible than is generally appreciated. Such flexibility might be related to a flexible form of social organization with several levels of affiliation (Chap. 7).

1.8 *Papio* Societies

The paleoanthropological record provides little information about social behavior and social organization, and none for a long period before the first archeological remains were deposited. However, the archeological evidence for later societies, as well as the historical record, suggest that the foundation for human social life might go back to early hominin societies comparable to those of baboons.

Baboons live in complex societies that may contain hundreds of individuals. On the basis of social organization, baboon species can be considered in two clusters. The so-called COKY baboons (Jolly 2020) are the chacma, olive, kinda, and yellow baboon species. These baboons live in *troops*, stable and independent communities that consist of multiple males and females (Chap. 7). The COKY classification is a convenient basis for some generalizations, but there are important distinctions within this cluster.

Olive and yellow baboons are the most similar in behavior. They are distinct from chacma baboons in ways that seem consistent with the view that chacmas represent the original baboon stock in southern Africa. For example, males in both of the more northerly species can form coalitions to achieve mating success (Chap. 7), while this pattern is completely absent from the chacma repertoire (Alberts and Altmann 2007). Kinda baboon troops are unusually large, often including more than

200 members, and male–female relations display interesting distinctions from other COKY species (Petersdorf et al. 2019; Chap. 8 in this book).

Two species of baboons live in multilevel societies in which smaller social units are nested in more comprehensive units, up to four levels of organization. Affiliation at the different levels serves different biological and ecological functions, including foraging, predator defense, and shared sleeping locations (Fischer et al. 2019; Jolly 2020). The most basic group in both species, often called a *one-male unit*, consists of a single adult male with one or more adult females and their offspring. There is significant variation between hamadryas and Guinea baboons with regard to male-female relations (Chap. 8).

1.9 Summary and Discussion

This chapter introduced the two main subjects of the book, early hominins and baboons. Coverage was somewhat unbalanced because of the abundance of new evidence and resultant controversies about early hominins. As living species, baboons lend themselves to more comprehensive research with a somewhat greater chance of agreement among the researchers.

Early hominins were defined as human ancestors and close relatives from the beginning of the lineage to about 1.7 million years ago, when major environmental and behavioral shifts took place. Hominins constitute an evolutionary lineage that separated from a common ancestor with the genus *Pan* (chimpanzees and bonobos) probably about 6 or 7 mya (though some measures place that date earlier). The most important early hominin genera are *Ardipithecus*, *Australopithecus*, (early) *Homo*, and perhaps *Kenyanthropus* (sunk into *Australopithecus* by some experts).

Baboons were defined as monkeys of the genus *Papio*. There is a long history of comparing this particular genus to early hominins as one way to elaborate on the archeological record and to gain insights into periods of hominin evolution for which there are gaps in the archeological record. This especially applies to the long time span before the archeological record began.

The phrase "last common ancestor" (referring to the common ancestor shared by *Pan* and *Homo*) is often capitalized and abbreviated as LCA (sometimes the more precise but unwieldy PHLCA is used). There is no fossil genus or species that is generally recognized as the LCA. Given the odds against finding that particular animal, there probably never will be. Instead, there are attempts to infer its characteristics from indirect sources, including extant apes (such as chimpanzees) and the more numerous ape species that preceded the LCA in the Miocene period. One such approach postulates an animal similar to the living chimpanzee; another reconstructs a more generalized kind of ape; a third argues that the LCA must have been unlike any living species.

Little information is added by three genera, dated to between 7 and 5 mya, that are purported to represent the earliest hominins. The fossil evidence is sparse and controversial. However, *Ardipithecus ramidus* provides more evidence and is widely accepted to be a hominin. It was probably bipedal, though in a way unlike modern humans. This was in part due to a foot with a grasping hallux (big toe) that facilitated arboreal movement.

Ardipithecus was followed chronologically by *Australopithecus*, first known from about 4 mya. Numerous species of *Australopithecus* have been identified, generating a great deal of controversy. Some can be placed in a chronological sequence, but others may have been contemporaries for significant periods of time. *A. afarensis* is especially important because the fossil evidence is relatively abundant and some of its anatomical features seem to be more modern than the corresponding features of other species in the genus. This is especially significant with regard to bipedal locomotion. The roughly contemporary taxon called *Kenyanthropus* was claimed to be more modern in its craniodental features, but the facial form has been attributed to distortion while the fossil was in the ground. The genus *Homo* seems to have begun to emerge at about 2.8 mya, but only gradually differentiated itself from its predecessors.

Reference to particular hominin taxa is convenient or necessary for some comparative purposes. However, it is often simpler to avoid some of the controversies and generalize about features common to all the undoubted hominins. All early hominins were bipedal, even though the details of gait may have varied. Despite a theory that bipedalism originated in the trees, the evolution of bipedal anatomy shows that hominins made an early commitment to spending a great deal of time on the ground.

Although there was a wide range of variation in body mass, most early hominins seem to fall into a range of about 25–40 kilograms. Part of this variation was sexual dimorphism, with males being larger than females, though the quantitative extent of the difference is debated. Very early in hominin evolution, the canine teeth were reduced in both males and females to the point where there was little or no projection above or below the rest of the tooth rows. Finally, the brains of early hominins were small, with a median size 1/3 that of modern brains or less in almost all species. This suggests significant limits on mental capabilities and justifies the use of models from other animals as one way to understand early hominins.

Among such models, baboons (monkeys of the genus *Papio*) are one of the most important. Six species of extant baboons live in different parts of Africa, but overlap and hybridize. They all have the same basic physical features, but vary significantly in ecology and social behavior. As with the early hominins, it is sometimes profitable to look at a particular species for comparative purposes. However, stronger arguments can often be made by considering characteristics that all baboons have in common.

Baboons are quadrupedal animals that are mostly terrestrial during the day. However, they are also agile climbers that use this ability to access food and to take refuge from predators in trees. For sleeping places, they climb trees or cliffs and they seem to prefer the latter when both are available. This suggests that the misleading term arbo-terrestrial should be replaced with *scando-terrestrial*, that is, climbing and ground walking.

Baboons are the largest monkeys other than mandrills, males weighing as much as 35 kg, but females are much smaller than males. The canine teeth of males are about three times longer than those of females. All baboons live in complex societies, some in troops and some in multilevel societies. The genus offers many possibilities for comparison with early hominins, leading to a variety of hypotheses about ecology and behavior.

Baboons are obviously different from early hominins in some important ways, but the gaps can be bridged by functional analogies (e.g., hominin weapons as substitutes for baboon canine teeth). Some features that seem very different lead to useful analogies (e.g., bipedalism as an indicator of terrestriality). Some features that are only somewhat similar may still make for useful comparisons (e.g., relatively large size compared to many other mammals). The next chapter begins to make comparisons directly and subject them to some analysis.

References

Alberts SC, Altmann J. The evolutionary past and the research future: environmental variation and life history flexibility in a primate lineage. In: Swedell L, Leigh SR, editors. Reproduction and fitness in baboons: Behavioral, ecological, and life history perspectives. New York: Springer; 2007. p. 277–303.

Alemseged Z. Reappraising the palaeobiology of *Australopithecus*. Nature. 2023;617:45–54. https://doi.org/10.1038/s41586-023-05957-1.

Almécija S, Hammond AS, Thompson NE, et al. Fossil apes and human evolution. Science. 2021;372 https://doi.org/10.1126/science.abb4363.

Andrews P. Last common ancestor of apes and humans: morphology and environment. Folia Primatol. 2020;91:122–48. https://doi.org/10.1159/000501557.

Antón SC, Middleton ER. Making meaning from fragmentary fossils: Early *Homo* in the Early to early Middle Pleistocene. J Hum Evol. 2023;179:103307. https://doi.org/10.1016/j.jhevol.2022.103307.

Arias-Martorell. The morphology and evolutionary history of the glenohumeral joint of hominoids: a review. Ecol Evol. 2018;9(1) https://doi.org/10.1002/ece3.4392.

Arias-Martorell J, Tallman M, Potau JM, et al. Shape analysis of the proximal humerus in orthograde and semi-orthograde primates: correlates of suspensory behavior. Am J Primatol. 2015;77(1):1–19. https://doi.org/10.1002/ajp.22306.

Bergman T, Haile AA, Beehner J. What is (not) a Baboon? Int J Primatol. 2018;39(6) https://doi.org/10.1007/s10764-018-0063-5.

Bobe R, Wood B. Estimating origination times from the early hominin fossil record. Evol Anthropol. 2021; https://doi.org/10.1002/evan.21928.

Bobe R, Manthi FK, Ward CV, et al. The ecology of *Australopithecus anamensis* in the early Pliocene of Kanapoi, Kenya. J Hum Evol. 2020;140 https://doi.org/10.1016/j.jhevol.2019.102717.

Boyle EK, Wood B. Human evolutionary history. In: Kaas J, editor. Evolution of nervous systems. New York: Academic Press; 2017. p. 19–36.

Casenave M, Kivell TL. Challenges and perspectives on functional interpretations of australopith postcrania and the reconstruction of hominin locomotion. J Hum Evol. 2023;175:103304. https://doi.org/10.1016/j.jhevol.2022.103304.

Chintalapati M, Moorjani P. Evolution of the mutation rate across primates. Curr Opin Genet Dev. 2020;62:58–64. https://doi.org/10.1016/j.gde.2020.05.028.

Clarke RJ, Kuman K. The skull of StW 573, a 3.67 Ma *Australopithecus prometheus* skeleton from Sterkfontein Caves, South Africa. J Hum Evol. 2019;134(2):102634. https://doi.org/10.1016/j.jhevol.2019.06.005.

Clarke RJ, Pickering TR, Heaton JL, Kuman K. The earliest South African hominids. Annu Rev Anthropol. 2021;50:125–43. https://doi.org/10.1146/annurev-anthro-091619-124837.

Crompton RH, Sellers WI, Thorpe SKS. Arboreality, terrestriality and bipedalism. Phil Trans R Soc Lond B Biol Sci. 2010;365:3301–14.

DeSilva JM, Gill CM, Cody T, et al. A nearly complete foot from Dikika, Ethiopia and its implications for the ontogeny and function of *Australopithecus afarensis*. Sci Adv. 2018;4(7):eaar7723. https://doi.org/10.1126/sciadv.aar7723.

DeSilva JM, McNutt E, Zipfel B, et al. Associated *Australopithecus afarensis* second and third metatarsals (A.L. 333-133) from Hadar, Ethiopia. J Hum Evol. 2020;146:102848. https://doi.org/10.1016/j.jhevol.2020.102848.

Du A, Rowan J, Wang SC, et al. Statistical estimates of hominin origination and extinction dates: a case study examining the *Australopithecus anamensis–afarensis* lineage. J Hum Evol. 2020;138 https://doi.org/10.1016/j.jhevol.2019.102688.

Elton S, Dunn J. Baboon biogeography, divergence, and evolution: morphological and paleoecological perspectives. J Hum Evol. 2020;145:102799. https://doi.org/10.1016/j.jhevol.2020.102799.

Fischer J, Zinner D. Introduction to special issue: Frontiers in baboon research. J Hum Evol. 2020;146 https://doi.org/10.1016/j.jhevol.2020.102822.

Fischer J, Higham JP, Alberts S, et al. The natural history of model organisms: insights into the evolution of social systems and species from baboon studies. elife. 2019;8:e50989. https://doi.org/10.7554/eLife.50989.

Foley RA, Lahr MM. Ghosts of extinct apes: genomic insights into African hominid evolution. Trends Eco Evol in press. 2023; https://doi.org/10.1016/j.tree.2023.12.009.

Frost SR, White FJ, Reda HG, Gilbert CC. Biochronology of South African hominin-bearing sites: a reassessment using cercopithecid primates. Proc Natl Acad Sci USA. 2022;119(45) https://doi.org/10.1073/pnas.2210627119.

Fuchs AJ, Gilbert CC, Kamilar JM. Ecological niche modeling of the genus *Papio*. Am J Phys Anthropol. 2018;166:812–23. https://doi.org/10.1002/ajpa.23470.

Gilbert CC, Frost SR, Pugh KD, et al. Evolution of the modern baboon (*Papio hamadryas*): a reassessment of the African Plio-Pleistocene record. J Hum Evol. 2018;122:38–69. https://doi.org/10.1016/j.jhevol.2018.04.012.

Granger DE, Stratford D, Bruxelles L, et al. Cosmogenic nuclide dating of *Australopithecus* at Sterkfontein, South Africa. Proc Natl Acad Sci. 2022;119(27):e2123516119. https://doi.org/10.1073/pnas.2123516119.

Grine FE, Leakey MG, Gathago PN, et al. Complete permanent mandibular dentition of early *Homo* from the upper Burgi Member of the Koobi Fora Formation, Ileret, Kenya. J Hum Evol. 2019;131:152–75. https://doi.org/10.1016/j.jhevol.2019.03.017.

Haile-Selassie Y, Beotas LG, Melillo S, et al. New species from Ethiopia further expands Middle Pliocene hominid diversity. Nature. 2015;521(14448):483–8. https://doi.org/10.1038/nature14448.

Hammond P, Lewis-Bevan L, Biro D, Carvalho S. Risk perception and terrestriality in primates: a quasi-experiment through habituation of chacma baboons (*Papio ursinus*) in Gorongosa National Park, Mozambique. Am J Biol Anthropol. 2022;179(1):48–59. https://doi.org/10.1002/ajpa.24567.

Heaton JS, Pickering T, Carlson KJ, et al. The long limb bones of the StW 573 *Australopithecus* skeleton from Sterkfontein Member 2: descriptions and proportions. J Hum Evol. 2019;133:167–97. https://doi.org/10.1016/j.jhevol.2019.05.015.

Higham JP, Warren Y, Adanu J, et al. Living on the edge: life-history of olive baboons at Gashaka-Gumti National Park, Nigeria. Am J Primatol. 2009;71:293–304.

Hunt KD. Why are there apes? Evidence for the co-evolution of ape and monkey ecomorphology. J Anat. 2016;228(4):630–85. https://doi.org/10.1111/joa.12454.

Johannsen L, Coward SRL, Martin GR, et al. Human bipedal instability in tree canopy environments is reduced by "light touch" fingertip support. Sci Rep. 2017;7:1135. https://doi.org/10.1038/s41598-017-01265-7.

Jolly CJ. Analogies and models in the study of the early hominins. In: Sponheimer M, Lee-Thorp JA, Reed KE, Ungar PS, editors. Early hominin paleoecology. Denver: University of Colorado; 2013. p. 437–55.

Jolly CJ. Philopatry at the frontier: a demographically driven scenario for the evolution of multilevel societies in baboons (*Papio*). J Hum Evol. 2020;146:102819. https://doi.org/10.1016/j.jhevol.2020.102819.

Jungers WL, Grabowski M, Hatala KG, Richmond BG. The evolution of body size and shape in the human career. Phil Trans R Soc B Biol Sci. 2016;371:20150247. https://doi.org/10.1098/rstb.2015.0247.

Kamilar JM. Geographic variation in savanna baboon (*Papio*) ecology and its taxonomic and evolutionary implications. In: Lehman SM, Fleagle JG, editors. Primate biogeography. Developments in primatology. New York: Springer; 2006. p. 169–200.

Kimbel WH, Villmoare B. From *Australopithecus* to *Homo*: the transition that wasn't. Phil Trans R Soc Lond B. 2016;371(1698):20150248. https://doi.org/10.1098/rstb.2015.0248.

King GE. Baboon perspectives on the ecology and behavior of early human ancestors. Proc Nat Acad Sci. 2022;119(45):e2116182119. https://doi.org/10.1073/pnas.2116182119.

Kopp GH, Sithaldeen R, Trede F, et al. A comprehensive overview of baboon phylogenetic history. Genes. 2023;14:614 ff. https://doi.org/10.3390/genes14030614.

Leakey MG, et al. New hominin genus from eastern Africa shows diverse middle Pliocene lineages. Nature. 2001;410:433–40.

Macchiarelli R, Bergeret-Medina A, Marchi D, Wood B. Nature and relationships of *Sahelanthropus tchadensis*. J Hum Evol. 2020;149 https://doi.org/10.1016/j.jhevol.2020.102898.

Magill CR, Ashley GM, Dominguez-Rodrigo M, et al. Dietary options and behavior suggested by plant biomarker evidence in an early human habitat. Proc Natl Acad Sci USA. 2016;113(11):2874–9. https://doi.org/10.1073/pnas.1507055113.

Manthi F, Plavcan JM, Ward CV. Introduction to special issue. Kanapoi: Paleobiology of a Pliocene site in Kenya. J Hum Evol. 2020;140:102718. https://doi.org/10.1016/j.jhevol.2019.102718.

Martin JE, Tacail T, Braga J, et al. Calcium isotopic ecology of Turkana Basin hominins. Nat Commun. 2020;11:3587. https://doi.org/10.1038/s41467-020-17427-7.

Martinez FE, Capelli C, Ferreira da Silva MJ, et al. A missing piece of the *Papio* puzzle: Gorongosa baboon phenostructure and intrageneric relationships. J Hum Evol. 2019;130:1–20. https://doi.org/10.1016/j.jhevol.2019.01.007.

McNutt EJ, Zipfel B, DeSilva JM. The evolution of the human foot. Evol Anthropol. 2018;27:197–217.

Mittermeier RA, Rylands AB, Wilson DE, editors. Handbook of the mammals of the world – volume 3. Primates. Barcelona: Lynx Edicions; 2013. p. 660–5.

Mongle C, Strait DS, Grine F. Expanded character sampling underscores phylogenetic stability of *Ardipithecus ramidus* as a basal hominin. J Hum Evol. 2019;131:28–39. https://doi.org/10.1016/j.jhevol.2019.03.006.

Moorjani P, Amorim CEG, Arndt PF, Przeworski M. Variation in the molecular clock of primates. Proc Natl Acad Sci USA. 2016;113(38):10607–12. https://doi.org/10.1073/pnas.1600374113.

Patterson DB, Braun DR, Allen K, et al. Comparative isotopic evidence from East Turkana supports a dietary shift within the genus *Homo*. Nat Ecol Evol. 2019;3:1048–56. https://doi.org/10.1038/s41559-019-0916-0.

Petersdorf M, Weyher A, Kamilar JM, et al. Sexual selection in the Kinda baboon. J Hum Evol. 2019;135(S37) https://doi.org/10.1016/j.jhevol.2019.06.006.

Plavcan JM. Sexual dimorphism in hominin ancestors. International Encyclopedia of Anthropology. Wiley; 2018.

Plavcan JM, Ruff CB. Canine size, shape, and bending strength in primates and carnivores. Am J Phys Anthropol. 2008;136:65–84. https://doi.org/10.1002/ajpa.20779.

Plummer TW, Ditchfield PW, Bishop LC, et al. Oldest evidence of tool making hominins in a grassland-dominated ecosystem. PLoS One. 2009;4(9):e7199. https://doi.org/10.1371/journal.pone.0007199.

Prang TC. The African ape-like foot of *Ardipithecus ramidus* and its implications for the origin of bipedalism. eLife Sci. 2019;8:e44433. https://doi.org/10.7554/elife.44433.

Prang TC, Ramirez K, Grabowski M, Williams SA. *Ardipithecus* hand provides evidence that humans and chimpanzees evolved from an ancestor with suspensory adaptations. Sci Adv. 2021;7:eabf2474.

Rak Y, Geffen E, Hylander W, et al. One hominin taxon or two at Malapa Cave? Implications for the origins of *Homo*. S Afr J Sci. 2021;117(5/6) https://doi.org/10.17159/sajs.2021/8747.

Reno PL, Lovejoy CO. From Lucy to Kadanuumuu: balanced analyses of *Australopithecus afarensis* assemblages confirm only moderate skeletal dimorphism. PeerJ. 2015;3(Suppl 49):e925. https://doi.org/10.7717/peerj.925.

Rose MD. Positional behavior in olive baboons (*Papio anubis*) and its relationship to maintenance and social activities. Primates. 1976;18(1):59–116.

Ruff CB, Squyres N, Junno JA. Body mass estimation in hominins from humeral articular dimensions. Am J Phys Anthropol. 2020;173(2) https://doi.org/10.1002/ajpa.24090.

Schroeder L, Ackerman R. Evolutionary processes shaping diversity across the *Homo* lineage. bioRxiv. 2017; https://doi.org/10.1101/136507.

Selby MS, Lovejoy O. Evolution of the hominoid scapula and its implications for earliest hominid locomotion. Am J Phys Anthropol. 2017;162(4) https://doi.org/10.1002/ajpa.23158.

Simpson SW, Levin NE, Quade J, et al. *Ardipithecus ramidus* postcrania from the Gona Project area, Afar Regional State, Ethiopia. J Hum Evol. 2019;129:1–45. https://doi.org/10.1016/j.jhevol.2018.12.005.

Spoor F, Leakey MG, O'Higgins P. Middle Pliocene hominin diversity: *Australopithecus deyiremeda* and *Kenyanthropus platyops*. Phil Trans R Soc B Biol Sci. 2016;371(1698):20150231. https://doi.org/10.1098/rstb.2015.0231.

Stamos PA, Alemseged Z. Hominin locomotion and evolution in the Late Miocene to Late Pliocene. J Hum Evol. 2023;178:103332. https://doi.org/10.1016/j.jhevol.2023.103332.

Strum SC. Why natural history is important to (primate) science: a baboon case study. Int J Primatol. 2019;40:596–612.

Su DF, Haile-Selassie Y. Mosaic habitats at Woranso-Mille (Ethiopia) during the Pliocene and implications for *Australopithecus* paleoecology and taxonomic diversity. J Hum Evol. 2022;163:103076. https://doi.org/10.1016/j.jhevol.2021.103076.

Su DF, Harrison T. The paleoecology of the Upper Laetolil Beds, Laetoli Tanzania: a review and synthesis. J Afr Earth Sci. 2015;101:405–19. https://doi.org/10.1016/j.jafrearsci.2014.09.019.

Suwa G, Sasaki T, Semaw S, et al. Canine sexual dimorphism in *Ardipithecus ramidus* was nearly human-like. Proc Natl Acad Sci. 2021;118(49):e2116630118. https://doi.org/10.1073/pnas.2116630118.

Swedell L. African papionins, diversity of social organization and ecological flexibility. In: Campbell CJ, Fuentes A, MacKinnon KC, Bearder SK, Stumpf RM, editors. Primates in perspective. 2nd ed. Oxford University Press; 2011. p. 241–77.

Swedell L, Plummer T. Social evolution in Plio-Pleistocene hominins: insights from hamadryas baboons and paleoecology. J Hum Evol. 2019;137:1–15. https://doi.org/10.1016/j.jhevol.2019.102667.

Vaneechoutte M, Mansfield F, Munro S, Verhaegen M. Have we been barking up the wrong ancestral tree? Australopithecines are probably not our ancestors. SCIEPublish; 2023. https://doi.org/10.35534/natanthropol.2023.10007.

Villmoare B. Early Homo and the role of the genus in paleoanthropology. Am J Phys Anthropol. 2018;165:72–89. https://doi.org/10.1002/ajpa.23387.

Villasenor A, Bobe R, Behrensmeyer AK. Middle Pliocene hominin distribution patterns in Eastern Africa. J Hum Evol. 2020;147:102856. https://doi.org/10.1016/j.jhevol.2020.102856.

Ward CV, Plavcan JM, Manthi FK. New fossils of *Australopithecus anamensis* from Kanapoi, West Turkana, Kenya (2012–2015). J Hum Evol. 2020;140 https://doi.org/10.1016/j.jhevol.2017.07.008.

Washburn SL, DeVore I. The social life of baboons. Sci Am. 1961;204(6):62–71. https://doi.org/10.1038/scientificamerican0661-62.

White T. Early Hominids diversity or distortion? Science. 2003;299:1994–7. https://doi.org/10.1126/science.1078294.

Will M, Pablos A, Stock JT. Long-term patterns of body mass and stature evolution within the hominin lineage. R Soc Open Sci. 2017;4:171339. https://doi.org/10.1098/rsos.171339.

Winder I. The biogeography of the *Papio* baboons: a GIS-based analysis of range characteristics and variability. Folia Primatol. 2014;85(5):292–318. https://doi.org/10.1159/000362545.

Yaxley K, Foley R. Reconstructing the ancestral phenotypes of great apes and humans (Homininae) using subspecies-level phylogenies. Biol J Linnean Soc. 2019;128(4) https://doi.org/10.1093/biolinnean/blz140.

Chapter 2
Why Compare Early Hominins to Baboons?

2.1 Introduction

The comparative approach discussed here entails three crucial questions: Why compare early hominins to any living species? Why compare them to baboons in particular? Why are baboons one of the most important primates for this purpose?

Direct evidence for early hominin ecology and behavior comes from the paleoanthropological record that is provided by archeology and paleontology (e.g., Clarke et al. 2021). This information has been augmented by reference to living species for several reasons (King 2001, 2022). First, because the record only addresses behavior that leaves physical remains, living species contribute to a more rounded interpretation of the past. Second, the living species suggest hypotheses about behavior and ecology where direct evidence is meager or lacking. Third, such evidence is extremely meager for millions of years of earlier hominin evolution. There is no archeological record from this time and the fossils are scarce, sparsely preserved, and highly controversial.

A variety of living species have been compared to early hominins, including social carnivores (Smith et al. 2012) and cetaceans (Yamagiwa and Karczmarski 2014). These are far beyond the scope of this book. In any event our primate relatives have provided most of the comparative material for early hominins, with chimpanzees receiving the most attention (Hopper and Ross 2020; Muller et al. 2017). Chimpanzees (and bonobos) have a unique phylogenetic relationship to hominins (Chap. 1); however, the further evolution of early hominins displays many similarities to the evolution of baboons. These parallels can be summed up as geographic distribution, environmental diversity, arboreal and terrestrial locomotion, medium size relative to other mammals, and sexual dimorphism in body mass.

G. E. King, *Baboon Perspectives on Early Human Ancestors*,
https://doi.org/10.1007/978-3-031-36769-4_2

2.2 Environment

As early hominins and baboons evolved, they spread across most of Africa and encountered a wide variety of environments. Both were challenged by long-term aridification.

2.2.1 Environmental Diversity

"Baboons evolved and radiated in parallel with hominins within a similar landscape and time frame, the savannahs and woodlands of Plio-Pleistocene Africa. It is therefore highly likely that they experienced similar selection pressures and evolutionary processes as hominins" (Fischer and Zinner 2020). By the same token, hominins experienced selection pressures and evolutionary processes similar to those that affected baboons.

Throughout the early paleoecological record, hominins are found in association with evidence for complex habitats with diverse environmental components (Chap. 1). There is widespread agreement among paleoanthropologists that these habitats can be characterized as *mosaic environments*. That is, they consisted of interspersed components that commonly included deciduous tropical woodlands and various more open areas, as well as a substantial body of water (lake, river, or floodplain), at least on a seasonal basis (Andrews 2020). This describes the habitat of *Ardipithecus* and most of the hominins that followed.

There has been some concern about the precise definition of "mosaic" (Reynolds et al. 2015). Some experts have claimed that mosaics are chimeras caused by hydraulic mixing of evidence from different places and levels (White et al. 2009). However, paleoecological reconstructions continue to refer to the mosaic concept (Su and Haile-Selassie 2022) or some equivalent descriptor such as "patchwork" (Magill et al. 2016).

Hominins varied through time and space in the ways that they exploited the mosaics. *Ardipithecus*, the earliest genus generally accepted as hominin, is considered to have been basically a woodland animal despite access to some more open areas (Simpson et al. 2019). *Australopithecus anamensis* apparently continued the open woodland existence (Fig. 2.1). Comparison of their numbers at different sites indicates that they thrived in the mixed environment, as opposed to wetter and more enclosed habitats (Bobe et al. 2020a, b; Manthi et al. 2020). As hominins evolved, they extended the adaptive envelope. Some continued in mosaics like their predecessors (Curran and Haile-Selassie 2016), but in varied temperature regimes (Su and Harrison 2015). Some, including *A. afarensis* and *Kenyanthropus*, seem to have ventured farther into open areas (Martin et al. 2020; Villasenor et al. 2020).

Early *Homo* continued the expansion into diverse environments, as demonstrated by two sites with archeological remains. The environment of one group, at Olduvai Gorge in Tanzania, dated to about 1.84 mya, was heterogeneous woodland.

Fig. 2.1 Contemporary woodland in a baboon habitat. (Photo by Glenn King. Manyara. Tanzania)

Biomarkers in the soil revealed a "patchwork" landscape in which hominins had access to woodland, spring-fed wetland, and adjacent grassland (Magill et al. 2016) or shrubland (Arráiz et al. 2017). A similar mixture of features, including riverine forest, continued down to about 1.3 mya (Dominguez-Rodrigo and Cobo-Sanchez 2017). Roughly contemporaneous with the Olduvai site, hominins lived in open grassland at Kanjera in Kenya (Plummer and Bishop 2016). The soil chemistry associated with archaeological remains provided information about photosynthesis systems, distinguishing grasses from other plants. The values indicate that the plant cover was more than 75% grass (Plummer et al. 2009). Antelope fauna indicate that grassy habitats were well represented throughout the region and not just located around the archeological site.

Baboon occupation of mosaic environments is illustrated by Gorongosa National Park in West Africa, which hosts interspersed forests, woodlands, grasslands, swamps, rivers, and a major lake that fluctuates with seasonal cycles. The park is home to 219 troops that seem to be a mix of chacma baboons and yellow baboons. For paleoanthropologist Rene Bobe, Gorongosa "brings to mind" the vegetation mosaics in which Pliocene and Pleistocene hominins evolved (Bobe et al. 2020a, b).

Probably much like *Ardipithecus*, some baboons live in wooded areas within mixed habitats. In Gombe National Park of Tanzania, for example, they coexist with the chimpanzees made famous by Jane Goodall. The Gombe baboons described by Ransom (1981) lived in a habitat with an abundance of trees, high food density, and streams for drinking water. Baboons studied by Rowell in Uganda (1966) similarly spent most of their time in forest. On the other hand, the grassland habitats that

Fig. 2.2 Baboon troop foraging in the open. (Photo by Curt Busse. Okavango, Botswana)

hominins eventually occupied are like those in which baboons were originally studied and in which the genus has had great success (Altmann and Altmann 1970; DeVore and Washburn 1963; Fig. 2.2).

Baboons match all of the environments that early hominins occupied, and perhaps more (Table 2.1). The genus has survived in virtually every kind of environment other than the central rainforest and the Sahara Desert (Altmann and Altmann 1970). This includes semidesert habitats in northeastern and southern Africa (Aldrich-Blake et al. 1971; Hamilton 1985; Kummer 1968) and montane habitats comparable to highland sites recently documented for *Australopithecus* (Mbua et al. 2016).

2.2.2 Environmental Changes

Comparison of hominin and baboon environments is not just a matter of static distribution. Hominins underwent two kinds of major long-term changes in their environment, giving rise to two major themes of explanation for early hominin adaptations: *variability selection* and *aridification* (Maslin et al. 2015). The former focuses on the ability to adjust to continuous or recurring changes in environmental conditions. The latter emphasizes the effects of hotter and drier habitats, varying from sparse woodlands to treeless grasslands. Variability and aridification are not mutually exclusive concepts, especially when applied to different time scales.

Table 2.1 Baboon environments

	Kinda baboons	Guinea baboons	Hamadryas baboons	Yellow baboons	Chacma baboons	Olive baboons
Rainforest						X
Guinean forest		X				
Dry forest		X				
Secondary forest		X				
Mangrove forest		X		X		
Gallery forest	X	X				X
Miombo woodland	X			X	X	X
Light woodland	X	X			X	X
Dense woodland			X		X	X
Montane forest			X			X
Grass meadow			X			
Swamp					X	X
Bush				X		
Shrub		X				
Scrub					X	X
Savanna		X	X		X	X
Steppe		X		X	X	
Semi-desert			X		X	X
Rocky desert			X			
Coastal			X	X	X	

Environments occupied by the six generally accepted baboon species, derived from descriptions in Mittermeier et al. (2013). Unlikely to be exhaustive

Eastern and southern Africa, the areas that have yielded the greatest quantities of evidence for hominin evolution, have seen "extreme" environmental variability during the last ten million years. This was caused by global climate shifts, local effects such as vulcanism, and lake basin dynamics resulting from such changes (Rocatti and Perez 2019). Hominin evolution began and continued through a period of complex environmental variation. African tropical regions alternated between forest/woodland habitats with high tree density and the expansion of savannas with varying degrees of tree cover. Variations occurred between regions and between localities.

Recognition of these changes led to an important theory of hominin evolution—that environmental variation was more important than any particular type of habitat (Maslin et al. 2015). It was posited that hominins probably underwent "variability selection." That is, natural selection favored the ability to respond to short-term and long-term environmental fluctuations, rather than adaptation to any particular environment (Potts 2013; Potts and Faith 2015).

Though fluctuation was clearly of great importance in hominin evolution, there is reason to think that drying trends and their vegetational consequences played an important role. Early hominins underwent cyclical aridification on local and regional scales and general aridification on a continental scale (Rocatti and Perez 2019). This resulted in larger areas of grassland within and adjacent to hominin habitats. Genus *Homo* emerged in East Africa at about 2.8 mya during a major drying trend and displayed increasing association with grassland habitats (Plummer and Bishop 2016; Robinson et al. 2017). Hominins adapted to the savannas that expanded and surrounded them over the course of millions of years (Plummer et al. 2009), though some continued to occupy woodland mosaics where such conditions were available (Magill et al. 2016). Hominins did not transition from woodland to savanna; they expanded into the savannas.

Baboons experienced the same kind of long-term Plio-Pleistocene changes as hominins, while taxa closely related to each of them went extinct. Extant baboons undergo short-term challenges comparable to those of early hominins, such as seasonal stress (Chowdhury et al. 2021; Johannes-Boyau et al. 2019). The yellow baboons of Amboseli in Kenya, for example, have survived "environmental change of a type and magnitude typical of … East African paleoenvironments" (Alberts and Altmann 2007: 282).

Alberts and Altmann (2007) considered baboons in general a "good model" for exploring the consequences of variability selection. Baboons display the responses predicted by the hypotheses. They are flexible in their locomotor system and social systems; they have a large brain in proportion to body size; and they are like humans and unlike most other primates in having adapted to a wide range of environments with little or no seasonality in their reproductive behavior.

Comparison among baboon species and populations can suggest which hominin behaviors could have stayed the same over the long term, and which had to change as they experienced environmental fluctuations and expanded into drier and more open habitats. Baboons, whether we consider one or a few species or the entire genus, are pertinent to the behavior of early hominins in any environment that they encountered.

2.3 Positional Behavior

Bipedal locomotion was a crucial adaptation at the beginning of hominin evolution. However, hominins retained considerable ability for arboreal movement for millions of years. Even modern humans can develop great facility in climbing.

2.3.1 Bipedalism and Terrestriality

Orrorin and *Sahelanthropus*, early possible hominins, display some debatable evidence of bipedalism (Boyle and Wood 2017). Stamos and Alemseged (2023) supported the bipedalism inference for these possible hominins, but characterized this aspect of their behavior as "primitive" in form and facultative in the sense that it was possible but relatively rare. These authors placed the same interpretation on the greater amount of evidence for *Ardipithecus*. Casenave and Kivell (2023) saw evidence for facultative bipedalism in the pelvis and legs of *Ardipithecus*. The rigidity of the midfoot and flexibility at the toes also signaled bipedalism. Prang (2019) viewed the *Ardipithecus* foot as similar to that of African apes, but considered the lengthened midfoot and reduced toes as indicating "propulsive capabilities associated with an early form of bipedalism."

Kozma et al. (2018) compared motion in humans, apes, and other primates to assess the functions of pelvic morphology and hip movement. They concluded that ape pelves permit enhanced climbing capability, but limitation of hip extension results in a crouched gait. Human pelves permit a greater degree of hip extension, which greatly improves walking economy, that is, distance traveled in relation to energy consumed. Application of these findings to fossil pelves led to the conclusion that *Australopithecus afarensis* and *A. africanus* had human-like hip extension and *Ardipithecus* was nearly human-like. They must have spent much of their time moving on the ground.

Since baboons are quadrupeds, the analogy with hominins in this case is primarily functional/ecological rather than anatomical/behavioral. Both taxa adapted to extensive activity on the ground. Baboons differ from most other primates in this regard, although there is a parallel with chimpanzees. The behavioral implications are profound, including foraging patterns (Chaps. 4 and 5) and responses to predators (Chap. 6).

It may also be significant that baboons are capable of limited bipedal posture and locomotion (Fig. 2.3). The reasons for this behavior in baboons might be pertinent to the origin of bipedal evolution in hominins, during a time long before *Ardipithecus*. Baboons may stand erect to gather food (Fig. 2.4) and they may walk or run bipedally to carry food . Early hominins might also have used erect posture to gather food and might have carried food to avoid competition with each other or danger from predators. Increased selection pressure along these lines could have favored anatomical changes.

On the premise that infant primates have a more diverse locomotor repertoire than adults, Druelle et al. (2017) theorized that locomotor development is a source of variation subject to natural selection. They observed six infant baboons at two different stages of development. During the same stage of development, the infants improved significantly in coordination between the hind limbs in spontaneous bipedal walking and in interlimb coordination in quadrupedal walking. The researchers hypothesized that neural networks underlying quadrupedal locomotion might also be employed to perform occasional bipedal walking. They inferred that a secondary locomotor mode experienced during infancy, as a byproduct of locomotor development, may lead to evolutionary innovation under appropriate selective pressures. Thus, a baboon analogy suggests an evolutionary pathway from quadrupedalism to bipedalism for hominins.

Fig. 2.3 Young chacma baboon in bipedal stance. (Photo by Curt Busse. Okavango, Botswana)

Fig. 2.4 Foraging chacma baboon stands bipedally to reach for tree branch. (Photo by Glenn King. On the road near Mkhuzi, South Africa)

2.3.2 Arboreality

While evolving bipedal capability for ground locomotion, hominins probably retained considerable ability for arboreal behavior. There is wide agreement on this point, but some debate as to what form(s) of behavior were involved. Table 2.1 summarizes some of the evidence and interpretations. Two obvious categories are vertical climbing and suspension. Selby and Lovejoy (2017) added the concept of clambering. Based on comparative evidence from gorillas and some New World monkeys, clambering was described as cautious movement through trees that makes equal use of all four limbs. All of these modes of locomotion have been attributed to *Ardipithecus*. The evidence from various species of *Australopithecus* suggests that substantial arboreal activity continued, but the balance between effective climbing and bipedal movement on the ground was shifting.

Some scientists have argued that arboreal traits in hominins later than *Ardipithecus* were relicts with no functional significance, because those hominins had become obligate or near-obligate terrestrial bipeds (e.g., Lovejoy 2009). Stamos and Alemseged (2023) responded that the system of postcranial traits in *Australopithecus* seems to have been relatively stable over the course of about two million years, contrasting with significant changes in craniodental morphology. They argued that the stability of climbing morphology in *A. afarensis* over such a long period is consistent with stabilizing natural selection, which indicates that arboreal behavior contributed to the fitness of these early hominins. Paradoxically, one case of individual fitness loss also provides evidence for arboreality in *A. afarensis*: perimortem fractures throughout the skeleton of "Lucy" are consistent with a vertical fall from a tree (Kappelman et al. 2016).

After a review, Casenave and Kivell (2023) concluded that *A. afarensis*, *A. prometheus*, and *A. sediba* present suites of anatomical features that differ from one another and suggest (subtly for some features and dramatically for others) that locomotor biomechanics were different for each taxon. Whatever the details, however, these and other hominins seem to have retained arboreal features for millions of years (although it is not clear just how frequent and significant arboreal behaviors were in any particular species) (Casenave and Kivell 2023).

Baboons are comparable to this varied hominin pattern of positional behavior. Although spending much of the day on the ground, they are agile in trees and on cliffs. These are the places where they find relative safety when confronted with predators during the day and when sleeping at night (Chap. 6). The fact that baboons often climb cliffs for sleeping, at least in some habitats, suggests that the term "arboreality" should be used loosely in relation to early hominins and perhaps replaced by a term such as *scando-terrestrial* (Chap. 1).

It is intriguing that Pavia's (2020) reconstruction of a South African hominin paleoenvironment included a possible cliff at the edge of an open grassland. Baboon behavior suggests the speculation that this provided local early *Homo* with a refuge, especially for sleeping at night. Chacma baboons in South Africa frequently use such cliffs for sleep. (Table 2.2)

Table 2.2 Arboreal features in early hominins

Ardipithecus
Anatomical features
Long forelimbs (Selby and Lovejoy 2017)
Long phalanges on hands and feet (Selby and Lovejoy 2017)
Long, curved manual and pedal phalanges (Stamos and Alemseged 2023)
Ape-like hip extension (Kozma et al. 2018)
Origin of hamstring muscles and related features of pelvis (Selby and Lovejoy 2017)
Locomotor interpretations
Suspensory behavior and vertical climbing (Prang et al. 2021)
Vertical climbing (Kozma et al. 2018)
Greater arboreal efficiency than *Australopithecus* (Stamos and Alemseged 2023)
Clambering (Simpson et al. 2019; Selby and Lovejoy 2017)
***Australopithecus* (various species)**
Humerus with mixture of orangutan-like and monkey-like features:
significant amount of suspensory behavior (Arias-Martorell 2018) and climbing (Melillo et al. 2021).
Hip extension different from *Ardipithecus*: reduced power in climbing (Kozma et al. 2018)
ape-like features in cochlea and semicircular canals: sense of balance needed for some arboreality (Beaudet et al. 2019)
Atlas bone (the first cervical vertebra) with "substantial similarities" to same feature in living apes.
Consistent with greater head mobility than in modern humans
Three-dimensional visual field adapted to arboreal behavior, especially vertical climbing.

2.4 Body Mass

Body size, usually considered in terms of weight or mass, is a basic feature of animals that affects almost every aspect of their biology, including (for example) locomotion, diet, energy requirements, social organization, and life history (Jungers et al. 2016; Grabowski and Jungers 2017). Variation in research methods has produced uncertainty as to the body mass of various early hominins (Grabowski et al. 2015; Ruff et al. 2020; Will et al. 2017). Differing approaches use different fossil bones as reference points. Material from the legs is preferable because they are weight-bearing. Commonly used are the talus from the ankle and the head of the femur, but these are not always available. No matter which fossil bones are used, extant models are necessary to formulate equations that relate bones to the mass of a living body. The usual models, chimpanzees and humans, can lead to significantly different results. The equations themselves represent differing mathematical approaches that can yield significantly different results.

The weights for hominin species presented in the first chapter (Table 1.1) were drawn from a survey of research available to Boyle and Wood (2017), with the intention of providing a comprehensive result as a basis for further discussion. More recent studies, summarized in Table 2.3, have more limited coverage. It is immediately apparent that the results from Jungers et al. (2016) and Will et al. (2017) are in close agreement with each other and with Boyle and Wood (2017). They suggest a range roughly 25–65 kg for early hominins. Ruff et al. (2020) produced outliers for several taxa by analyzing humeri rather than lower limb bones. However, the analysis of one individual by Simpson et al. (2019) obtained estimates ranging from about 50 to 63 kg from the femur and about 56 kg from the talus.

Recent discoveries have expanded the sample of *A. anamensis* to 74 individuals, including the first known postcranial remains of a small individual (Ward et al. 2020). Based on limb bones (humerus, radius, tibia, capitate, manual phalanx) and dental evidence, the largest individuals of this species were about the same size as the largest members of *A. afarensis*, that is, up to about 70 kg. The figures for *Homo habilis* fall within the loose parameters established for *Australopithecus*.

In relation to any of the varied assessments of hominin body mass, baboons are on the whole significantly smaller. The weight of most baboons falls into the range of 15–35 kilograms (Fischer et al. 2019). There is an overlap at the lower end, but the largest early hominins seem to have been twice the size of the largest extant baboons.

Aside from the overlap, there are reasons to regard early hominins and baboons as comparable in size. First, despite the differences between them, baboons and early hominins are medium-sized mammals. This makes for similar relations to other mammals. For two simple examples, both the hominins and the baboons were/are capable of killing hares for food and susceptible to being killed and eaten by leopards (Chaps. 5 and 6). Another reason for comparison is that both the hominins and baboons are large enough to cope with some of the same problems, such as making successful defenses against leopards.

Finally, baboons are among the very few primates that approach the size of the hominins and also occupy a comparable range of environments. Chimpanzees and orangutans are better size matches for early hominins. However, orangutans live largely arboreal lives in the tropical forests of Southeast Asia. Chimpanzees are

Table 2.3 Recent body mass estimates of hominin taxa

Taxon	Jungers et al. (2016)	Will et al. (2017)	Simpson et al. (2019)	Ruff et al. (2020)*
Ardipithecus	32 (*N* = 1)	32 (*N* = 1)	50–63	50, 45 (*N* = 1)
A. anamensis	46 (*N* = 1)	–	–	68, 55 (*N* = 1)
A. afarensis	41 (25–64)	40 (25–64)	–	30–75
A. africanus	31 (23–43)	31 (23–43)	–	40–60
A. sediba	26 (23–29)	27 (23–30)	–	40–41
H. habilis	34 (27–38)	48 (38–65)	–	–
H. ergaster	49 (29–64)	51 (32–68)	–	52–53

These are recent estimates that were not used in Table 1.1. Fewer taxa have been reexamined. The results overlap with the earlier work, but provide higher figures for all taxa except *Homo ergaster*. *All studies reference lower limbs, except for the use of humeri by Ruff et al. (2020). All measurements in kilograms, rounded to the nearest whole number. Mean given first, followed by range in parentheses, except where *N* = 1. *Australopithecus africanus* includes *A. "prometheus"* more like hominins and baboons in that they are largely terrestrial African primates that live in a wide range of environments. Nevertheless, they have not penetrated the full range of environments occupied by baboons and have had limited success in the more demanding ones. The maximum size of male mandrill monkeys is similar to orangutans and they are more terrestrial; however, they are largely limited to the tropical forests of Africa.

2.5 Sexual Dimorphism

Like most monkeys and apes, early hominins were sexually dimorphic in body mass. In some hominin species, the degree of sexual dimorphism in this trait may have approached that of baboons. Sexual dimorphism in the size of the canine teeth is also important in many primate species, including baboons. In contrast, sexual dimorphism in the canine teeth seems to have been unimportant in early hominins.

2.5.1 Sexual Dimorphism in Body Mass

Establishing the range of weight variation within a species is bound up with disputes about sexual dimorphism. One way to interpret a wide range of variation within a group of related fossils is that males and females are significantly different in size (gorillas are a familiar living example of such a species). Alternatively, the material in question may represent more than one species; this interpretation requires that other markers of species distinction be present.

Sex difference in early hominin size is the subject of a long-running and intense debate because sexual dimorphism has important social and ecological implications (Cassini 2020; Plavcan 2018). One common explanation is aggressive competition among males for mates, which in turn has implications for social organization. An important alternative is niche partitioning between males and females for access to resources, especially food, which alleviates conflict. A long-standing hypothesis that combines ecological and social factors is that males defend females and young against predators (Washburn and DeVore 1961). Whatever hypotheses are favored, there is general agreement that the issue of sexual dimorphism in size is important.

There is considerable disagreement about the degree of dimorphism in early hominins. Some researchers have concluded that early hominins in general were little different from modern humans in sexual dimorphism. Reno and Lovejoy (2015) argued that small sample sizes had exaggerated previous results and that their analysis of *A. afarensis* fossils demonstrated moderate dimorphism on the order of chimpanzees and living humans (see also Reno et al. 2010). They suggested that other studies indicated similar results for several other early hominin species.

However, numerous studies using various criteria have concluded that sexual dimorphism in early hominins was significantly greater than in modern humans and that our ancestors did not approach the modern condition until after the genus *Homo* had appeared. Recent discoveries of *A. anamensis* fossils seem to support the sexual dimorphism interpretation. A partial tibia, undoubtedly from an adult, is only 75% of the size of previously known one. Other than size, the fossils exhibit similar morphology, indicating membership in the same species. A similar range of variation exists between some of the largest and smallest *A. afarensis* tibias. Together with the dental data, the fossil sample suggests a similarly wide range of body sizes in these two *Australopithecus* species, possibly due to similar levels of sexual dimorphism (Ward et al. 2020). Using fossil footprints to reconstruct foot size and comparative data to infer body mass, Villmoare et al. (2019) concluded that *A. afarensis* sexual dimorphism was comparable to that of gorillas, that is, males about twice the size of females (see also Kimbel and Delezene 2009; Masao et al. 2016).

Reconstruction of hominin sex differences in body mass by Grabowski et al. (2015) suggested a steady decline in size dimorphism from *A. afarensis* to *H. erectus* and ultimately to *H. sapiens*. In their study of fossil footprints, Villmoare et al. (2019) concluded that sexual dimorphism was significantly reduced in *H. erectus* compared to earlier hominins, but was still greater than in modern humans. The drastically changing environmental and social circumstances of early hominins may well have selected for a drastic reduction in sexual dimorphism. Interpretation may be complicated further by variation across the fossil record. *Ardipithecus* seems to show minimal size dimorphism while dimorphism in *Australopithecus* species varies from "modest to strong" (Plavcan 2018).

Sexual dimorphism in body mass seems to have been substantial for at least some early hominin species. In some cases, this feature may have reached the point of males being twice the size of females. If this is correct, then baboons may be a better match for early hominins than are chimpanzees. Chimpanzee sexual dimorphism approximates that of modern humans while the rate of male/female mass in baboons varies from 1.55 to 2.20 among the six species (Fischer et al. 2019; see also Table 2.3). Variation among baboons is significant and seems to be related to important aspects of behavior (Petersdorf et al. 2019). Comparison among baboons species may contribute to an understanding of variations among hominins.

2.5.2 Sexual Dimorphism in Canine Teeth

Sexual dimorphism in the canine teeth is a vital difference between baboons and hominins because canine size has profound implications for ecology and social behavior. Suwa et al. (2021) emphasized the idea that reduction of the canine teeth "indicates a profound behavioral shift associated with comparatively weak levels of male aggression." The phenomenon also raises questions about the ability of early hominins to protect themselves against predators.

In brief, a reduction in canine size can be interpreted as a reduction in fighting ability. However, an alternative to large canine teeth for fighting is the use of *extrasomatic* (non-anatomical) weapons, that is, artifacts. This point will be explored further in relation to predator defense (Chap. 6). For now, the main implication is that large canine teeth do not eliminate baboons from consideration as models for early hominin ecology and behavior.

2.6 Summary and Discussion

Paleoanthropology provides direct evidence for the ecology and behavior of early hominins. However, it is limited to phenomena that leave physical remains. Living species augment this record. They also provide hypotheses pertaining to gaps in the record and to the long period before hominin behavior began to provide archeological remains.

The main rationale for comparison of early hominins with baboons is provided by several major functional similarities. First of all, both taxa are unusual among primates in their wide distribution across diverse habitats from woodlands to hot grasslands and cold highlands. As they spread across these habitats, both taxa adapted to drastic cycles of climate change with great flexibility in their physiology and behavior. Of very broad significance, flexibility in positional behavior allowed hominins and baboons to travel on the ground during most of each day while retaining the ability to climb trees and cliffs when necessary, especially for safer sleeping.

Body mass is a somewhat more problematic point of comparison. Early hominins were significantly larger than baboons, with a possible range of 25–75 kg. However, the baboon range of 15–35 kg overlaps with early hominins. Both taxa are medium sized compared to mammals in general and are relatively large compared to most other primates. Although a few other primates match the hominins in size more closely than do baboons, none of these species come close to occupying the range of environments shared by hominins and baboons. Baboons can suggest answers to questions about crucial features of hominin life, such as subsistence and danger, in any environment that the hominins encountered.

A controversial aspect of body mass is the degree of difference between males and females in a species. This sexual dimorphism varies across baboon species but is relatively large in all of them. Most analyses of early hominin fossils have concluded that sexual dimorphism in body size was substantial, perhaps even as great as in baboons. If this is correct, then baboons may provide better perspectives on a variety of issues than less dimorphic species such as chimpanzees. Size dimorphism decreased during the course of hominin evolution. Variation among baboon species in this characteristic may provide clues about the social and/or ecological factors involved.

A crucial difference between early hominins and baboons is the size and shape of the canine teeth. The canines of both sexes in hominins were hardly different from incisors in size and shape. In baboons, the canines in both sexes are long and sharp and this is especially the case in males. One interpretation is that canine reduction in hominins accompanied a reduction in aggressive behavior. An alternative is that hominins came to rely on extrasomatic weapons at an early date.

References

Alberts SC, Altmann J. The evolutionary past and the research future: environmental variation and life history flexibility in a primate lineage. In: Swedell L, Leigh SR, editors. Reproduction and fitness in baboons: Behavioral, ecological, and life history perspectives. New York: Springer; 2007. p. 277–303.

Aldrich-Blake FPG, Bunn TK, Dunbar RIM, Headley PM. Observations on baboons, *Papio anubis*, in an arid region in Ethiopia. Folia Primatol. 1971;15:1–35.

Altmann SA, Altmann J. Baboon ecology: African field research. Chicago: University of Chicago Press; 1970.

Andrews P. Last common ancestor of apes and humans: morphology and environment. Folia Primatol. 2020;91:122–48. https://doi.org/10.1159/000501557.

Arias-Martorell. The morphology and evolutionary history of the glenohumeral joint of hominoids: a review. Ecol Evol. 2018;9(1) https://doi.org/10.1002/ece3.4392.

Arráiz H, Barboni D, Ashley GM, et al. The FLK Zinj paleolandscape: reconstruction of a 1.84 Ma wooded habitat in the FLK Zinj-AMK-PTK-DS archaeological complex, Middle Bed I (Olduvai Gorge, Tanzania). Palaeogeog Palaeoclim Palaeoecol. 2017;488:9–21. https://doi.org/10.1016/j.palaeo.2017.04.025.

Beaudet A, Clarke RJ, Bruxelles L, et al. The bony labyrinth of StW 573 ("Little Foot"): implications for early hominin evolution and paleobiology. J Hum Evol. 2019;127(8) https://doi.org/10.1016/j.jhevol.2018.12.002.

Bobe R, Manthi FK, Ward CV, et al. The ecology of *Australopithecus anamensis* in the early Pliocene of Kanapoi, Kenya. J Hum Evol. 2020a;140 https://doi.org/10.1016/j.jhevol.2019.102717.

Bobe R, Martinez FI, Carvalho S. Primate adaptations and evolution in the Southern African Rift Valley. Evol Anthropol. 2020b;29(3) https://doi.org/10.1002/evan.21826.

Boyle EK, Wood B. Human evolutionary history. In: Kaas J, editor. Evolution of nervous systems. New York: Academic Press; 2017. p. 19–36.

Casenave M, Kivell TL. Challenges and perspectives on functional interpretations of australopith postcrania and the reconstruction of hominin locomotion. J Hum Evol. 2023;175:103304. https://doi.org/10.1016/j.jhevol.2022.103304.

Cassini MH. Sexual size dimorphism and sexual selection in primates: primate sexual size dimorphism and sexual selection. Mamm Rev. 2020;50(3) https://doi.org/10.1111/mam.12191.

Chowdhury S, Brown JL, Swedell L, et al. Costs of seasonality at a southern latitude: Behavioral endocrinology of female baboons in the Cape Peninsula of South Africa. Horm Behav. 2021;134(1):105020. https://doi.org/10.1016/j.yhbeh.2021.105020.

Clarke RJ, Pickering TR, Heaton JL, Kuman K. The earliest South African hominids. Annu Rev Anthropol. 2021;50:125–43. https://doi.org/10.1146/annurev-anthro-091619-124837.

Curran S, Haile-Selassie Y. Paleoecological reconstruction of hominin-bearing middle Pliocene localities at Woranso-Mille, Ethiopia. J Hum Evol. 2016;96:97–112. https://doi.org/10.1016/j.jhevol.2016.04.002.

DeVore I, Washburn SL. Baboon ecology and human evolution. In: Howell FC, Bourliere F, editors. African ecology and human evolution. Chicago: Aldine; 1963. p. 335–67.

Dominguez-Rodrigo M, Cobo-Sanchez L. The spatial patterning of the social organization of modern foraging *Homo sapiens*: a methodological approach for understanding social organization in prehistoric foragers. Palaeogeog Palaeoclim Palaeoecol. 2017;488(S7) https://doi.org/10.1016/j.palaeo.2017.06.008.

Druelle F, Aerts P, Berillon G. The origin of bipedality as the result of a developmental by-product: the case study of the olive baboon (*Papio anubis*). J Hum Evol. 2017;113:155–61. https://doi.org/10.1016/j.jhevol.2017.07.010.

Fischer J, Zinner D. Introduction to special issue: Frontiers in baboon research. J Hum Evol. 2020;146 https://doi.org/10.1016/j.jhevol.2020.102822.

Fischer J, Higham JP, Alberts S, et al. The natural history of model organisms: insights into the evolution of social systems and species from baboon studies. eLife Sci. 2019;8:e50989. https://doi.org/10.7554/eLife.50989.

Grabowski M, Jungers WL. Evidence of a chimpanzee-sized ancestor of humans but a gibbon-sized ancestor of apes. Nat Commun. 2017;8(1) https://doi.org/10.1038/s41467-017-00997-4.

Grabowski M, Hatala KG, Jungers WL, Richmond BG. Body mass estimates of hominin fossils and the evolution of human body size. J Hum Evol. 2015;85:75–93. https://doi.org/10.1016/j.jhevol.2015.05.005.

Hamilton WJ III. Demographic consequences of a food and water shortage to desert Chacma baboons, *Papio ursinus*. Intl J Primatol. 1985;6(5):451–62. https://doi.org/10.1007/BF02735570.

Hopper LM, Ross SR, editors. Chimpanzees in context. A comparative perspective on chimpanzee behavior, cognition, conservation, and welfare. Chicago: University Chicago Press; 2020.

Joannes-Boyau R, Adams JW, Austin C. Elemental signatures of *Australopithecus africanus* teeth reveal seasonal dietary stress. Nature. 2019;572(7767) https://doi.org/10.1038/s41586-019-1370-5.

Jungers WL, Grabowski M, Hatala KG, Richmond BG. The evolution of body size and shape in the human career. Phil Trans R Soc B Biol Sci. 2016;371:20150247. https://doi.org/10.1098/rstb.2015.0247.

Kappelman J, Ketcham RA, Pearce S, et al. Perimortem fractures in Lucy suggest mortality from fall out of tall tree. Nature. 2016;537:503e507.

Kimbel WH, Delezene L. "Lucy" redux: a review of research on *Australopithecus afarensis*. Am J Phys Anthropol. 2009;140(S49):2–8. https://doi.org/10.1002/ajpa.21183.

King GE. The once and future baboon: a source of analogies for earliest hominid adaptations. Conference. 2001; Am Assn Phys Anthropol.

King GE. Baboon perspectives on the ecology and behavior of early human ancestors. Proc Nat Acad Sci. 2022;119(45):e2116182119. https://doi.org/10.1073/pnas.2116182119.

Kozma EE, Webb NM, Harcourt-Smith WEH, et al. Hip extensor mechanics and the evolution of walking and climbing capabilities in humans, apes, and fossil hominins. Proc Natl Acad Sci USA. 2018;115(16):4134–9. https://doi.org/10.1073/pnas.1715120115.

Kummer H. Social organization of hamadryas baboons, a field study. Chicago: University Chicago Press; 1968.

Lovejoy O. Reexamining human origins in light of *Ardipithecus ramidus*. Science. 2009;327(5967):781.

Magill CR, Ashley GM, Dominguez-Rodrigo M, et al. Dietary options and behavior suggested by plant biomarker evidence in an early human habitat. Proc Natl Acad Sci USA. 2016;113(11):2874–9. https://doi.org/10.1073/pnas.1507055113.

Manthi F, Plavcan JM, Ward CV. Introduction to special issue. Kanapoi: Paleobiology of a Pliocene site in Kenya. J Hum Evol. 2020;140:102718. https://doi.org/10.1016/j.jhevol.2019.102718.

Martin JE, Tacail T, Braga J, et al. Calcium isotopic ecology of Turkana Basin hominins. Nat Commun. 2020;11:3587. https://doi.org/10.1038/s41467-020-17427-7.

Masao FT, Ichumbaki EB, Cherin M, et al. New footprints from Laetoli (Tanzania) provide evidence for marked body size variation in early hominins. eLife Sci. 2016;5:1–29. https://doi.org/10.7554/eLife.19568.

Maslin MA, Shultz S, Trauth MH. A synthesis of the theories and concepts of early human evolution. Phil Trans R Soc Lond B Biol Sci. 2015;370:20140064.

Mbua E, Kusaka S, Kunimatsu Y, et al. Kantis: a new *Australopithecus* site on the shoulders of the Rift Valley near Nairobi, Kenya. J Hum Evol. 2016;94:28–44. https://doi.org/10.1016/j.jhevol.2016.01.006.

Melillo S, Beotas LG, Saylor BZ, et al. New Pliocene hominin remains from the Leado Dido'a area of Woranso-Mille, Ethiopia. J Hum Evol. 2021;153:102956. https://doi.org/10.1016/j.jhevol.2021.102956.

Mittermeier RA, Rylands AB, Wilson DE, editors. Handbook of the mammals of the world – volume 3. Primates. Barcelona: Lynx Edicions; 2013. p. 660–5.

Muller MN, Wrangham RW, Pilbeam PR, editors. Chimpanzees and human evolution. Cambridge MA: Belknap Press (Harvard); 2017.

Pavia M. Palaeoenvironmental reconstruction of the Cradle of Humankind during the Plio-Pleistocene transition, inferred from the analysis of fossil birds from Member 2 of the hominin-bearing site of Kromdraai (Gauteng, South Africa). Quater Sci Rev. 2020;248:106532. https://doi.org/10.1016/j.quascirev.2020.106532.

Petersdorf M, Weyher A, Kamilar JM, et al. Sexual selection in the Kinda baboon. J Hum Evol. 2019;135(S37). https://doi.org/10.1016/j.jhevol.2019.06.006.

Plavcan JM. Sexual dimorphism in hominin ancestors. International Encyclopedia of Anthropology. Wiley; 2018.

Plummer TW, Bishop L. Oldowan hominin behavior and ecology at Kanjera South, Kenya. J Anthropol Sci. 2016;94:29–40. https://doi.org/10.4436/JASS.94033.

Plummer TW, Ditchfield PW, Bishop LC, et al. Oldest evidence of tool making hominins in a grassland-dominated ecosystem. PLoS One. 2009;4(9):e7199. https://doi.org/10.1371/journal.pone.0007199.

Potts R. Hominin evolution in settings of strong environmental variability. Quater Sci Rev. 2013;73:1–13. https://doi.org/10.1016/j.quascirev.2013.04.003.

Potts R, Faith JT. Alternating high and low climate variability: the context of natural selection and speciation in Plio-Pleistocene hominin evolution. J Hum Evol. 2015;87:5–20. https://doi.org/10.1016/j.jhevol.2015.06.014.

Prang TC. The African ape-like foot of *Ardipithecus ramidus* and its implications for the origin of bipedalism. eLife Sci. 2019;8:e44433. https://doi.org/10.7554/elife.44433.

Prang TC, Ramirez K, Grabowski M, Williams SA. *Ardipithecus* hand provides evidence that humans and chimpanzees evolved from an ancestor with suspensory adaptations. Sci Adv. 2021;7:eabf2474.

Ransom TW. Beach troop of the Gombe. Lewisburg PA: Bucknell University Press; 1981.

Reno PL, Lovejoy CO. From Lucy to Kadanuumuu: balanced analyses of *Australopithecus afarensis* assemblages confirm only moderate skeletal dimorphism. PeerJ. 2015;3(Suppl 49):e925. https://doi.org/10.7717/peerj.925.

Reno PL, McCollum MA, Meindi RS, Lovejoy O. An enlarged postcranial sample confirms *Australopithecus afarensis* dimorphism was similar to modern humans. Phil Trans R Soc B Biol Sci. 2010;365(1556):3355–63. https://doi.org/10.1098/rstb.2010.0086.

Reynolds SC, Wilkinson DM, Marston CG, O'Regan HJ. The 'mosaic habitat' concept in human evolution: past and present. Trans R Soc S Afr. 2015;70(1):57–69. https://doi.org/10.1080/0035919X.2015.

Robinson JR, Rowan J, Campisano CJ, et al. Late Pliocene environmental change during the transition from *Australopithecus* to *Homo*. Nat Ecol Evol. 2017;1(6):159. https://doi.org/10.1038/s41559-017-0159.

Rocatti G, Perez SI. The evolutionary radiation of hominids: a phylogenetic comparative study. Sci Rep. 2019;9:15267. https://doi.org/10.1038/s41598-019-51685-w.

Rowell T. Forest living baboons in Uganda. J Zool Lond. 1966;149:344–64.

Ruff CB, Squyres N, Junno JA. Body mass estimation in hominins from humeral articular dimensions. Am J Phys Anthropol. 2020;173(2) https://doi.org/10.1002/ajpa.24090.

Selby MS, Lovejoy O. Evolution of the hominoid scapula and its implications for earliest hominid locomotion. Am J Phys Anthropol. 2017;162(4) https://doi.org/10.1002/ajpa.23158.

Simpson SW, Levin NE, Quade J, et al. *Ardipithecus ramidus* postcrania from the Gona Project area, Afar Regional State, Ethiopia. J Hum Evol. 2019;129:1–45. https://doi.org/10.1016/j.jhevol.2018.12.005.

Smith JE, Swanson EM, Reed D, Holekamp KE. Evolution of cooperation among mammalian carnivores and its relevance to hominin evolution. Curr Anthropol. 2012;53(S6) https://doi.org/10.1086/667653.

Stamos PA, Alemseged Z. Hominin locomotion and evolution in the Late Miocene to Late Pliocene. J Hum Evol. 2023;178:103332. https://doi.org/10.1016/j.jhevol.2023.103332.

Su DF, Haile-Selassie Y. Mosaic habitats at Woranso-Mille (Ethiopia) during the Pliocene and implications for *Australopithecus* paleoecology and taxonomic diversity. J Hum Evol. 2022;163:103076. https://doi.org/10.1016/j.jhevol.2021.103076.

Su DF, Harrison T. The paleoecology of the Upper Laetolil Beds, Laetoli Tanzania: a review and synthesis. J Afr Earth Sci. 2015;101:405–19. https://doi.org/10.1016/j.jafrearsci.2014.09.019.

Suwa G, Sasaki T, Semaw S, et al. Canine sexual dimorphism in *Ardipithecus ramidus* was nearly human-like. Proc Natl Acad Sci. 2021;118(49):e2116630118. https://doi.org/10.1073/pnas.2116630118.

Villasenor A, Bobe R, Behrensmeyer AK. Middle Pliocene hominin distribution patterns in Eastern Africa. J Hum Evol. 2020;147:102856. https://doi.org/10.1016/j.jhevol.2020.102856.

Villmoare B, Hatala K, Jungers W. Sexual dimorphism in *Homo erectus* inferred from 1.5 Ma footprints near Ileret, Kenya. Sci Rep. 2019;9(1):7687. https://doi.org/10.1038/s41598-019-44060-2.

Ward CV, Plavcan JM, Manthi FK. New fossils of *Australopithecus anamensis* from Kanapoi, West Turkana, Kenya (2012–2015). J Hum Evol. 2020;140 https://doi.org/10.1016/j.jhevol.2017.07.008.

Washburn SL, DeVore I. The social life of baboons. Sci Am. 1961;204(6):62–71. https://doi.org/10.1038/scientificamerican0661-62.

White TD, Asfaw B, Beyene Y, et al. *Ardipithecus ramidus* and the paleobiology of early hominids. Science. 2009;326(5949):64–86. https://doi.org/10.1126/science.1175802.

Will M, Pablos A, Stock JT. Long-term patterns of body mass and stature evolution within the hominin lineage. R Soc Open Sci. 2017;4:171339. https://doi.org/10.1098/rsos.171339.

Yamagiwa J, Karczmarski L, editors. Primates and cetaceans, field research and conservation of complex mammalian societies. New York: Springer; 2014.

Chapter 3
Conceptual Frameworks and Criticisms

3.1 Introduction

Various attempts have been made to introduce intellectual concepts or systems that provide the hominin–baboon comparison with a rationale and consistency of approach. Most or all of these can be subsumed by the concepts of *model*, *analogy*, and *scenario*. Each of these will be examined in the following sections as to its applicability to hominin–baboon comparison.

Some critics have completely rejected baboon perspectives while others have argued that other primates are better sources. Discussion of these criticisms leads into a closer comparison of baboons with other primates with regard to reaching a better understanding of early hominins. The final part of the chapter discusses strengths and shortcomings in the criticisms and asserts that baboons offer more to reconstruction of early hominin behavior than any other primate except for chimpanzees. In some ways baboons are even more significant for that purpose than chimpanzees.

3.2 Models

The word "model" has been used widely but loosely, creating uncertainty about its meaning. Moore (1996) listed and discussed numerous synonyms, many of which have been employed in the study of primate behavior (Table 3.1). He distinguished between those that are potentially useful, if applied carefully, and those that contribute little or actually mislead. Boldfaced terms in Table 3.1 are those that are treated as useful in the present discussion. On the negative side, "just-so story" is a term frequently used by critics of models for behavior in extinct species. Derived from the title of children's stories by Rudyard Kipling, it implies that the model it maligns

G. E. King, *Baboon Perspectives on Early Human Ancestors*,
https://doi.org/10.1007/978-3-031-36769-4_3

Table 3.1 Meanings of the word "Model" in comparative studies (Based on Moore 1996)

Meanings for "Model"	Explanation/Example
Analogy	A relation based on similarity between attributes of two things
Model (noun)	Known member of an analogous pair
Model (verb)	Process of constructing an aspect of the unknown member of an analogous pair
Scenario	A broad, internally consistent "story" about a specific extinct taxon
A set of formal instructions	For example, mathematical representation of an idea or system
Sketch	Deliberately self-effacing term (intended to provoke thought without criticism[a])
Overview	Similar to sketch
Hypothesis	An idea amenable to testing; narrower than a scenario
Reconstruction (noun)	Synonymous with scenario, but sounds more precise (misleading, to be avoided)
Reconstruction (verb)	Process of using contemporary sources to hypothesize about aspects of the past[a]
Metaphor	Figure of speech intended to evoke a feeling of similarity; not analytically useful
"Theory"	Often, a formal-sounding synonym for "idea" or "suggestion"
"Just-so story"	A fairy tale; someone else's model with which one disagrees
A body of theory	For example, natural selection
A set of ideas with a central focus	For example, role of carnivory in human evolution
A direct analogy	For example, arborealism in hominins and apes
A world view	For example, scientific knowledge

[a]My interpretation of Moore and/or of the concept

is a fairy tale. Moore's addition of "someone else's model" to the definition is an apt dig at the rancor in some discussions of the topic. Perhaps the greatest problem with models is the word itself, because it seems to imply a comprehensive picture of the referent. This can lead to unrealistic expectations of numerous and detailed similarities.

Two kinds of models have been presented as relevant to reconstructing early hominin behavior through comparison with extant primates: *referential* and *conceptual*. Mathematical models (e.g., Dunbar 2001) can be considered a form of conceptual model because they conform to Moore's definition of a "set of formal instructions." Such models, however, are beyond the scope of this book.

3.2.1 Referential Models

Moore (1996) described referential models as using "real phenomena" to illuminate others that are more difficult to study. In this context, the "real phenomena" are extant primates; early hominins are the problematic target. Moore was most interested in chimpanzees as the extant model for hominins, but his general concepts apply equally well to baboon perspectives. This includes the ways in which referential models can be used:

1. Demonstration of behaviors that might have been possible for the problematic subjects, such as tool use by chimpanzees with materials that would leave no archeological remains.
2. Facilitating the incorporation of demographic factors into reconstruction. An example is the dispersal (migration) of individuals from one community to another, typically one sex or the other.
3. Suggesting important new ideas. For instance, Moore hypothesized that male–female differences in chimpanzee insectivory might be a model for the evolution of the sexual division of labor in hominins.
4. Generating detailed scenarios as a basis for testable hypotheses. This is illustrated by a scenario in which hominin females form kinship-based units for shared protection, child care, and food-getting (Swedell and Plummer 2019; Chap. 9 in this book).

I would add that referential models can provide evidence for both general patterns and particular activities. An example of such a general pattern is the scenario of an omnivorous diet, based on dental traits of extant species that might be found in the teeth of fossils. An example of a particular activity is digging up roots for food, which is performed by both chimpanzees and baboons (Chap. 4).

Following Susman (1987) and Potts (1987), Moore (1996) advocated a referential approach that is based on intraspecific variability in order to model diachronic change in the hominin lineage. He compared "forest" and "savanna" chimpanzees for this purpose. At the time he wrote, this was largely limited to framing questions to be answered. Much more is known now, including information about limits on chimpanzee exploitation of savanna biomes (Marchant et al. 2020). Baboons, in contrast, encompass the full range of environments experienced by hominins (this important point has been noted previously and will be noted again). As for using variation, as in Moore's modeling, this can be derived from the differences among the six baboon species as well as variation within each of the species. Even Guinea baboons, found in a relatively small region, have populations in forests and in the arid Sahel (Zinner et al. 2021; Chap. 1 in this book).

There is no single baboon model (King 2022), except in the most general terms (e.g., terrestriality, sexual dimorphism). This became apparent early in the development of field research, with reports of baboons in gallery forests and woodlands rather than savannas (Ransom 1981; Rowell 1966). Now we have six different species of baboons living in an even wider range of environments, including, for example, chacma baboons in the swampy Okavango Delta in Botswana (Cheney and Seyfarth 2007) and hamadryas baboons in the semideserts of lowland Ethiopia (Schreier and Swedell 2012).

Whatever referential model one may bring to bear, it is vital to specify its characteristics and to be clear about the specific behavior pattern(s) under study (cf. Moore 1996). Given the variation among baboons, it is also important to specify the particular baboon species or group of species that is the basis for the model. Throughout this book, an attempt has been made to name the baboon species that were the subjects of each cited study or set of studies.

3.2.2 Conceptual Models

Conceptual models are constructed very differently from referential models and are held by some to be superior (Tooby and DeVore 1987). They are based on theoretical analysis rather than specific real-life subjects. Conceptual models draw on general principles from behavioral ecology and other fields in order to specify theoretical relationships pertinent to a particular situation. The process is based on the following premises: (1) the gene is the unit of selection, and (2) animals act like strategists promoting their *inclusive fitness* (Tooby and DeVore 1987). This concept refers to the idea that animals behave in ways that support their relatives, a phenomenon caused by the probability that relatives share the same genes for the same kind of "altruistic" behavior.

Conceptual models tend to be very speculative, which heightens the risk of being refuted by new evidence (Stanford 2012). Tooby and DeVore (1987), for example, made the theory-based assumption that social animals must emigrate from their natal groups in order to avoid inbreeding depression. Subsequent fieldwork demonstrated that neither sex consistently emigrated in a number of primate species (Moore 1996).

Though placed in opposition by some scientists, the conceptual and referential approaches overlap and can be complementary (Stanford 2012). Strum and Mitchell (1987) created a conceptual model for early hominins that concluded with a description that sounded very much like baboon behavior (King 2001). They acknowledged that looking at baboons was a good "first step" toward reconstructing early hominin behavior. An approach that includes both referential and conceptual models is likely to be the most effective, because reciprocal refinement can result in a better understanding of the problem at hand (Moore 1996).

Some primatologists and others continue to denigrate and reject referential models, including baboons. Others continue to use them effectively, as did many of the sources for this book. Referential models have derived new life from recently developed knowledge of diversity in the genus *Papio*. Recent findings about baboons and hominins present us with a series of challenging choices rather than a unitary model.

3.3 Analogies

Analogies provide a flexible framework for dealing with specific behavior patterns that have been chosen for investigation. Though analogies in a general sense are sometimes considered a kind of model (Moore 1996; Table 3.1), it is best to distinguish them because analogies can be defined in relatively rigorous way. Analogy in the strict sense refers to a likeness of relationship rather than a simple resemblance (Jolly 2013). It postulates that a functional connection in one case parallels a functional connection in the other. In reconstructing the prehistoric past the extant case is used to formulate a predictive hypothesis about the past case.

The functional connection between predation and escape is a simple but important example. The presence of dangerous predators (A) leads extant primates (e.g., baboons) to climb trees to escape them (functional connection B), from which we hypothesize that the presence of dangerous predators in the paleontological record (C) caused prehistoric primates (e.g., hominins) to seek refuge in trees (functional connection D).

This example illustrates another aspect of strict analogies: they delimit the boundaries of comparison (Jolly 2013). The analogy in question is limited to the predator–prey relationship. It can be narrowed further to relations with particular types of predators, such as terrestrial mammals or raptors (Chap. 6). This use of analogies falls into the "referential" category, but it does not exclude amplification by the principles associated with conceptual models.

The concept of strict analogy provides useful guidelines for hominin-baboon comparisons, but there is no reason to be rigidly bound by it. More general analogies (the looser, more common usage of the term) can also provide useful ideas. For example, baboons and hominins are distinctive among primates in combining the status of medium-sized mammals and a high level of sociality with life in diverse habitats in Africa. Given these parallels, diversity in the social organization of baboons suggests alternative possibilities for patterns of social organization in early hominins, such as the nature of long-term relationships between particular males and females (Chap. 8). Such loose analogies are heuristic and do not in themselves produce conclusions with any degree of probability. However, they can stimulate the formation of testable hypotheses.

The emphasis in this book is on carefully selected analogies rather than comprehensive models. Increasing knowledge of variation among baboons (Fischer and Zinner 2020) makes this approach both necessary and profitable. The basis for selection of particular analogies is the need to address particular problems in understanding early hominin ecology and behavior.

3.3.1 Scenarios

Productive thought can also come from broader and more dynamic settings of the kind often termed *scenarios*. These might be considered extended analogies of the loose variety (i.e., without precisely delimited boundaries). In the explication by Greene (2017), scenarios summarize evolutionary patterns and processes in a phylogenetic context. They are constructed by (1) describing *phenotypes*, the physical characteristics of an organism, *including behavior*; (2) inferring the evolutionary processes that produced those characteristics; and (3) integrating those components with ecological and other ancillary data.

At their best, such evolutionary scenarios are "factually dense narratives" with the following characteristics: they (1) contain no known falsehoods; (2) are transparent as to their empirical and methodological shortcomings; (3) can be rejected based on new discoveries, and (4) their potential ideological pitfalls are flagged for scrutiny. If constructed carefully and regarded skeptically, such scenarios can synthesize knowledge, inspire research (Greene 2017), and lead to testable hypotheses (Moore 1996).

Strum (2019) described a scenario for the social behavior of male troop-living olive baboons that she observed. They did not have a stable dominance hierarchy and some displayed social intelligence by using nonaggressive strategies in competition with other troop members. Bonds with females and infants were part of this suite of behaviors. Strum emphasized that recognition of this complex system (only partially described here) required the integration of her quantitative data with long-term qualitative observations. This is a "factually dense narrative" that can be expanded into a phylogenetic origin and applied to early hominins, taking account of "ancillary data" for baboon ecology and hominin paleoecology.

3.4 Criticism and Papiophobia

During the 1960s Sherwood Washburn inspired many people to study baboons as a source of hypotheses about hominin adaptations to living on the ground and in the relatively open habitats loosely called savannas (DeVore and Washburn 1963; Washburn and Devore 1961). The resultant "baboon model" once pervaded accounts of hominin evolution and then, subjected to severe criticism, its influence declined. Some pronounced its demise. By the year 2000 a popular textbook said that the model was seldom used and that baboons are only of interest for illustrating the variability of primate behavior. This conclusion has been overturned by abundant information from recent research on both baboons and early hominins, as well as increasingly incisive thought about how the baboon data pertain to early hominins. In the following discussion of criticisms, the term "model" will be used loosely because the subjects of the discussion have used the term in various ways.

Some objections to the baboon model were never valid, while others were resolved by increasing knowledge of both baboons and early hominins. It is worth considering some longstanding arguments for two reasons. One is to warn against perpetuation of past errors. Though overt criticism seems to have subsided, baboons are now ignored in some contexts where they are obviously relevant (e.g., Newson and Richerson 2021, discussion in Chap. 9). The other reason to consider criticisms is to distinguish those that raise valid issues and learn from them. These reasons correspond roughly to a division of critical arguments into two general categories: absolutist (baboons are irrelevant) and relativist (other species are better). Some absolutist arguments are so strained that I think of them as *papiophobia*. The relativist category is the one that contains useful critiques.

3.4.1 Absolutist Arguments

According to these arguments, baboon perspectives are useless or misleading and should be entirely dismissed. Some absolutist arguments were based on logical confusions of one kind or another, such as the unwarranted alignment of baboons with other models for human evolution. In one version of this fallacy, baboons were conflated with human hunter-gatherers. A writer described the "basic formula" of the baboon model as a "baboon type" society with hunting added (Fedigan 1982). The author then listed and criticized 15 alleged traits of the "hunting way of life" before returning to baboons. Although baboon and hunter-gatherer models have sometimes been used in conjunction, there is no logical necessity for doing so. The two models have distinct sources and apply to different stages of hominin evolution.

Some critiques of baboon–hominin comparison confused models with inferences: baboon analogies were rejected because perceived conclusions were considered wrong. One such text asserted that the baboon model should be dismissed because there was no evidence that a terrestrial primate in open country is subject to greater predator pressure. Assessment of predator pressure in open country (or any other habitat) is an inference drawn from a model; it is not part of the method itself. The fact is that ongoing research shows that open country is more dangerous and that baboons respond in highly patterned ways (Chap. 6).

Another example of a misused inference is the idea that any baboon model is wrong because males do not always defend the troop. Specifically, the report by Rowell (1966) was cited as to the fact that the presence of predators resulted in male baboons outrunning others for the trees. Not mentioned by the critics was the fact, also reported by Rowell, that the only large predators that approached baboons in her study area were humans. Most baboons will run from humans, who have hunted them for decades or centuries or millennia (depending on the area in question). Male baboons, however, often confront other large predators such as leopards and cheetahs (Altmann and Altmann 1970; Chap. 6 in this book). The baboon model required modification, not rejection.

According to Tanner (1981:19–20), "the earlier picture of savanna baboon social life was so incomplete as to be misleading." A footnote informed the reader that this was because the importance of females was neglected. Some absolutist objections were more overtly based on ideology. For example, the baboon model was criticized as "androcentric" (Martin and Voorhies 1975). If androcentrism means a form of prejudice that distorts the data, then the problem is in the researchers rather than the subjects. The remedy is better research rather than rejection of the model. If objections are raised because of distaste for established facts (such as male dominance or male aggression against females), this is not science.

The criticism about neglecting females had some validity with regard to early research on baboon behavior. However, decades of fieldwork by more diverse investigators have provided us with an enriched understanding of baboons that accords females a key role in social life and ecology (e.g., Strum 1987). This has led to more sophisticated models of early hominin behavior (e.g., Strum 2019; Swedell and Plummer 2019).

In some cases, critics contradicted each other. The first comparisons between baboons and early hominins emphasized savanna adaptations, so some critics portrayed baboons as forest animals that occasionally ventured into grasslands (Martin and Voorhies 1975). Put simply, hominins were savanna animals while baboons were forest animals. When paleoanthropology provided evidence that early hominins engaged in arboreal locomotion in wooded habitats, baboon analogies were dismissed because they could only tell us about "adapting to open country" (Susman 1987). Hominins were forest animals while baboons were savanna animals. In short, the baboons were too forest for some critics and too savanna for others.

Both formulations were simplistic with regard to both baboons and early hominins (Chap. 2). The earliest hominins may have lived in relatively closed woodlands, but never in dense rainforest or "jungle." Their successors occupied mixed habitats that included deciduous woodlands and more open areas. Expanding across Africa, hominins encountered major geographic variation as well as diverse local habitats. During millions of years of existence, they were subjected to major environmental fluctuations and a general drying trend. Baboons, also living in Africa, though later in time, experienced much the same conditions.

Some critics contradicted themselves. Tanner (1981) rejected the baboon model because it had allegedly not been compared to other "savanna" monkeys. Her example was patas monkeys, which deal with predators by hiding or running. This is the opposite of the chimpanzee model that she favored, from which she inferred that both male and female hominins deterred predators by throwing things at them. Baboons, like chimpanzees, are capable of active defense against predators. Baboons, like chimpanzees, live in relatively large groups that make this possible. Patas monkeys live in smaller groups with a single adult male in each one. Furthermore, while Tanner rejected the baboon model because of "extreme" sexual dimorphism, the same degree of sexual dimorphism is found in patas monkeys.

The argument about sexual dimorphism, setting aside spurious comparisons, actually had some potential validity. As discussed in Chap. 2, sexual dimorphism in body mass is substantial in all baboon species, and males can be twice the size of females in one or two species. Human size dimorphism, on the other hand, is relatively minor. Some researchers consider early hominins to be comparable to modern humans in this regard, but many view the fossil record as showing much greater dimorphism in early hominins than in modern humans.

It is also relevant that there is significant variation in sexual dimorphism among baboon species. The ratio of male to female mass can be as little as 1.5 (for modern humans the figure is about 1.15). Explanations for variation among baboon species may shed light on factors in early hominin evolution. Evolutionary *changes* in hominin sexual dimorphism might be modeled by variation across baboon species (cf. Moore 1996).

Finally, some absolutists belabored the obvious—that baboons are not close relatives of the hominin lineage. Such assertions refuse to recognize the importance of both homology and analogy in evolution. Without going into theoretical complexities, these concepts refer to the fact that some similarities between species are related to their common ancestry and other similarities are due to adaptation to similar circumstances. A classic example of the latter is the structurally different wings that allow both bats and birds to fly.

3.4.2 Relativist Arguments

Relativist critiques are more important than absolutist because they are more likely to lead to constructive discussions of the best strategies for understanding early hominins. Relativists argued that other primates were equal to or better than baboons for the reconstruction of early hominin behavior. In examining these alternatives, it should be borne in mind that inferences from different primates are not always in conflict with one another. In some cases, they are mutually reinforcing, pointing to the same or similar conclusions. In other instances, they are complementary, providing different components of the answer to a question about early hominins.

Chimpanzees

Many referential models and analogies for early hominins are based on chimpanzees, because of the common ancestry of the two lineages. This includes two approaches. One considers the possibility of homologies shared by chimpanzees and humans, derived from the LCA. Chimpanzees, for instance, provide important suggestions regarding the origin of vital human features such as tool use and culture (Boesch 2012). Such models postulate that hominins and chimpanzees retained the same ancestral traits and/or evolved in the same direction because of those traits (Andrews 2020; Moore 1996). Chimpanzees are certainly of interest in the reconstruction of early hominin ecology due to the phylogenetic connection and also the similarity in brain size. This is a sound approach and baboon perspectives on hominins must continuously adjust to its findings.

Less compelling is the argument that some chimpanzee populations are models for early hominins because they are "savanna" dwellers (Moore 1996). The term is sometimes used loosely and may simply suggest that the habitat of a given population is not dominated by forest cover (Marchant et al. 2020). Recent studies and reviews have attempted to treat the subject with more rigor. Van Leeuwen et al. (2020) reviewed 43 chimpanzee field research sites across equatorial Africa. Nine of these were characterized as "savanna" sites, based on the terminology of the researchers or their description of the vegetation cover of the landscape. These habitats are described as "hot, dry and open, dominated by woodland and grassland vegetation types, and with minimal forest cover." The common quantified features that separated all of these sites from forests were annual rainfall of less than 1360 mm/year and forest cover of less than 12.5%.

As noted by Hunt and McGrew (2002), such habitats may include woodland with the grassland and are often labeled mosaic. Fongoli, for example, is a "savanna-mosaic" (Wessling et al. 2018b) or "savanna-woodland mosaic" (Wessling et al. 2018a). Semliki is simply a "mosaic" that includes forest, wooded grassland, and open grassland (Hamilton et al. 2019). Chimpanzees in these habitats display low population densities and may spend relatively little time in the most open parts of their range. The more open habitats represent some of the extremes of where chimpanzees can persist—occupying the biogeographical edges of the species' distribution (Wessling et al. 2020). Many of these groups show signs of environmental stress in relation to heat and dehydration, and at least some of these chimpanzees seem reluctant to exploit the grassland parts of their habitats. One indication of stress is the frequency of thermoregulatory behaviors such as cave use in the hottest and driest landscapes (Lindshield et al. 2021).

Observations at Semliki "indicate that even in open or mosaic habitats … chimpanzees remain forest primates, utilizing primarily forest resources and only a minor subset of savanna resources" (Hamilton et al. 2019: 560). These chimpanzees harvest fruits from open areas close to the forest edge, but they are "uneasy and fearful" when they do so. Olive baboons were also observed at Semliki, by the same researchers, and the baboons did not display the same "aversion" to the open parts of the environment as did other monkeys and the chimpanzees.

Baboons have gone beyond grasslands to survive in deserts and semideserts. Table 3.2 compares the extremes of chimpanzee "savanna" environments with some

Table 3.2 Chimpanzees and baboons in arid habitats

Species	Location	Rainfall	Dry season(s)	Temperature	Vegetation
Chimpanzees	9 locations	Lowest mean annual 750 mm	Longest 7 months	Highest mean/ann. 29 °C	Savanna-forest mosaic
Olive baboons	Laikipia Plateau Kenya	Mean annual ~42 mm	Two seasons total 7 months	Highest daily temp 34 °C	Dry savanna with gallery forests
Chacma baboons	Tsaobis Namibia	Mean annual ~123 mm	Approaches 8 months	Highest *shade* temp ~ 40 °C	Riparian woodland and desert scrub
Hamadryas baboons	Filoha Ethiopia	?	?	Mean daily (1y) 31.5 °C high ~35 °C	*Acacia* scrub, open grassland

This table compares the extremes of "savanna" chimpanzee habitats (Van Leeuwen et al. 2020) with figures from two studies of baboons in arid habitats: olive baboons (Musyoki and Strum 2016), chacma baboons (King et al. 2011), and hamadryas baboons (Schreier and Swedell 2012). The available figures are not strictly comparable, but they strongly suggest the adaptive gap between "savanna" chimpanzees and baboons that have penetrated more arid habitats. The text provides more detail selected examples of baboons in open country. A much more detailed comparison of baboon ecologies can be found in Zinner et al. (2021).

In relation to open country and early hominins, an important parallel between chimpanzees and baboons is susceptibility to predation despite their size. This issue provides an example of complementary hypotheses from the two sources. Male baboons have large canine teeth, which can be used to fight predators, but early hominins did not have such dentition. Nevertheless, our ancestors could have become effective defenders by substituting weapons for teeth. Chimpanzee evidence shows that they might have used clubs and spear-like sharpened sticks (Chap. 6).

This last point brings up the assertion by Tanner (1981:19) that "the transitional hominins of both sexes could doubtless throw branches and leaves at potential predators at least as effectively as chimpanzees today." First, this statement is unreferenced and unlikely, especially as regards deterring a predator like a leopard by throwing leaves. Second, this is an ideological response to the idea that males defend females. Female chimpanzees may be just as good at throwing things as males, but the idea that this applies to early hominins depends on the assumption that hominin sexual dimorphism in body mass was no greater than the relatively small difference in chimpanzees. To most paleoanthropologists the fossil record tells a different story (Chap. 2).

Other Apes

Several scientists have suggested analogies for early hominins from gorillas. Macho (2018), for example, drew on contrasts between gorilla populations. The reliable vegetation available to mountain gorillas allows for accelerated development of the young, which seems to be related to relatively small brains. Lowland gorillas feed mainly on fruit, which is a rich but unreliable source nutrition. A slower rate of growth reduces the risk of starvation (presumably because the smaller body requires less). The brain continues to grow after weaning and is ultimately larger than that of a mountain gorilla. Gorillas are even more terrestrial than chimpanzees and more like hominins in that respect. However, as discussed earlier, body mass has a powerful effect on many aspects of life and gorillas are far larger than chimpanzees or early hominins. The size of gorillas imposes very different limits on their biology.

Following Elton (2006), Macho acknowledged that baboons are good ecological models for early hominins because they are eclectic omnivores that live in large groups, some of which are multilevel societies. Along with hominins, snub-nosed monkeys, and Japanese macaques, baboons are the only (relatively) large-bodied primates to survive at higher latitudes where environments are more seasonal and demanding (northern Asia in the case of macaques).

Orangutan facility with tools may represent homologies among the great apes (including humans). However, ecological relevance to early hominins is limited because these abilities are mainly displayed in captive and rehabilitant settings. In their natural lives, most orangutans are more arboreal than early hominins are likely to have been. They also tend toward a relatively solitary existence, especially adult males. Orangutans are pertinent to the hypothesis of arboreal bipedalism in hominins (Crompton et al. 2010), but this tells us little or nothing about the functions and/or selective influences on terrestrial bipedalism. The other Asian apes, the gibbons, live in small groups that have been compared to human families. However, these groups tend to be territorial and lack the larger community that is characteristic of humans, presumably early hominins, and baboons (Chap. 7).

Monkeys Other Than Baboons

Monkeys other than baboons have also been proposed as models for early hominins. The case of the patas monkeys was discussed above. Geladas (*Theropithecus gelada*) are a more appropriate model for hominins (e.g., Jolly 1970). They are relatively large, ground-foraging primates that associate in large groups. However, in contrast to baboons, geladas are limited to highland regions of northeastern Africa. They provide some analogies for the reconstruction of early hominin traits; however, many of these, such as multilevel societies and consumption of grasses, are also found in baboons.

Another genus that has been compared to hominins is *Chlorocebus*, which includes the well-known vervet and closely related species. Baboons and early hominins both differ from *Chlorocebus* in spanning diverse environments, including very open ones. Vervets and their kin are creatures of the forest and woodland fringes. According to Altmann (1998), baboons differ from these taxa in that baboons "stride unhesitatingly into the open from the cover of the woodland." They walk long distances, often without shade or water, with "relative immunity" to predators. As one result of these capabilities, they consume a much wider variety of foods than vervets. These and other characteristics of baboons are related to their size. While baboons fall short of hominin size, vervets are far behind both with a maximum weight of about 6 kg. Alberts and Altmann (2007) reported a crisis at Amboseli, Kenya, in which baboons responded well to deteriorating conditions while vervets experienced a substantial population decline.

Comparison with All Primates

The order Primates, containing hundreds of species, offers many possibilities for comparison with hominins. For some purposes, the whole order can be considered (Strier 2001). Common features at this level are very generalized, but nonetheless important. For example, hominins probably lived in social groups from the beginning of their evolution as almost all primate species do today.

On the other hand, generalizations that encompass the entire primate order can tell us little or nothing about the consequences of living on the ground in relatively open habitats as our ancestors eventually did and baboons do. This is because the great majority of primates are forest dwellers and primarily arboreal.

3.5 Summary and Discussion

Baboons have often been characterized as a "model" for early hominin behavior, a term that has caused some confusion because of multiple definitions and theoretical disagreements. The most commonly used kind of model has been, and continues to be, the referential model. It uses a real phenomenon (in this case, baboon behavior) to illuminate another phenomenon that is more difficult to study (early hominins). Referential models can demonstrate the possibility of certain behaviors, facilitate the use of demographic factors in reconstruction, suggest important new ideas, and generate scenarios.

Conceptual models are more theoretical and complex. They postulate that animals behave like strategists to maximize their inclusive fitness. Inclusive fitness refers to behavior that promotes the survival and reproductive success of relatives as well as the individual. Conceptual models produce more sophisticated results, but tend to be more speculative than referential models and more subject to empirical contradictions. The two types of models are not mutually exclusive and can be used to check on one another. There is no single baboon model. Six different species, and variation within the species, provide many possibilities.

Analogies provide a flexible framework for dealing with specific behavior patterns that are chosen for investigation. Analogy can be defined rigorously as a likeness of relationship, that is, parallel functional connections in two different cases. Here, the analogies have to do with understanding the factors connected with a selected behavior pattern in extant primates and the possible application of these relationships to particular early hominins. More loosely formulated similarities can be useful for heuristic purposes and may be the first step toward constructing testable hypotheses.

Scenarios are broader than analogies but more constrained than comprehensive models. If constructed carefully, with assumptions made explicit, they can be useful for the organization and synthesis of knowledge and as an inspiration for further research. They should integrate behavior with underlying evolutionary processes and with ecological data.

Criticisms of the hominin–baboon comparison can be roughly divided into two categories. Absolutists have completely rejected the idea, while relativists have postulated that other primates are better sources of information and hypotheses. Some absolutist arguments were based on logical confusions and misused inferences. In some instances, the critics contradicted each other or themselves.

Many relativist critiques offered constructive criticism of baboon perspectives and the possibility of coordinated effort. Chimpanzees are often cited as the best model for early hominins, especially with reference to their common ancestry. This is a sound approach and baboon perspectives must be adjusted to its findings. Ecological comparison of chimpanzees with early hominins is also of value, but more limited in possibilities. This is because "savanna" chimpanzees are mostly occupants of mosaic habitats in which they may minimize their use of open areas and may suffer stresses not found in other chimpanzees. In contrast, baboons have successfully colonized the full range of environments that are manifested in the early hominin record. This includes habitats that are hotter and drier than any that chimpanzees are known to occupy.

Other apes are also close relatives of hominins and can provide some useful analogies. However, the possibilities are more limited. Gorillas are huge, and body mass has a powerful effect on many aspects of life. Orangutans are found only in Asian biomes, tend to be more arboreal than baboons or early hominins, and tend to be solitary rather than social (especially adult males). Gibbons are highly arboreal Asian apes. They live in small groups with a superficial resemblance to human families, but which lack the community context that is characteristic of baboons, chimpanzees, and hominins.

Monkeys other than baboons are sometimes compared to early hominins. The most noteworthy are geladas and the *Chlorocebus* group (vervets and closely related species). However, geladas are limited to the highlands of northeastern Africa. Vervets and their relatives are much smaller than baboons and, like chimpanzees, are only tentative users of open areas.

Chimpanzees are the only primates that rival baboons with regard to providing information relevant to reconstruction of early hominin behavior. Chimpanzees have the vital status of having the last common ancestor with hominins. However, baboons have many advantages over chimpanzees in helping us to understand the problems and possible adaptations of early hominins as they expanded into hotter, drier, and increasingly open habitats.

References

Alberts SC, Altmann J. The evolutionary past and the research future: environmental variation and life history flexibility in a primate lineage. In: Swedell L, Leigh SR, editors. Reproduction and fitness in baboons: Behavioral, ecological, and life history perspectives. New York: Springer; 2007. p. 277–303.

Altmann S. Foraging for survival, yearling baboons in Africa. Chicago: University Chicago Press; 1998.

Altmann SA, Altmann J. Baboon ecology: African field research. Chicago: University of Chicago Press; 1970.

Andrews P. Last common ancestor of apes and humans: morphology and environment. Folia Primatol. 2020;91:122–48. https://doi.org/10.1159/000501557.

Boesch C. Wild cultures: a comparison between chimpanzee and human cultures. Cambridge, UK: Cambridge University Press; 2012.

Cheney DL, Seyfarth RM. Baboon metaphysics: the evolution of a social mind. Chicago: University of Chicago Press; 2007.

Crompton RH, Sellers WI, Thorpe SKS. Arboreality, terrestriality and bipedalism. Phil Trans R Soc Lond B Biol Sci. 2010;365:3301–14.

DeVore I, Washburn SL. Baboon ecology and human evolution. In: Howell FC, Bourliere F, editors. African ecology and human evolution. Chicago: Aldine; 1963. p. 335–67.

Dunbar R. Brains on two legs: Group size and the evolution of intelligence. IN de Waal FBM, editor. Tree of origin, what primate behavior can tell us about human social evolution. Harvard University Press; 2001. p. 173–191.

Elton S. Forty years on and still going strong: the use of hominoid-cercopithecoid comparisons in paleoanthropology. J R Anthropol Inst. 2006;12:19–38.

Fedigan LM. Primate paradigms: sex roles and social bonds. Montreal: Eden Press; 1982.

Fischer J, Zinner D. Introduction to special issue: Frontiers in baboon research. J Hum Evol. 2020;146 https://doi.org/10.1016/j.jhevol.2020.102822.

Greene HW. Evolutionary scenarios and primate natural history. Am Nat. 2017;190(S):S69.

Hamilton M, Nelson SV, Fernandez DP, Hunt KD. Detecting riparian habitat preferences in "savanna" chimpanzees and associated fauna with strontium isotope ratios: implications for reconstructing habitat use by the chimpanzee-human last common ancestor. Am J Phys Anthropol. 2019;170:551–64. https://doi.org/10.1002/ajpa.23932.

Hunt KD, McGrew WC. Chimpanzees in the dry habitats at Assirik, Senegal, and at Semliki Wildlife Reserve, Uganda. In: Boesch C, Hohmann G, Marchant LF, editors. Behavioral diversity in chimpanzees and bonobos. Cambridge UK: Cambridge University Press; 2002. p. 35–51

Jolly C. The seed-eaters: a new model of hominin differentiation based on a baboon analogy. Man, NS. 1970;5(1):5–26. https://doi.org/10.2307/2798801.

Jolly CJ. Analogies and models in the study of the early hominins. In: Sponheimer M, Lee-Thorp JA, Reed KE, Ungar PS, editors. Early hominin paleoecology. Denver: University of Colorado; 2013. p. 437–55.

King GE. The once and future baboon: a source of analogies for earliest hominid adaptations. Conference (Invited paper for SL Washburn memoriam): Am Assn Phys Anthropol. 2001.

King GE. Baboon perspectives on the ecology and behavior of early human ancestors. Proc Nat Acad Sci. 2022;119(45):e2116182119. https://doi.org/10.1073/pnas.2116182119.

King A, Clark FE, Cowlishaw G. The dining etiquette of desert baboons: the roles of social bonds, kinship, and dominance in co-feeding networks. Am J Primatol. 2011;73:768–74.

Lindshield S, Hernandez-Aguilar RA, Korstjens AH. Chimpanzees (*Pan troglodytes*) in savanna landscapes. Evol Anthropol. 2021;30(1) https://doi.org/10.1002/evan.21924.

Macho GA. Referential models for the study of hominin evolution. In: Schwartz JH, editor. Rethinking human evolution. Cambridge MA: MIT Press; 2018. p. 252–65.

Marchant LF, Wessling E, Lindshield S. Introduction to the special issue on savanna chimpanzees. Int J Primatol. 2020;71(1) https://doi.org/10.1007/s10764-020-00188-x.

Martin MK, Voorhies B. Female of the species. New York: Columbia University Press; 1975.

Moore J. Savanna chimpanzees, referential models and the last common ancestor. In: McGrew WC, Marchant L, Nishida T, editors. Great ape societies. Cambridge UK: Cambridge University Press; 1996. p. 275–92.

Musyoki CM, Strum SC. Spatial and temporal patterns of home range use by olive baboons (*Papio anubis*) in eastern Laikipia, Kenya. Afr J Ecol. 2016:8. https://doi.org/10.1111/aje.12294.

Newson L, Richerson PJ. A story of us, a new look at human evolution. New York: Oxford University Press; 2021.

Potts R. Transportation of resources: reconstructions of early hominid socioecology: a critique of primate models. In: Kinzey WG, editor. The evolution of human behavior: primate models. Albany: SUNY Press; 1987. p. 28–48.

Ransom TW. Beach troop of the Gombe. Lewisburg PA: Bucknell University Press; 1981.

Rowell T. Forest living baboons in Uganda. J Zool Lond. 1966;149:344–64.

Schreier AL, Swedell L. Ecology and sociality in a multi-level society: ecological determinants of spatial cohesion in hamadryas baboons. Am J Phys Anthropol. 2012;148(4):580–8. https://doi.org/10.1002/ajpa.22076.

Stanford CB. Chimpanzees and the behavior of *Ardipithecus ramidus*. Annu Rev Anthropol. 2012;41:139–49. https://doi.org/10.1146/annurev-anthro-092611-145724.

Strier KB. Beyond the apes: reasons to consider the entire primate order. In: de Waal F, editor. Tree of origin, what primate behavior can tell us about human social evolution. Cambridge MA: Harvard University Press; 2001. p. 69–93.

Strum SC. Almost human. New York: WW Norton; 1987.

Strum SC. Why natural history is important to (primate) science: a baboon case study. Int J Primatol. 2019; https://doi.org/10.1007/s10764-019-00117-7.

Strum SC, Mitchell W. Baboon models and muddles. In: Kinzey WG, editor. Evolution of human behavior: primate models. Albany: SUNY Press; 1987. p. 87–104.

Susman R. Pygmy chimpanzees and common chimpanzees: models for the behavioral ecology of the earliest hominids. In: Kinzey WG, editor. The evolution of human behavior: primate models. Albany: SUNY Press; 1987. p. 72–86.

Swedell L, Plummer T. Social evolution in Plio-Pleistocene hominins: insights from hamadryas baboons and paleoecology. J Hum Evol. 2019;137:1–15. https://doi.org/10.1016/j.jhevol.2019.102667.

Tanner NM. On becoming human. New York: Cambridge University Press; 1981.

Tooby J, DeVore I. The reconstruction of hominid behavioral evolution through strategic modeling. In: Kinzey WG, editor. Evolution of human behavior: primate models. Albany: SUNY Press; 1987. p. 183–238.

van Leeuwen KL, Hill RA, Korstjens AJ. Classifying chimpanzee (*Pan troglodytes*) landscapes across large-scale environmental gradients in Africa. Int J Primatol. 2020; https://doi.org/10.1007/s10764-020-00164-5.

Washburn SL, DeVore I. The social life of baboons. Sci Am. 1961;204(6):62–71. https://doi.org/10.1038/scientificamerican0661-62.

Wessling EG, Kühl H, Mundry R, et al. The costs of living at the edge: seasonal stress in wild savanna-dwelling chimpanzees. J Hum Evol. 2018a;121:1–11. https://doi.org/10.1016/j.jhevol.2018.03.001.

Wessling EG, Deschner T, Mundry R, et al. Seasonal variation in physiology challenges the notion of chimpanzees (*Pan troglodytes verus*) as a forest-adapted species. Front Ecol Evol. 2018b;6(a60):1–21. https://doi.org/10.3389/fevo.2018.00060.

Wessling E, Gonzalez PD, Llana M, et al. Chimpanzee (*Pan troglodytes verus*) density and environmental gradients at their biogeographical range edge. Intl J Primatol. 2020;41(3) https://doi.org/10.1007/s10764-020-00182-3.

Zinner D, Klapproth M, Schell A. Comparative ecology of Guinea baboons (*Papio papio*). Primate Biol. 2021;8:19–35. https://doi.org/10.5194/pb-8-19-2021.

Chapter 4
Subsistence and Stress

4.1 Introduction

Long standing methods for reconstructing the feeding habits of fossil species include the study of jaw function along with tooth size, shape, and structure. However, this work is largely limited to foods that *could* have been eaten. More recently developed methods indicate the kinds of food that were *actually* eaten (Teaford et al. 2023). One such method is the study of dental microwear—microscopic scratches and pits caused by the foods consumed. Stable isotopes, extracted from teeth and bones, provide additional information about food, as well as inferences about the environment that provided the food.

The vital element carbon illustrates stable isotope analysis. Three isotopes of carbon vary in atomic weight due to different numbers of neutrons in the nucleus of the atom. Carbon-14 (used in dating fossils and other items) is unstable, resulting in radioactive decay. Carbon-12 and carbon-13 are stable and maintain a constant ratio in the atmosphere, which is absorbed by plants. That ratio is changed within different plants by alternative processes of photosynthesis. Plants using these processes are grouped as C_3, C_4, and CAM (the last stands for crassulacean acid metabolism). The altered ratios are preserved in animals (bones, teeth, hair) and provide information about what the animals ate. That in turn implies the kind of environment that the animals were exploiting.

Each of these methods entails complexities and ambiguities, which argues for a "multi-proxy" approach (Teaford et al. 2023). Addition of baboon analogies to this comprehensive approach can provide more detailed hypotheses concerning issues such as which plants were eaten, which parts of the plants were eaten, and what foraging behaviors were needed to obtain them. In discussing models for hominin dietary reconstruction, Paine and Daegling (2023:5) noted that extant papionins occur in a variety of habitats "that are generally associated with ancient hominins"

G. E. King, *Baboon Perspectives on Early Human Ancestors*,
https://doi.org/10.1007/978-3-031-36769-4_4

and that both taxa "have been faced with the specific ecological challenges these habitats present."

4.2 Dietary Variation and Trends

A major feature of early hominin evolution was a series of dietary shifts, apparently related to both variation across habitats and long-term changes in the environment. Paleoanthropology follows several lines of evidence that trace these variations in broad terms. Baboon analogies suggest important details.

4.2.1 Hominin Dietary Shifts

The earliest hominins, represented by *Ardipithecus*, have been characterized as "nearly pure" C_3 feeders. This indicates that they subsisted on the products of trees and shrubs in relatively closed woodlands (Martin et al. 2020). *Australopithecus anamensis* may have undergone relatively little change from *Ardipithecus*, adding some variety to the diet of C_3 foods with occasional forays into open areas (Bobe and Reynolds 2022; Bobe et al. 2020; Bobe et al. 2022; Martin et al. 2020). However, Quinn (2019) recalculated isotope values for some *A. anamensis* and postulated that the species added a substantial amount of C_4 food by foraging across diverse habitats in a mosaic setting. She argued this view to be more consistent with dental features (tooth structure and microwear) that indicate consumption of tough and/or hard foods. Roughly contemporary with *A. anamensis*, an *Australopithecus* species in Chad (classification still somewhat controversial) had a diet that was "predominantly" C_4 (Bobe and Reynolds 2022).

Whatever the earlier situation, there is general agreement that the C_4 component was significantly expanded in later hominin stages. After about 3.5 mya, multiple hominin taxa in eastern Africa (Martin et al. 2020; Quinn et al. 2021) and southern Africa (Grine et al. 2020; Sewell et al. 2022) began incorporating more C_4 foods into their diets. Isotopic evidence and tooth microwear suggested that *A. africanus* was a mixed feeder that may have "effectively straddled different ecotones" (Grine et al. 2020). Carbon isotope analysis from Sterkfontein indicated "a high proportion" of C_4 foods in the diet by 2.6 mya if not earlier (Sewell et al. 2022). Phytoliths extracted from the teeth of *A. sediba*, dating to about 2 mya, also provided evidence for a significant amount of C_4 food (Reynolds 2022). These developments are thought to be associated with increasing exploitation of "savannas" (recall that savanna is a broad term that can include any habitat with significantly reduced tree cover).

The diet of earliest *Homo* in South Africa resembled that of *A. anamensis* until a significant increase in C_4 took place after 1.7 mya, which was about a million years after the genus first appeared (Bobe and Reynolds 2022; Patterson et al. 2019). It seems that subsistence analogies from baboons can be useful in reconstructing early *Homo*, as well as its predecessors. This view is consistent with a gradual transition from *Australopithecus* to *Homo*, as postulated by Kimbel and Villmoare (2016; see Chap. 1). The shift after 1.7 mya was probably influenced by increased consumption of animal foods, which will be discussed at length in Chap. 5.

4.2.2 Generalists and Specialists

All of the hominin situations postulated in the preceding section are paralleled by variations in behavioral ecology across baboon populations. One of the most important points that emerges from comparative study is that baboons tend to favor forest and woodland habitats like those that have been postulated for *Ardipithecus* and its predecessors. The yellow baboons of Tana River in Kenya provide an example. The forest there covers less than 9% of the area that the baboons regularly use, but 42% of food consumption observations were made in the forest. Similarly, forest components of habitat were used more than predicted by chance by olive baboons in Ivory Coast (Kunz and Linsenmair 2008) and Guinea baboons in Senegal (Zinner et al. 2021). Some baboons are described simply as "forest dwelling" (Johnson et al. 2012). Forest and woodland baboons make extensive use of C_3 foods such as fruits and flowers, but most of them spend a significant amount of time foraging in various kinds of savannas. Thus, baboons can provide analogies and scenarios for various early hominins that pertain to details of their subsistence behavior.

Paleoanthropology demonstrates hominin expansion into more open habitats with different food resources. Baboon analogies suggest how the hominins managed the change. According to researchers who carried out a 5-year study of yellow baboons in Mikumi National Park, Tanzania, "Eclectic feeding and selectivity are a powerful combination; together they probably go a long way toward explaining the baboons' success over a large part of a continent with diverse ecosystems" (Norton et al. 1987:115). This is a scenario that is readily transferred to the expansion of the hominins.

Codron et al. (2008) analyzed the feeding ecology of chacma baboons in South African savannas and found that they were like "australopiths" in consuming substantial amounts of C_4 foods. Attributing this to parallel shifts from wooded habitats to open savannas, they cautiously inferred that "there is some benefit in an analogous comparison." However, they posited that differences between hominins and baboons, rather than similarities, might provide valuable information. For example, they found baboons to be less isotopically variable than *Australopithecus*, which suggested to them that the hominins were even more generalist than the baboons.

This is an important difference with regard to the *total* diets of hominins and baboons. However, it allows for useful similarities in their *basic* foods or food categories. Furthermore, many or most modern baboons are probably at least as generalist as "nearly pure" C_3 feeders such as *Ardipithecus* and perhaps *A. anamensis*. Comparative study of extant baboons can probably help us to understand the stages by which earlier hominins increased the breadth of their diets. Different extant baboon populations suggest the changes that took place as hominins expanded from woodlands into more open areas or were confronted by more open habitats resulting from aridification.

Norton et al. (1987) pointed out that the breadth of baboon diet was easily underestimated because of studies that were too short or focused on other matters. For example, a distinguished primatologist listed 65 plant species eaten by the baboons that he studied; then increased the number to 94 after 1 more year of research. Norton and colleagues identified 185 species of baboon foods in one locality after 5 years of study. Long-term fieldwork at Amboseli has led to the conclusion that baboons can subsist on almost any plants (Alberts and Altmann 2007). This kind of flexibility must have been necessary for hominins to survive in diverse habitats from eastern to southern Africa (Paine et al. 2019) and in diverse localities within those regions (Henry et al. 2019).

Selectivity is the complementary dimension of subsistence in both baboons and early hominins. Paine et al. (2019) posited that many early hominins were probably selective feeders, eating particular plants and plant parts depending on habitat and season. The baboon evidence strongly supports this view. In the study by Norton et al. (1987), 185 food species exploited by the baboons were selected from at least 700 possibilities. Furthermore, in consuming approximately 80% of their food species, the baboons ate only one or two parts of the plant. They ignored other parts or removed them by stripping or peeling. In the Budongo Forest of Uganda, olive baboons strongly emphasized a few species out of the 51 food plants identified in a 6-month study (Okecha and Newton-Fisher 2006). Forest dwelling baboons at Kibale in Uganda ate some parts and discarded others (Johnson et al. 2012).

Norton et al. (1987) summed up the baboon pattern in this way: baboons are adapted to consuming a wide array of foods and make choices among them in the context of "an ever-changing mosaic" of possible foods in combinations that satisfy their energetic and nutritional needs. Such a suite of adaptations could well have propelled early hominins on the way to their expansion throughout most of Africa in the context of diverse and changing environments. Within this broad framework, extant baboons provide more specific analogies for early hominin selection and treatment of particular food species.

4.3 Food Sources

Stable isotopes, carbon and others, provide only a first approximation to the ecology and behavior of subsistence. A variety of foods can account for each of the isotopes. Some reasonable hypotheses and guesses can be provided by paleoenvironments and chimpanzee analogies. Baboon evidence can add more detail for a wider variety of contexts.

4.3.1 Fruit

Hominins apparently originated as C_3 feeders in woodlands and forests, where fruits can provide a large portion of such a diet. Given the hominin phylogenetic relationship with chimpanzees, the latter often termed "fruit specialists," hominins probably derived an emphasis on this source from the Common Ancestor. Most baboon populations are consistent with this scenario (Fig. 4.1, Table 4.1). The chacma exception in the table is probably due to the colder and drier conditions at the southern tip of Africa.

In an example from West Africa (Ivory Coast), two troops of olive baboons of different sizes were observed in a savanna-forest mosaic (Kunz and Linsenmair 2008). Population density was lower than expected for this kind of habitat (1.2 baboons per km^2), which was hypothesized to be related to the "highly frugivorous" diet. The baboons spent about 50% of their feeding time on the fruits and seeds of many plant species and devoted relatively little time to more abundant but lower quality foods. Both troops, though very different in size, had comparatively large home ranges. This allowed them to cope with the irregular availability of fruit as compared with other plant foods. They used the forest more often than expected by chance, presumably to pursue the fruit diet.

Across the continent, in Ethiopia, another population of forest-living olive baboons also demonstrated a preference for fruit (Mullu and Solomon 2016). Fruit was the largest portion of their diet (about 28%), followed by leaves (about 20%). Figs of a sweet tasting variety provided a variety of nutrients. In contrast to

Fig. 4.1 Chacma baboon eating *Cordia* fruit. (Photo by Curt Busse. Okavango, Botswana)

Table 4.1 Food preferences of forest/woodland baboons

Species	Olive baboons *Papio anubis*	Olive baboons	Olive baboons	Olive baboons	Hamadryas baboons	Chacma baboons *P. ursinus*
Location	Comoé Park Ivory Coast	Kibale Park Uganda	Arba Minch Ethiopia	Budongo Forest Uganda	Borena-Sayint Ethiopia	Tokai Forest South Africa
References	Kunz and Linsenmair (2008)	Johnson et al. (2012)	Mullu and Solomon (2016)	Okecha and Newton-Fisher (2006)	Ibrahim et al. (2023)	Johnson et al. (2013)
Method	Feeding time	Feeding events	Total scans	Feeding time	Scan of activities	Focal subject
Favored foods	Fruit 29% Grass 29% Seeds 18% Flowers 6% Leaves 5%	Fruit 46% Stems 33% Tubers 7% Leaves 7% Seeds 4%	Fruit 28% Leaves 20.5% Grass 15.5% Bark 15% Roots 9.7%	Fruit 34% Leaves/Grass 17% Seeds 13% Garbage 17% Misc. 19%	Fruit 32% Graminoid blades[a] 21% Young leaves 13% Cereals 9% Flowers 6%	Leaves 58% Seeds and nuts 15% Mushrooms 15% USOs 9% Fruit 2%

The preponderance of olive baboons is an artifact of the literature that lists food preferences. Relevant material from other baboon species is discussed in the text

[a]Graminoids are herbaceous plants with grass-like leaves

hamadryas baboons in arid lowland habitats, a hamadryas population with access to "large areas of natural forest" in the Ethiopian highlands had a diet in which fruit was the leading component (Ibrahim et al. 2023).

Fruit trees were of special significance in a study of competition between humans and kinda baboons on the fringe of a national park in the D.R. Congo (Kazaba et al. 2020). The kinda baboons ate fruit in all nine of the tree species that they exploited, especially the wild loquat tree (*Uapaca kirkiana*) and the monkey orange tree (*Strychnos cocculoides*). The researchers recommended domestication of these species in order to alleviate the competition.

The taste for fruit in wild baboons is illustrated by their response to the intrusive prickly pear cactus (*Opuntia stricta*). The fruits, with sweet low-acid pulp, became a baboon favorite (Schoeninger et al. 2019). Experiments with olive baboons demonstrated that they were "among the most sugar-sensitive nonhuman primates tested so far" (Laska et al. 1999:25). The researchers inferred that this might be a criterion for food selection. Since baboons share this taste with humans, it seems likely that their responses are analogous to those of early hominins.

An important point related to this food choice is that baboons generally prefer forest habitats when available. This bears on an old controversy about whether hominins expanded into savannas to take advantage of new opportunities or were forced to do so by changing conditions. The fact that baboons prefer forests despite their success in savannas suggests that early hominins were forced to adapt to savannas.

4.3.2 High-Protein Plants

Early hominins in mosaic environments, despite a preference for forest or woodland, probably used the savanna components to add C_4 foods to the diet (Bobe et al. 2022; Martin et al. 2020; Quinn 2019). Baboons are consistent with this scenario and indicate that proteinaceous plants were an important factor. Bentley-Condit and Power (2018) reported on the food of yellow baboons in a partially forested habitat. Comparing forest to savanna samples, they found the savanna items to be higher in crude protein and minerals. Applying a narrower focus to the typically eclectic diet, they examined the top 15 foods. The savanna items had significantly higher crude protein, ash (an indicator of mineral content), magnesium, and manganese. Forest items were higher in gross energy.

Johnson et al. (2017) studied olive baboons in a "heterogeneous forested habitat" and found a pattern of "protein prioritization." They followed individual females and quantified nutrient intake from both feeding observations and nutritional analysis. The baboons maintained a 5:1 kilocalorie ratio of nonprotein energy to protein, based on annual intake. As food availability changed seasonally, the baboons maintained protein intake while exploiting energy sources where possible. The savanna baboon diet contains a greater proportion of protein-rich foods than is the case of sympatric ungulates (Codron et al. 2006).

Protein prioritization is also displayed by humans, with kilocalorie/protein ratios comparable to the baboons. This suggests an evolved ecological strategy that allowed both lineages to exploit energy in multiple environmental contexts (Johnson et al. 2017). In modern human populations, with a different pattern of food availability, protein prioritization seems to cause some humans to overconsume energy as they try to maintain protein intake (Johnson et al. 2017). It is considered a central factor in obesity and associated cardiometabolic disease.

4.3.3 Savanna Plants

Another evaluation of potential hominin foods was developed with reference to chimpanzees, but is also relevant to baboons. Copeland (2009) noted the evidence that early hominins occupied relatively more open and arid savannas than those in which any chimpanzees live. She compared plant foods at savanna chimpanzee sites with those in semiarid savannas that are not inhabited by chimpanzees but are potentially similar to environments occupied by some early hominins. In the semiarid savanna, trees that produce fleshy fruits are manifested in fewer species and lower densities. The most abundant potential hominin plant foods are seasonally available *Acacia* flowers and seeds/pods, grass seeds, and the underground parts of marsh plants. Copeland explicitly cited baboon diets as demonstrating the value of these foods.

4.3.4 Underground Storage Organs

Underground Storage Organs (USOs) are starchy underground organs in which many plants retain carbohydrates and water during periods that are unfavorable for growth (Dominy et al. 2008). USOs are often classified and named according to their structure and orientation in the ground, which can be confusing. Most are stems of some kind. *Corms* are swollen and compacted stems. *Rhizomes* are horizontal stems that can produce roots and shoots. *Tubers* are generally thickened roots or swellings on roots. *Bulbs* are modified shoots. Some plants with USOs use the C_3 pathway and others use the C_4 pathway. Regardless of category, the USOs of many plants are rich in starches that the body can convert into sugars that are essential for brain development and maintenance (Macho 2016).

Dominy et al. (2008) evaluated the various USOs as to their suitability for consumption by early hominins. Rhizomes are too tough to have been a useful food for hominins under most circumstances, although the rhizomes of some aquatic plants are an exception. Tubers are more plausible, but few tuberous plants use the C_4 pathway. Corms and bulbs are the most likely hominin foods and conform to previous authors' hypotheses in this respect. They are widespread, low in fiber, more likely to be found in C_4 plants, and are sought by very few other animals. They are also gritty, which is consistent with wear on hominin teeth. Bulbs are the tougher and more elastic of these two forms, leaving corms as the most valuable for the subsistence of most hominins.

Direct evidence for USO consumption by early hominins comes from Kanjera South, where use-wear on the tools of early *Homo* in a savanna environment indicate the processing of tubers. Dense patches of tubers in this location offered a significant source of food (Bishop et al. 2022). Preparation with tools explains the use of tubers, which are tougher than corms or bulbs.

USOs occur in the diets of forest and woodland baboons, but with limited significance. They occur among the top listed foods for three of the five populations in Table 4.1, but as relatively minor components. In the savannas of eastern Africa, USOs were more important. Grass USOs were a rich source of nutrients, as well as water, during the dry seasons (Altmann and Altmann 1970; DeVore and Washburn 1963).

Macho (2016) emphasized the significance of USOs for the spread of *Australopithecus* into more open and fragmented habitats with greater seasonality. USOs made up for the decreased supply of fruit by offering "energy-dense" foods that were also high in nutrients and water content. She compared hominins with baboons as to the benefit of this source: eclectic feeders in a similar range of environments, needing USOs to compensate for less fruit as a supply of glucose necessary for the growth and maintenance of relatively large brains. Most animals break down starches through the action of salivary amylase and baboons have a concentration of the enzyme that is greater than twice that of modern humans (Macho 2016). Hominins might have evolved to a similar level if they had not begun to use tools and fire to process USOs.

4.3.5 Grasses and Sedges

Fruit may not have been the only attraction in forests and woodlands. Henry et al. (2019) collected plant samples from several localities in South Africa in order to explore the implications for early hominin use of similar habitats. One of their important findings was that woodland habitats were more nutritionally valuable than anticipated. Grasses from these wooded environments were generally good resources with relatively high protein content and low in chemical and mechanical contents that interfered with feeding. This could have been particularly important in the dry season, when protein content among all of the sampled plants was lower. Despite strong temperature and rainfall variation between seasons, most nutritional properties other than protein remained relatively constant.

Hominins in East Africa around 3 mya inhabited a vegetational mosaic in which grasslands were becoming more common (Bobe et al. 2022). By 2 mya some early *Homo* were living in open savannas where grass was the predominant vegetation (Bishop et al. 2022). At about the same time, South Africa underwent increasing aridity that led to a regional expansion of grasslands. Strontium/carbon ratios suggest a higher level of either insectivory or “grazing” (Sewell et al. 2022). Baboon analogies support both insects (Chap. 5) and grasses as hominin foods (Figs. 4.2 and 4.3).

Paine et al. (2018) agreed with previous writers that early hominins almost certainly paralleled baboons in making grass seeds part of their broad dietary repertoire. However, noting the limited seasonal availability of grass seeds, they focused

Fig. 4.2 Yellow baboon foraging in grass. (Photo by Oliver Paine. Amboseli National Park, Kenya)

Fig. 4.3 Yellow baboon eating the base of a grass stem. (Photo by Oliver Paine. Amboseli National Park, Kenya)

on grass *leaves* as potential hominin food. Grass leaves are the single most abundant C_4 resource in modern African savannas. Analysis of savanna grass leaves for their nutrient value and physical characteristics showed that species differed significantly with regard to traits that would have been favorable for the hominin diet. Some grass blades would have been good food sources because they were less tough, lower in fiber, and higher in protein.

Paine et al. (2018) were explicitly encouraged in their work by numerous studies showing that grasses are a major source of food for baboons in savanna environments. Grass is the most important single food for at least some savanna baboons with bulk intake up to 40% and as high as 90% during dry seasons for some populations (DeVore and Washburn 1963). Baboons demonstrate how grasses can be an important year-around food. Depending on the season they shift their attention to different parts of the grasses: seeds, thick lower stems, and rootlike underground rhizomes (Macho 2015).

The grass-like sedges that grow in watered localities have also been proposed as important food sources for hominins (Lombard 2022; Paine et al. 2018). At Kanjera South in Kenya, early *Homo* lived in an arid savanna and the use-wear on their tools indicates the processing of herbaceous plants such as sedges (Bishop et al. 2022). The high protein content of grasses and sedges in the wetlands of Amboseli in Kenya during the dry season may be one reason that baboons have done relatively well there, even during some periods of severe drought (Alberts and Altmann 2007; Paine et al. 2018). Grasses with bulbs and sedges with corms are likely to use the C_4 pathway. They are common over much of eastern and southern Africa, the areas where most hominin fossils have been found (Dominy et al. 2008).

Schoeninger et al. (2019) doubted the significance of grasses for baboons and, therefore, the relevance of baboons to early hominins in this context. They cited references compiled by Crowley et al. (2010) for isotope analysis of baboon teeth that showed little or no C_4 intake. Schoeninger and colleagues accounted for the small amounts of C_4 evidence as coming from intrusive sources: consumption of crops or fruit of the intrusive prickly pear cactus. On the other hand, isotype analyses by Codron et al. (2008) found that baboons eat substantial amounts of C_4 foods, comparable to the C_4 proportion of the *Australopithecus africanus* diet.

Most behavioral studies, especially in eastern Africa, have reported grasses to be an important part of the baboon diet (e.g., Altmann and Altmann 1970; DeVore and Washburn 1963). They stripped the seeds with their fingers or mouths, exploiting a highly nutritious food during dry seasons. Baboons ate grass blades after wet seasons, when the blades were young and green.

4.3.6 More About Plant Foods

The varied ecologies of baboon populations suggest ways in which hominins might have augmented the basic foods. As eclectic feeders, baboons are able to include valuable but rarer food sources when they become available. Seeds from fruits and grasses are important in the diet of baboons (Table 4.1) and presumably in the diet of early hominins. However, some are more valuable than others. On the Laikipia Plateau in Kenya, baboons consumed *Acacia* seeds with protein/fiber ratios almost five times higher than those assessed for other plants by Paine et al. (2019). That is, they were highly nutritious and very easy to digest. These seeds constituted only about 3% of the diet, but this may underrate their seasonal nutritional value.

Yellow baboons in Amboseli National Park spent long periods of time feeding on exudates from fever trees (*Acacia xanthophloea*) and, exemplifying baboon selectivity, paid little attention to the umbrella trees of the same genus (*A. tortilis*) (Hausfater and Bearce 1976). Fever trees produce greater quantities of exudate, and chemical analysis indicated that it is probably composed of a complex polysaccharide that contains at least four sugar constituents—potentially nutritious for baboons and other animals. In contrast, low solubility and apparently low carbohydrate content suggest that umbrella tree exudate has little nutritional value. Furthermore, its taste (which is unpleasant to humans) suggests that it contains toxic compounds. Fever tree gum is tasteless and odorless for humans. Early hominins could presumably have distinguished between these two potential foods as readily as baboons and humans.

Bark stripping by primates and other animals was investigated with regard to its commercial significance, but the results have implications for baboon ecology and its relationship to early hominin subsistence behavior. Thirteen primate species have been recorded stripping bark from trees of commercial value worldwide, but three species are responsible for most of the damage in large-scale plantations (Di Bitetti 2019). The two Old World species are Sykes monkeys (*Cercopithecus albogularis*) and chacma baboons. Yellow baboons are also implicated, but in fewer reports. Actions to mitigate this problem, including the massive killing of primates, have proven ineffective in the long term, which seems to suggest that there is some compelling reason for the behavior.

The most frequently cited hypothesis for primate bark stripping is that they eat soft bark when or where their natural food is scarce. However, this hypothesis is not generally supported by empirical evidence. Instead, *Eucalyptus* bark may be sought for its high sodium content. Pines are apparently bark stripped to consume the sugary phloem during their growing season (when bark is presumably more easily peeled off) (Di Bitetti 2019). This evidence suggests that bark stripping can be valuable for subsistence whether other foods are scarce or not.

4.3.7 Toxic Plants

In concluding their comparative analysis of protein/fiber ratios, Paine et al. (2019) noted that hominins might have been influenced by other important nutrients, including lipids and water. Codron et al. (2006) reported chacma baboon consumption of the succulent but toxic species *Euphorbia ingens* and they cited other accounts of baboons in various localities consuming the same species or at least the same genus. The baboons might be immune, but this would raise the question of why immunity would have evolved. Codron and colleagues noted that these succulents are high in lipids and water. Since water was plentiful for the baboons in question (as it would have been for many woodland hominins), they concluded that lipids were the most plausible factor for consuming these toxic plants.

Yellow baboons in Mikumi Park, Tanzania, occasionally fed from the cassod tree (*Senna siamea*), which contains a variety of toxic compounds, including a trypsin inhibitor that causes severe digestive distress in humans and other mammals. The baboons ate seeds from the tree, but no other parts. The quantity was small, being consumed by individuals during less than 1% of follows (Kitegile 2022). However, consumption was concentrated during the latter part of a long dry season (5–6 months). The tree increased seed production during this time when other foods became scarce, that is, a fallback food.

To interpret the behavior of these baboons, Kitegile referred to the work of Altmann (1999) on yellow baboon consumption of *Vachellia tortilis*, another tree that contains a trypsin inhibitor. In both cases adult females and subadults processed the seeds before eating, while adult males rarely did. Removal of the seed coat reduced the toxin but also reduced the available protein. Males were able to tolerate the toxin because of their greater body size and longer digestive tract. The molecule of the trypsin inhibitor, which is a protein, may split during a long enough period in the digestive tract and allow the absorption of more amino acid and less toxin. Early hominins were larger than extant baboons and might have been more able to take advantage of such fallback foods.

4.4 Ingestion of Other Substances

Baboons ingest substances other than plants or animals, perhaps for certain nutrients and perhaps for other health reasons. Early hominins, living in similar environments, are likely to have had similar problems and might have responded in the same way.

4.4.1 Geophagy

Geophagy, the consumption of soils and clays, is common among humans and other primates (Pebsworth et al. 2018). Pebsworth et al. (2011) continually monitored the largest and most frequently visited geophagy sites in their study area with camera traps and recorded visits by baboons on almost half of 545 days. During 3500 baboon visits, camera traps captured almost 60 h of geophagy. The baboons spent more time ingesting white alkaline soils with high percentages of clay and fine silt, which contained higher concentrations of sodium, than acidic soils that contained higher concentrations of iron. Pregnant females spent more time in consuming soil than baboons of any other age class, sex, or reproductive state. The soils in question would be effective at alleviating gastrointestinal distress and possibly supplementing minerals for all age/sex classes, but might also have met more specific age/sex requirements. The full distribution of geophagy in baboons is unknown but reports include other chacma populations as well as olive and yellow baboons (Pebsworth et al. 2018).

4.4.2 Medicinal Plants

Chimpanzee consumption of medicinal plants is well known. Some baboon evidence augments the hypothesis that such behavior may have benefitted early hominins. Scientists studying hamadryas baboons at the Awash Falls in Ethiopia noted that although the desert date tree (*Balanites aegyptiaca*) grew all around the falls, only the baboons living below the falls ate the tree's fruit. These baboons were exposed to a parasitic worm found in water snails and the desert date fruit can repel these snails. Baboons living above the falls were not in contact with the water snails and therefore had no need of the presumably medicinal fruit (Admassu et al. 2013).

4.5 Foraging and Stress

Life in forests and woodlands might have been relatively easy for hominins, though there must have been some stresses, including seasonal scarcity of some foods (and danger from predators, see Chap. 6). To the extent that they foraged in savannas, hominins would have encountered new sources of stress resulting from changes in diet and more exposure to the sun. Baboons demonstrate solutions that hominins might have found.

4.5.1 Dietary Stress

Paine et al. (2019) reviewed a history of studies leading to the conclusion that many hominin habitats were "characterized by stark seasonality and mosaic habitats, just as modern African savannas are today." They followed Copeland's (2009) comparative study of chimpanzee habitats to the conclusion that early hominins could not have depended on fruit to the extent that their hominoid ancestors did, and probably shifted to a diet more like that of most baboons. This led Paine and colleagues to perform nutritional analyses of plants growing in eastern and southern savannas across both wet and dry seasons. They discovered a general decline in protein/fiber ratios during dry seasons. Most important for this discussion, they found dry season increases in the protein levels of foods like those that are often heavily consumed by baboons: sedges, grasses, and underground plant organs.

The role of feeding problems in stress is suggested by a seminatural experiment resulting from food made available by humans. Lodge et al. (2013) measured hormonal levels in two troops of olive baboons in the same population. The troops faced similar ecological challenges, but differed in that one troop augmented its diet by raiding crops. Based on detailed feeding observations and nutritional analysis of food samples, the crop-raiding troop experienced 50% higher energy intake rates and 50% lower glucocorticoid levels (indicating stress), compared to the troop that did not exploit crops. Energetic stress in the non-raiders was associated with elevated progesterone levels and lower reproductive output. Variation from richer to poorer natural food sources, either geographic or temporal, might have had similar effects on hominin populations.

4.5.2 Heat Stress

To the extent that early hominins spent time in open areas rather than shaded woodland, they would have been subjected to heat stress. Hominins evolved distinctive adaptations for coping with this problem, the most obvious being perspiration. Humans have by far the highest eccrine sweat gland density among primates, resulting from multiple interacting mutations (Aldea et al. 2021).

How might early hominins have coped with heat stress before distinctively human adaptations appeared? Wheeler (1991) posited that bipedalism provided relief from heat. The major benefit would have been reduction in the surface area of the body directly exposed to the sun. In addition, windspeeds are higher and air temperatures lower away from the ground, both increasing the rate of heat dissipation. Wheeler argued that the drastic adaptation of bipedalism would have been essential for a large-brained primate under such conditions.

Baboons remained quadrupedal (and furred) while becoming the most successful primates in colonizing the African savannas (Mitchell et al. 2009), which casts some doubt on Wheeler's hypothesis. The baboon analogy can be questioned on the basis of the fact that baboon brains are much smaller than those of early hominins. On the other hand, a key feature in favor of the comparison is that baboon brains are like human brains in that they lack special mechanisms (such as are found in other animals) to cool them (Hill 2006; Maloney et al. 2007).

The presence of heat stress, and successful adaptation to it, are both represented in a study by Gesquiere et al. (2008). Measurement of hormone levels in female baboons confirmed seasonal heat stress, yet they were able to reproduce throughout the year. Baboons cope with heat by means of behavior patterns that are simple and could easily have been discovered and adopted by early hominins. They rest in shade as much as possible (cf. Wheeler 1994), especially during midday , when they also engage in a drinking session that facilitates evaporative cooling (Hill 2006). Even in savanna habitats, some shade trees are available and are incorporated into the ranging pattern of a troop (Fig. 4.4). Baboons occasionally sleep in caves (Hamilton 1982; McGrew et al. 2003), which may provide thermoregulation. Measured by scientists, the temperature in one cave remained steady and mild

Fig. 4.4 Chacma baboons resting in the shade under a tree at midday. (Photo by Glenn King. On the road near Mkhuzi, South Africa)

compared to outside conditions (Barrett et al. 2004). Chacma baboons seem to be more prone than other *Papio* species to enter caves, which may be due to more extreme temperature variation in southern Africa.

4.5.3 Dehydration

Humans and other terrestrial animals must maintain a tight balance of water gain and water loss each day in order to survive (Pontzer et al. 2021). Pontzer and colleagues designed a comparative study to investigate the phenomenon in humans. It measured water turnover in five human populations (including hunter-gatherers in semiarid savanna) and all three genera of great apes (housed in zoos and rainforest sanctuaries). Controlling for external factors, the researchers found that water turnover was 30–50% lower in humans than in apes despite the loss of water through sweating. Water stress apparently led to previously unknown water conservation adaptations in hominin physiology.

Pontzer and colleagues turned to baboons to expand their perspective with naturalistic information. They characterized water sources as "an ecological tether" for savanna baboons that constrains daily travel (they noted that the same seems true of "savanna" chimpanzees). This implies that at some point hominins began to evolve mechanism(s) for reduced water demand. The researchers related this to the appearance of the external nose in early *Homo* at about 2 mya. This structure reduces water loss and has been proposed as a water conservation adaptation. The hypothesis is consistent with increasing association between early *Homo* fossils and dry, open paleoenvironments.

The hypothesis of Pontzer and colleagues suggests two phases in the comparison of hominins with baboons. The first phase is defined by four million years or more (during which external noses had not yet evolved) when hominins were presumably "tethered" to water sources. In the second phase, beginning about 2 mya, hominins evolved one or more unique mechanisms for water retention as they penetrated farther into open environments and evolved a human-like hunting and gathering adaptation.

Like baboons (and chimpanzees) in woodland habitats such as Gombe National Park in Tanzania, early hominins could drink water from a variety of sources while moving around in response to food availability (Ransom 1981). With expansion into more open areas, water must have become a more limiting resource. However, the "tether" would vary in strength from time to time and place to place. Yellow baboons at Amboseli, for example, went without drinking on 3 of 19 days and drank only once on 11 days (Altmann and Altmann 1970).

Variations in water availability are associated with variations in baboon behavior that suggest further analogies with early hominins. Where water is scarce, for example, troops may converge on water sources and conflicts sometimes occur (Altmann and Altmann 1970). Another response to scarcity (and perhaps to pollution) is to dig holes to obtain water (also done by some chimpanzees).

In more recent work at Amboseli (Paietta et al. 2022), the baboon response to artificial wells indicates an intense need for water. They accessed the wells during dry periods and extended their range to encompass new wells when older ones had dried up. The baboons continued use of the wells, despite danger from pastoralists and their dogs (over a period of years, one-third of all adult deaths were linked to these encounters).

4.6 Summary and Conclusions

Feeding habits of fossil species can be reconstructed in several ways. Visible features of the teeth and chewing apparatus show what kind of foods the animals *could* have eaten. Two approaches provide evidence for the general types of food that they *actually* ate. Microwear analysis examines the marks that food left on fossil teeth. Stable isotope analysis reveals the chemical evidence of food types left in the teeth and bones. The latter depends on stable ratios between variants of elements such as carbon. These are absorbed by plants and altered by the processes of photosynthesis. For example, the C_3 photosynthetic pathway usually represents foods such as fruits. Grasses are often indicated by the C_4 pathway. However, there are overlaps (e.g., grasses may be C_4 or C_3) and these are still rather broad categories of food. Baboon analogies can suggest specifics about the particular foods that were exploited and how they were obtained.

The earliest hominins, inhabiting forests and woodlands, depended almost entirely on C_3 foods. Later hominins, such as *Australopithecus*, incorporated more C_4 foods into the diet as they expanded into savannas, or the environment around them changed into savanna. Paleoanthropologists have postulated that this transition was possible because hominins could choose from a wide variety of foods and could select the most nutritious parts of those items. Some have alluded to the fact that baboons, spanning very similar environments, are also eclectic and selective feeders. Baboons are like humans in that they tend to adjust their food intake to maximize protein, so this almost certainly was a key feature of early hominin diets as well.

Fruit must have played a large role in the diets of early hominins. This inference from chimpanzee ecology is strongly supported by baboon analogy. Fruit is the preferred type of food for all baboons wherever it is available in any significant quantity. However, the baboon evidence indicates important changes that took place as hominins expanded into more open and fragmented environments. Grass probably became more important. Baboons eat grass seeds and blades. They also eat the underground storage organs (USOs) of grasses and other plants. These are structures in which plants retain water, starch, and other nutrients when conditions are not right for the use of these substances for growth. Hominins were probably like baboons in excavating these plant organs, especially during dry season. The starch from USOs can turn into sugars that support the growth and maintenance of large brains.

Some baboon populations consume less common foods, and early hominins might have augmented their diets with comparable items. These include acacia seeds with unusually high protein content; tree exudates that may contain polysaccharides with diverse sugar components; and soft tree bark that might be high in sodium. Some baboons eat toxic items that may supply them with important nutrients such as lipids.

Baboons, other primates, and many extant human populations ingest various clays and soils. These items might provide sodium and/or some essential minerals. Some might also have medicinal properties. The latter may also be true of some of the plants that baboons consume. Possible functions of such ingesta include relief of gastric distress and resistance to parasites. Early hominins might not have eaten the same items, but might well have had comparable ones in their diets.

Foraging animals encounter various stresses as they seek food. Stresses would certainly have increased as hominins dealt with more open and fragmented habitats, culminating in savannas with only scattered trees. Baboons demonstrate the stresses in this range of environments and some of the ways in which early hominins might have coped with them.

Food is and was less readily available in savannas, especially during dry seasons. The flexibility of the eclectic/selective subsistence adaption, as seen in baboons, would have helped to alleviate the problem. One specific solution to dry seasons, underground storage organs, was probably available to most hominin populations.

With minimal tree cover, heat would have been a major stress, especially dangerous for brains without the defensive mechanisms evolved by other savanna mammals. Baboons adjust with behavioral patterns that could easily have been adopted by early hominins: use of shade whenever available, midday rest, and occasional sheltering in caves.

In dry seasons water would have been a "tether" for early hominins, as it is for baboons. Baboons illustrate the results of this constraint on behavior. One concomitant is reduced foraging due to limitations on travel. Another is the convergence of separate social groups at water sources, potentially resulting in complex social interactions, some hostile, that could be avoided during other times of the year.

Two vital topics have been left out of this chapter. One is the role of animal prey in the expansion of C_4 foods. This extensive subject is covered in the next chapter. The other topic is the stress caused by predators while baboons are foraging and even when they are sleeping above the ground. This is the subject of Chap. 6.

References

Admassu M, Bekele A, Kim JC. Nutritional composition of *Balanites aegyptiaca* (desert date) and *Hyphaene thebaica* (doum palm) fruits consumed by hamadryas baboons (*Papio hamadryas hamadryas*) in Awash National Park, Ethiopia. J Nutr Ecol Food Res. 2013;1:198–206. https://doi.org/10.1166/jnef.2013.1037.

Alberts SC, Altmann J. The evolutionary past and the research future: environmental variation and life history flexibility in a primate lineage. In: Swedell L, Leigh SR, editors. Reproduction and fitness in baboons: Behavioral, ecological, and life history perspectives. New York: Springer; 2007. p. 277–303.

Aldea D, Atsuta Y, Kokalari B, et al. Repeated mutation of a developmental enhancer contributed to human thermoregulatory evolution. Proc Natl Acad Sci USA. 2021;118(16):e2021722118. https://doi.org/10.1073/pnas.2021722118.

Altmann SA. Foraging for survival, yearling baboons in Africa. Chicago: University of Chicago Press; 1999.

Altmann SA, Altmann J. Baboon ecology: African field research. Chicago: University of Chicago Press; 1970.

Barrett L, Gaynor D, Rendall D, et al. Habitual cave use and thermoregulation in chacma baboons (*Papio hamadryas ursinus*). J Hum Evol. 2004;46(2):215–22. https://doi.org/10.1016/j.jhevol.2003.11.005.

Bentley-Condit VK, Power ML. The nutritional content of Tana River yellow baboon (*Papio cynocephalus*) foods in a partially forested habitat. PLoS One. 2018;13(11):e0207186. https://doi.org/10.1371/journal.pone.0207186.

Bishop LC, Plummer TW, Braun DR, et al. Fauna and paleoenvironments of the Homa Peninsula, western Kenya. In: Reynolds SC, Bobe R, editors. African paleoecology and human evolution. New York: Cambridge University Press; 2022. p. 360–75.

Bobe R, Reynolds SC. Hominid paleoenvironments in tropical Africa from the Late Miocene to the Early Pleistocene. In: Reynolds SC, Bobe R, editors. African paleoecology and human evolution. New York: Cambridge University Press; 2022. p. p161–86.

Bobe R, Manthi FK, Ward CV, et al. The ecology of *Australopithecus anamensis* in the early Pliocene of Kanapoi, Kenya. J Hum Evol. 2020;140 https://doi.org/10.1016/j.jhevol.2019.102717.

Bobe R, Geraads D, Wynn JG, et al. Fossil vertebrates and paleoecology of the Pliocene Hadar Formation at Dikika, Ethiopia. In: Reynolds SC, Bobe R, editors. African paleoecology and human evolution. New York: Cambridge University Press; 2022. p. 229–241.

Codron D, Lee-Thorp JA, Sponheimer M, et al. Inter- and intrahabitat dietary variability of chacma baboons (*Papio ursinus*) in South African savannas based on fecal δ13C, δ15N, and %N. Am J Phys Anthropol. 2006;129:204–14.

Codron D, Lee-Thorp JA, Sponheimer M, et al. What insights can baboon feeding ecology provide for early hominin niche differentiation? Intl J Primatol. 2008;29:757–72. https://doi.org/10.1007/s10764-008-9261-x.

Copeland SR. Potential hominin plant foods in northern Tanzania: semi-arid savannas versus savanna chimpanzee sites. J Hum Evol. 2009;57:365–78. https://doi.org/10.1016/j.jhevol.2009.06.007.

Crowley BE, Carter ML, Karpanty SM, et al. Stable carbon and nitrogen isotope enrichment in primate tissues. Oecologia. 2010;164(3):611–26. https://doi.org/10.1007/s00442-010-1701-6.

DeVore I, Washburn SL. Baboon ecology and human evolution. In: Howell FC, Bourliere F, editors. African ecology and human evolution. New York: Aldine; 1963. p. 335–67.

Di Bitetti MS. Primates bark-stripping trees in forest plantations – A review. Forest Ecol Mgt. 2019;449(1) https://doi.org/10.1016/j.foreco.2019.117482.

Dominy NJ, Vogel ER, Yeakel J. Mechanical properties of plant underground storage organs and implications for dietary models of early hominins. Evol Biol. 2008;35(3):159-175. https://doi.org/10.1007/s11692-008-9026-7

Gesquiere LR, Khan M, Shek L, et al. Coping with a challenging environment: effects of seasonal variability and reproductive status on glucocorticoid concentrations of female baboons (*Papio cynocephalus*). Hormones Behav. 2008;54(3):410–6.

Grine F, Lee-Thorp J, Blumenthal S, et al. Stable carbon isotope and molar microwear variability of South African australopiths in relation to paleohabitats and taxonomy. In: Schmidt CW, Watson JT, editors. Dental wear in evolutionary and biocultural contexts. New York: Academic Press; 2020. p. 187–223. https://doi.org/10.1016/B978-0-12-815599-8.00009-5.

Hamilton, WJ III. Baboon sleeping site preferences and relationships to primate grouping patterns. Am J Primatol. 1982;3:41–53.

Hausfater G, Bearce WH. Acacia tree exudates: their composition and use as a food source by baboons. E Afr Wildl J. 1976;14:241–3.

Henry AG, Hutschenreuther A, Paine OCC, et al. Influences on plant nutritional variation and their potential effects on hominin diet selection. Rev Palaeobot Palynol. 2019;261:18–30. https://doi.org/10.1016/j.revpalbo.2018.11.001.

Hill RA. Thermal constraints on activity scheduling and habitat choice in baboons. Am J Phys Anthropol. 2006;129:242–9.

Ibrahim H, Bekele A, Fashing PJ, et al. Feeding ecology of a highland population of hamadryas baboons (*Papio hamadryas*) at Borena-Sayint National Park, northern Ethiopia. Primates. 2023;64:513–26. https://doi.org/10.1007/s10329-023-01077-6.

Johnson C, Swedell L, Rothman JM. Feeding ecology of olive baboons (*Papio anubis*) in Kibale National Park, Uganda: preliminary results on diet and food selection. Afr J Ecol. 2012; https://doi.org/10.1111/j.1365-2028.2011.01316.x.

Johnson C, Raubenheimer D, Rothman JM. 30 days in the life: daily nutrient balancing in a wild chacma baboon. PLoS One. 2013;8(7):e70383. https://doi.org/10.1371/journal.pone.0070383.

Johnson C, Raubenheimer D, Swedell L, et al. Forest baboons maximize energy and prioritize protein: implications for evolution of human diet. Conference: American Association of Physical Anthropologists; 2017.

Kazaba PK, Ngoie CK, Mugaruka RK, et al. Ethnobotanical study of the competition between humans and baboons (*Papio kindae*) for wild fruit trees in the fringe of the Kundelungu National Park, D.R. Congo. Ethnobot Res Appl. 2020;19:1–8. https://doi.org/10.32859/era.19.08.1-11.

Kimbel WH, Villmoare B. From *Australopithecus* to *Homo*: the transition that wasn't. Philos Trans R Soc B. 2016;371(954):20150248. https://doi.org/10.1098/rstb.2015.0248.

Kitegile AS. New diet on the menu: yellow baboon foraging on cassod tree (*Senna siamea*), a trypsin inhibitor legume. Tanz J Forest Nat Conserv. 2022;91:151–62.

Kunz BK, Linsenmair KE. The disregarded West: diet and behavioural ecology of olive baboons in the Ivory Coast. Folia Primatol. 2008;79(1):31–51. https://doi.org/10.1159/000108384.

Laska M, Schüll E, Scheuber H-P. Taste preference thresholds for food-associated sugarss in baboons (*Papio hamadryas anubis*). Int J Primatol. 1999;20(1):25–34.

Lodge E, Ross C, Ortmann S, MacLarnon AM. Influence of diet and stress on reproductive hormones in Nigerian olive baboons. Gen Comp Endocrin. 2013;191:146–54. https://doi.org/10.1016/j.ygcen.2013.06.016.

Lombard M. Sedge foodplants growing in the Cradle of Humankind, South Africa, and *Cyperus Esculentus* tubers (*Patrysuintjies*) as a C4 superfood. Open Quater. 2022;8(5):1–21. https://doi.org/10.5334/oq.110.

Macho G. Pliocene hominin biogeography and ecology. J Hum Evol. 2015;87:78–86. https://doi.org/10.1016/j.jhevol.2015.06.009.

Macho G. The implications of morphology, mechanics, and microstructure of teeth for understanding dietary drivers in human evolution. In: Lee-Thorp J, Katzenberg AM, editors. The Oxford handbook of the archaeology of diet. Oxford University Press; 2016.

Maloney SK, Mitchell D, Mitchell G, Fuller A. Absence of selective brain cooling in unrestrained baboons exposed to heat. Am J Physiol Regul Integr Comp Physiol. 2007;292:R2059–67. https://doi.org/10.1152/ajpregu.00809.2006.

Martin JE, Tacail T, Braga J, et al. Calcium isotopic ecology of Turkana Basin hominins. Nat Comm. 2020;11:3587. https://doi.org/10.1038/s41467-020-17427-7.

McGrew WC, McKee JK, Tutin CEG. Primates in caves: two new reports of *Papio* spp. J Hum Evol. 2003;44:521–6. https://doi.org/10.1016/S0047-2484(03)00042-3.

Mitchell D, Fuller A, Maloney SK. Homeothermy and primate bipedalism: is water shortage or solar radiation the main threat to baboon (*Papio hamadryas*) homeothermy? J Hum Evol. 2009;56:439–46. https://doi.org/10.1016/j.jhevol.2009.03.003.

Mullu D, Solomon M. Feeding ecology of olive baboon (*Papio anubis*) in Arba Minch Forest, Arba Minch Ethiopia. J Nat Sci Res. 2016;6(21):1–5.

Norton GW, Rhine RJ, Wynn GW, Wynn RD. Baboon diet: A five-year study of stability and variability in the plant feeding and habitat of the yellow baboons (*Papio cynocephalus*) of Mikumi National Park, Tanzania. 1987;48(1–2):78–120. https://doi.org/10.1159/000156287

Okecha AA, Newton-Fisher NE. The diet of olive baboons (*Papio anubis*) in the Budongo Forest Reserve, Uganda. In: Newton-Fisher NE, Notman H, Reynolds V, Paterson JD, editors. Primates of western Uganda. New York: Springer; 2006. p. 61–73.

Paietta EN, Weibel CJ, Jansen DA, et al. Troubled waters: water availability drives human-baboon encounters in a protected, semi-arid landscape. Biol Conserv. 2022;274:109740. https://doi.org/10.1016/j.biocon.2022.109740.

Paine OCC, Daegling. The game of models: dietary reconstruction in human evolution. J Hum Evol. 2023;174:103295. https://doi.org/10.1016/j.jhevol.2022.103295.

Paine OCC, Koppa A, Henry AG, et al. Grass leaves as potential hominin dietary resources. J Hum Evol. 2018;117:44–52. https://doi.org/10.1016/j.jhevol.2017.10.013.

Paine OCC, Koppa A, Henry AG, et al. Seasonal and habitat effects on the nutritional properties of savanna vegetation: potential implications for early hominin dietary ecology. J Hum Evol. 2019;133:99–107. https://doi.org/10.1016/j.jhevol.2019.01.003.

Patterson DB, Braun DR, Allen K, et al. Comparative isotopic evidence from East Turkana supports a dietary shift within the genus *Homo*. Nat Ecol Evol. 2019;3:1048–56. https://doi.org/10.1038/s41559-019-0916-0.

Pebsworth PA, Bardi M, Huffman MA. Geophagy in chacma baboons: patterns of soil consumption by age class, sex, and reproductive state. Am J Primatol. 2011;73:1–10. https://doi.org/10.1002/ajp.21008.

Pebsworth P, Huffman MA, Lambert JE, Young SL. Geophagy among nonhuman primates: a systematic review of current knowledge and suggestions for future directions. Am J Phys Anthropol. 2018;168(S67) https://doi.org/10.1002/ajpa.23724.

Pontzer H, Brown MH, Wood BM, et al. Evolution of water conservation in humans. Curr Biol. 2021;31. (in press) https://doi.org/10.1016/j.cub.2021.02.045.

Quinn RL. Isotopic equifinality and rethinking the diet of *Australopithecus anamensis*. Am J Phys Anthropol. 2019;169:403–21. https://doi.org/10.1002/ajpa.23846.

Quinn RL, Lewis J, Brugal J-P, et al. Influences of dietary niche expansion and Pliocene environmental changes on the origins of stone tool making. Palaeogeog Palaeoclimatol Palaeoecol. 2021;562 https://doi.org/10.1016/j.palaeo.2020.110074.

Ransom TW (1981) Beach troop of the Gombe. Bucknell U Press, Lewisburg NY

Reynolds SC. The Southern African sites: Paleoenvironmental syntheses and future research prospects. In: Reynolds SC, Bobe R, editors. African paleoecology and human evolution. New York: Cambridge University Press; 2022. p. 49–65.

Schoeninger MJ, McGrew WC, Phillips CA. Evolutionary implications of non-human primate diets. In: Lee-Thorp IN, Katzenberg MA, editors. The Oxford handbook of the archaeology of diet. Oxford University Press; 2019. p. 1–24.

Sewell LC, Kibii JM, Reynolds SC. The paleoenvironments of Sterkfontein: old questions, new approaches. In: Reynolds SC, Bobe R, editors. African paleoecology and human evolution. New York: Cambridge University Press; 2022. p. 92–101.

Teaford MF, Ungar PS, Grine FE. Changing perspectives on early hominin diets. Proc Natl Acad Sci USA. 2023;120(7):e2201421120. https://doi.org/10.1073/pnas.2201421120.

Wheeler PE. The thermoregulatory advantages of hominid bipedalism in open equatorial environments: the contribution of increased convective heat loss and cutaneous evaporative cooling. J Hum Evol. 1991;21(2):107–15. https://doi.org/10.1016/0047-2484(91)90002-D.

Wheeler PE. The thermoregulatory advantages of heat storage and shade-seeking behavior to hominids foraging in equatorial savannah environments. J Hum Evol. 1994;26(4):339–50. https://doi.org/10.1006/jhev.1994.1021.

Zinner D, Klapproth M, Schell A. Comparative ecology of Guinea baboons (*Papio papio*). Primate Biol. 2021;8:19–35. https://doi.org/10.5194/pb-8-19-2021.

Chapter 5
Animal Foods and the Origins of Meat Eating

5.1 Introduction

Faunivory is a convenient term for the consumption of all forms of animal food, including invertebrates, vertebrates, and eggs. The expansion of C_4 resources in hominin diets was probably due in part to increased consumption of animals that fed on C_4 plants (Martin et al. 2020). Living primates provide clues to the particular kinds of prey animals involved and how they were obtained. Chimpanzee evidence is highly relevant, but has its limitations, especially due to the emphasis on arboreal prey. Baboon analogies augment the chimpanzee evidence in some cases and provide alternative possibilities in others.

Meat can denote the flesh and organs of vertebrates, though another common usage limits it to mammals. Meat, especially from mammals, has played a major role in theories of hominin evolution. Compared to other sources of animal food, many mammals are larger and/or more elusive. Like early hominins and unlike most chimpanzees' faunivory, baboons hunt mammalian prey on the ground.

5.2 Invertebrates

Baboons are like many other primates in consuming a variety of invertebrate animals. Most of them are terrestrial arthropods, including insects and spiders. Some baboons and a few other monkeys are able to access shellfish. These are all food sources that could have been used by early hominins, but would leave little or no evidence after millions of years. The great shell middens beloved by archeologists came later in hominin evolution.

G. E. King, *Baboon Perspectives on Early Human Ancestors*,
https://doi.org/10.1007/978-3-031-36769-4_5

5.2.1 Arthropods

Insects and other arthropods are consumed by most primates and in many human cultures around the world (Lesnik 2017, 2018). The use of arthropods for food is unlikely to be represented in the paleoanthropological record, but primate evidence suggests that this contributed to hominin subsistence. Chimpanzees are famous for their use of probing tools to obtain a variety of insects such as termites, ants, and bees. A baboon analogy suggests at least one source of insect food that did not necessitate the use of tools or require ground penetration of any kind. Baboons differ from many other primates in their exploitation of swarming insects such as locusts, dragonflies, and caterpillars, which are captured by hand. These sometimes provide a group with food for hours or even days (Altmann and Altmann 1970; Schreier et al. 2019).

Three chacma troops were under observation in the Namib desert when they were presented with an outbreak of large grasshoppers (Hamilton et al. 1978). Baboons in one troop ate these insects as an almost exclusive diet while they were available. Various vegetative food items, especially *Acacia* seeds, added some diversity, but troop movements were not oriented to these alternatives. Instead, the baboons remained in the vicinity of a waterhole and one nearby sleeping cliff and moved only a few hundred meters per day. In contrast, in the absence of grasshoppers, their day routes averaged several kilometers and ended at various alternative roosts. At the time of the grasshopper swarm, fecal material under the sleeping cliffs of the two baboon troops upriver established that they too were concentrating on those insects. Yellow baboons at Amboseli also fed on swarming grasshoppers with a capture rate of 75% (Altmann and Altmann 1970).

When parasitic scale insects attacked mopane trees in the home ranges of five chacma troops, the insects became their main food during the outbreak (Hamilton et al. 1978). One troop ignored abundant alternative foods such as seed pods and palm nuts, which were heavily utilized during other times. Baboons in a troop a few kilometers away, with only a few scattered mopane trees in their range, maintained a diet that was nearly animal-free.

Hamadryas baboons at Filoha ate locusts and dragonflies when they swarmed, chasing and jumping up to catch them (Schreier et al. 2019). The majority of attempts were successful and the baboons spent 30–60 min in these feeding sessions. Most of the 200 members of the study group participated. Monthly insect consumption rose to 2% of the monthly diet during peak periods. The researchers noted that this pattern was consistent with observations of other hamadryas baboons and other *Papio* species.

Baboons can also obtain non-swarming insects for food. In at least one population, chacma baboons overturned rocks to feed on underlying invertebrates (Mare et al. 2019). That prey were the goal of the behavior was indicated by the fact that the baboons selected some rocks to turn over while ignoring others. Concentration on medium-sized rocks maximized the balance between food energy obtained and effort expended. Early hominins may well have obtained insects from such terrestrial sources as well as from swarms.

5.2.2 *Aquatic Invertebrates*

From their earliest appearance, hominins were associated with bodies of water of various kinds: lakes, rivers, streams, and wetlands (Andrews 2020). This suggests that aquatic resources may have been of some importance to them. De Chevalier et al. (2022) maintained that exploitation of aquatic food sources is an issue of particular significance for understanding human evolution. Their thinking was that hominins are the primate group with the highest degree of adaptedness to aquatic environments and the only group in which true coastal and maritime adaptations have evolved.

Lakes in the time of *Australopithecus anamensis* were stocked with diverse mollusks and fish (Van Bocxlaer 2020). Baboons have obtained such foods from lakes and also from marine sites. Chacma baboons of Cape Peninsula in South Africa took advantage of intertidal organisms, such as shellfish and the remains of crabs. Crustaceans and mollusks are rich in protein and some provide energy (Lewis and O'Riain 2019). Congruence between observation-based estimates of diet composition and those generated by isotope-based models, using baboon hair and feces, confirmed that these chacmas consumed small amounts of marine foods on a seasonal basis (Lewis et al. 2017). Marine foods were also part of the diet in a population of yellow baboons in Somalia (P. Messeri 1978, cited by Lewis and O'Riain 2019).

5.3 Vertebrates: Fish

Stewart (2010) suggested that early hominins caught fish, a process that advanced from inadvertence to opportunistic hand-catching to deliberate hand-catching. Baboon evidence supports this scenario. Matsumoto-Oda and Collins (2016) reported two cases of fish eating by olive baboons, at sites in Kenya and Tanzania. In both instances the consumers were adults (rather than exploring juveniles) that ate recently dead or dying fish. The opportunity for fish eating occurred by chance, suggesting that it would be difficult for the baboons to eat fish regularly. These researchers noted comparative evidence from orangutans and Japanese macaques, indicating that fish-eating occurred when fruits or other foods were scarce. They viewed this as evidence that fish are favored as a fallback food. Chacma baboons observed in Namibia, like the olive baboons in East Africa, ate dead or dying fish in dry conditions (Hamilton and Tilson 1985).

Supporting Stewart's scenario, the Namibian chacma baboons also captured live fish by hand (Hamilton and Tilson 1985). A river at this desert site periodically dried and left waterholes containing fish. Baboons walked into pools and groped around to collect larger dead fish from the bottom. Live fish were captured as they floundered in shallows or when they surfaced. Occasionally baboons slapped the water surface, stunning fish that they then grabbed. Some individuals entered larger pools and seized active fish beneath the surface, especially under boulders. Large fish, some more than 30 cm long, were captured in this way. This baboon analogy suggests that hand capture of live fish was a possibility for early hominins.

Two factors suggest some doubts about the relevance of the Namibian case to early hominins. First, the fish were made available by unrelated human activities upstream. Second, the extent of baboon fishing may have been related to the absence of competition from piscivorous birds such as herons in this very dry environment. Acknowledging these caveats, the analogy still implies that early hominins might have added fish to their diet wherever and whenever conditions were favorable.

5.4 Terrestrial Vertebrates

Meat can be defined as the flesh and viscera of vertebrates, and may include eggs (Watts 2020). The emergence of regular meat consumption, especially the distinctive exploitation of meat from mammals, is regarded as a major development in hominin evolution. Three hypotheses have been proposed for the adaptive significance of meat eating in primates (Watts 2020). Compared to plant foods, meat (1) is denser in calories, (2) provides higher quality protein, and/or (3) provides essential micronutrients such as vitamins. These hypotheses apply as much to hominins and baboons as they do to other primates.

Baboon data are pertinent to several interrelated questions about meat eating in early hominins. How did the behavior begin? How did it develop to the point of leaving remains detectable by archeology? What kind of behavior was necessary to obtain meat? The sections below consider the evidence for meat eating in early hominins and baboons; baboon analogies for the methods by which early hominins might have obtained meat; and hints from baboons as to the motivational basis for meat eating and procurement in early hominins.

5.4.1 Meat Eating

On current evidence, consumption of mammals seems to have become well established by the time of early *Homo*. Ferraro et al. (2013) critically examined archeological sites dating to more than 3 mya that yielded remains of mammals that were apparently butchered with tools. The sites were isolated in time and space and the data limited, leaving open the question of how important meat was to hominins during that period. Comparison with evidence from three large and well-preserved faunal assemblages dating to about 2 mya led the researchers to infer a fundamental shift in subsistence patterns around that time. They concluded that these hominins transported and ate numerous individuals of small bovid species. They defined "small bovid" broadly as comparable to extant Grant's gazelles, which may weigh anywhere from 35 to 80 kg. Archeology shows that butchery became "more prevalent" in East Africa shortly after 2 mya (Patterson et al. 2019). Calcium isotope analysis supports the implication that a major expansion of faunivory took place after the appearance of *Homo* (Martin et al. 2020).

Comparative evidence from baboons and other primates strongly implies that consumption of animal matter was part of the hominin diet from the beginning of the lineage. It would have consisted mainly of invertebrates and small vertebrates such as birds and lizards. This tendency would have provided the platform for a more distinctive orientation to mammalian prey. Three groups of primates stand out as consumers of meat: chimpanzees, baboons, and the capuchin monkeys of the Americas (Watts 2020). Chimpanzees and baboons demonstrate vertebrate capture in African habitats similar to those occupied by early hominins. Together, they support the view that early hominins ate meat, probably from some early time in the evolution of the lineage.

More distinctively, and more pertinent to the major innovation in hominin subsistence, consumption of mammal flesh is relatively prominent in chimpanzee and baboon diets. The genus *Papio* stands apart from other monkey genera with respect to the breadth of vertebrate species that they exploit, and because mammals are eaten by every baboon species for which relevant data exist (Sommer et al. 2016; Table 5.1), except for the (so far) less-studied Kindas. Baboons eat 21 wild mammal species belonging to 5 different taxonomic orders, as well as immature domestic sheep and goats (Watts 2020).

Papio meat eating spans the geographic and environmental distribution of the genus from the chacma baboons of the Namibian desert (Davies and Cowlishaw 1996) to olive baboons in the Nigerian forest (Sommer et al. 2016) to Guinea baboons in Senegal (Goffe and Fischer 2016) to hamadryas baboons in Ethiopia (Schreier et al. 2019). Particulars differ significantly from one environment to another, with different implications for various aspects of hominin evolution.

Data from olive baboons were gathered in Nigeria at the wettest and most forested site studied to that date (Sommer et al. 2016). Despite abundant wildlife, meat eating was rare compared to many other baboon sites. Mammals were killed at a rate of one during 1291 h of observation. The rarity of meat eating in that location probably reflects the difficulty of acquiring prey animals when vegetation cover is dense. Olive baboons associated with a Ugandan forest ate a prey animal every 30 h (Rowell 1966). Early hominins may have begun to expand the exploitation of land mammals while still closely associated with some forests or woodlands.

Chacma baboons in a mountainous habitat ate vertebrates at a rate of one every 78.5 observation hours (Allan et al. 2022; Table 5.1). A key meat source was young antelopes, particularly bushbuck, which were consumed once every 115 h. The relatively high rate of meat eating suggests that the behavior may have been stimulated when early hominins penetrated the higher altitudes documented in recent archeological findings (Chap. 2). Anecdotal reports on chacma baboons note some of the same prey elsewhere (e.g. Davies and Cowlishaw 1996), but add domestic fowl and goats (Ade n.d.; Jackson 1978; McKee 1992).

Meat seemed to be virtually absent from the diet of hamadryas baboons until they were studied at Filoha, a somewhat richer habitat than most of those occupied by the species (Schreier et al. 2019; Table 5.1). These baboons maintained a small amount of animal matter in the diet throughout the year, alternating between insects and vertebrates. They were seen to eat vertebrates 38 times in a year of observation. The difference between Filoha and other hamadryas sites seems likely to be due to availability in a habitat that is slightly more abundant than the deserts occupied by other populations of the species. Prey availability and the availability of feeding time shaped vertebrate predation (Schreier et al. 2019).

Schreier et al. (2019) addressed a hypothesis of meat eating that may also be relevant to early hominins: Seasonal variation in plant food availability has been suggested to motivate dietary flexibility in a range of species and thus primates may seek more prey when key plant resources are unavailable. They found no relationship between faunivory (either insect or vertebrate) and either rainfall or the consumption of staple plant foods. Thus, in this case at least, faunivory seems to be independent of the supply of plant foods.

Table 5.1 Baboon prey

Prey type	Total (%)	Olive	Yellow	Chacma	Guinea	Hamadryas
Ungulates	135 (40%)	113	11	8 [+39][a]	1	1 [+15][b]
Hares	84 (25.5%)	57	20	1	2	4 [+7][b]
Birds	41 (12.5%)	16	17	7 [+7][a]	4	1 [+11][b]
Primates	30 (9.1%)	0	25	6	0	0
Rodents	18 (5.5%)	1	4	13 [+3][a]	0	0
Herpetofauna	15 (4.6%)	4	10	[4][a]	1	[3][a]
Mammals (?)	5 (1.5%)	3	2	0	0	0
Other	1 (0.3%)	1 bat	0	1 hyrax[a]	0	1 hyrax[b]
All	329 (100%) [+91]	196	88	35 [+54][a]	4	6 [+37][b]

Abridged with gratitude from Sommer et al. (2016), where references for all figures can be found. In a few cases the exact number of prey animals were unknown, so these were counted as one individual in the summaries. Herpetofauna included lizards, a snake, and frogs. In five cases the prey could be identified as a mammal, but not as any particular species. Data from more recent papers are added in brackets

[a]Allan et al. (2022): Chacma baboons in a montane environment. Two cases of scavenging were not counted as predation. Fifteen failed attempts were recorded

[b]Schrier et al. (2019): Hamadryas in scrubland/grassland typical of the species. Of 38 meat-eating episodes, 11 kills were observed. In the other instances, the meat was so fresh that the researchers inferred that it was a recent kill. These data were not broken down by taxa

5.4.2 Foraging and Hunting

Ferraro et al. (2013) concluded from their analysis of archeological sites (see above) that hominins (presumably early *Homo*) engaged in regular hunting of small bovids. This was based on the quantities of the animal bones and the evidence that had been transported to the places where they were discovered. Baboon behavior suggests earlier meat-getting patterns that provided the foundations for such archeologically visible sites.

Both chimpanzees and baboons provide hints as to the earlier faunivorous practices of hominins. Baboon analogies augment chimpanzee evidence in some respects, but provide alternative possibilities in others. Chimpanzees' favored prey are arboreal monkeys (Bugir et al. 2021) that they hunt with the kind of arboreal skill that may never have been in the repertoire of vertically climbing hominins (Chap. 1). Hominin climbing skills may have persisted for a long time, but any approximation to ape-like agility in the trees would have been greatly diminished by the time of early *Homo* (Chap. 1). Baboons find all of their animal foods on the ground (or in adjacent waters). In this respect baboon analogies are more relevant to hominin behavior than chimpanzee patterns.

The simplest and most common way that baboons obtain meat is to happen across a helpless animal, most often an infant antelope hidden in the grass (DeVore and Washburn 1963; Sommer et al. 2016). Such discoveries can be considered an aspect of foraging. However, baboons sometimes engage in higher levels of activity to obtain prey. They hunt some animals in the sense of stalking and/or pursuit. Yellow baboons chase hares and sometimes groups of almost all age/sex classes participate (Hausfater 1976). Also, adult males "definitely hunted or stalked young vervet monkeys in the strict sense of these words" (Hausfater 1976:48). Goffe and Fischer (2016) described Guinea baboons as stalking and chasing prey (although dense vegetation prevented any estimate of the distance covered by chases). Captures included a hare and seven antelopes (three of the latter identified as bushbuck).

A survey of baboon meat acquisition (Sommer et al. 2016) found that 60% of prey animals were simply "grabbed" in a chance encounter. However, 40% were "spotted and then chased." About 90% of the prey animals were immature and all of the ungulate prey were immature. However, immatures ranged in size, strength, and speed from neonates to subadults. Baboons at Gashaka struggled to maintain control of young antelopes (Sommer et al. 2016). Cases like these represent hunting, as opposed to foraging.

The most intense and elaborate hunting practices of any baboons were recorded at Gilgil, Kenya (beginning with Harding 1973; summarized by Strum 1981). Individual olive baboons altered their movements to engage in "active searching." They left the troop to walk through gazelle herds or detoured into thickets to find dik-diks. A pursuit could last for almost 10 min and take the baboon almost 300 m away from the troop.

"Complex" hunts, always directed at Thomson's gazelles (Fig. 5.1), involved more than one baboon and lasted more than 10 min. Searching, stalking, and pursuit

Fig. 5.1 Thomson's gazelle, a prey animal of some baboons. Fawns are easily killed, but hunting larger animals is difficult. (Photo by Glenn King, Ngorongoro Crater, Tanzania)

in some cases totaled as much as 2 h. The direct distance from start to capture could be as much as 1600 m, but the actual distance covered by an individual could be 4000 m. Some cases seemed to suggest intentional coordination of two types: one baboon chased a gazelle in the direction of another hunter, or several baboons pursued the same prey in sequence. A baboon's top speed seemed to be limited to a duration of about 5 min.

The significance of this behavior is subject to a major caveat: there were no other large predators at Gilgil to compete with or endanger the baboons (Strum 1981). This is a situation that would rarely if ever have been encountered by early hominins (Chap. 6). However, the behavior does suggest hunting capabilities available to early hominins when they found ways to deal with other predators, presumably with the development of weapons.

5.4.3 Hunting and Scavenging

Another way to obtain meat is by scavenging, that is, taking it from the carcasses of animals killed by predators or dead from other causes. Archeological evidence and primate analogies support the view that hunting was more common than scavenging in early hominins. Three archeological sites dated to about 2 mya were interpreted as the result of hunting rather than scavenging (Ferraro et al. 2013; see above).

Evidence of butchery indicated that carcasses were acquired in a relatively complete state, providing the hominins with primary access to the flesh and organs. Bovid fossils displayed few or no carnivore tooth marks (see also Parkinson et al. 2021).

Domínguez-Rodrigo et al. (2021) argued for hominin scavenging at a site that they investigated. Baboons scavenge, but the behavior is rare (Watts 2020). Several instances of scavenging were seen at Gilgil, where there were no other predators to compete with the baboons for carcasses (Strum 1981). One report from another site described a single instance of baboons appropriating a bushbuck that a leopard had killed and cached in a tree (Allan et al. 2022). On the other hand, the same study reported a high level of predation by baboons. It also alluded to the absence of baboon encounters with numerous potential prey species, suggesting that the baboons were a key component of several species' "landscapes of fear." Thus, baboon evidence indicates that early hominins obtained most of their meat by means other than scavenging, including foraging and hunting.

An argument that hominins depended on scavenging for meat is that, being bipedal, they could not have chased animals with the speed and agility of baboons. However, baboons demonstrate hunting practices that would have been possible for early hominins. One is the relay chase, such as was practiced by the olive baboons of Gilgil (Strum 1981). Bipedal hominins might have been successful with this tactic against juvenile animals that were mobile but lacked the speed and endurance of adults.

Another possibility is the surround. This has not been reliably reported for hunting by baboons, but the pattern occurs in the mobbing of leopards: if baboons are able to isolate a leopard, "they immediately surround it, alarm calling, and lunging at it, seemingly without fear" (Cheney and Seyfarth 2007:46). Leopards are sometimes injured or killed in such incidents (Altmann and Altmann 1970; see Chap. 6). Leopards are dangerous carnivores and most adult leopards weigh between 40 kg and 90 kg. If baboons can kill a leopard in a surround, early hominins could certainly have used the tactic to kill small or even medium-sized antelopes.

Oliver et al. (2019) added more archeological evidence by comparing bovid mortality profiles at sites located at Kanjera and Olduvai Gorge. The remains of prime adults dominated the bovid profile from the heterogeneous woodland habitat at Olduvai, leading to the inference that the hominins used cover to ambush their prey. On the other hand, juvenile bovids were predominant in the remains from the grassland of Kanjera. In those circumstances, limited cover presumably necessitated opportunistic emphasis on more vulnerable prey, some probably captured after short chases (Oliver et al. 2019). Acquisition of juveniles could have originated in baboon-like behavior of the earliest hominins. Even ambush (Bunn and Gurtov 2014) might have been anticipated by early hominins, according to a baboon analogy. In predation on sheep, goats, and domestic fowl, "the typical hunting strategy involves sitting still until the potential prey is close enough so that it can be leapt at and caught" (Sommer et al. 2016: 74). When two small antelopes ran past an adult male "sitting quietly on a log," he jumped at them (but missed) (Altmann and Altmann 1970).

5.4.4 Meat: Costs and Motivation

In many instances baboons procure meat with relatively little effort, especially in the fortuitous discovery of hidden antelope fawns. However, other lines of evidence indicate significant costs in hunting for meat. This in turn contributes to the inference that baboons are strongly motivated to obtain meat. Along with similar evidence from chimpanzees, this suggests the hominin foundation for elaboration of hunting and scavenging.

Chasing prey can cost baboons significantly in energy expenditure and time away from foraging for plant foods and engaging in social behavior. The ultimate examples are the complex hunts by olive baboons at Gilgil (Strum 1981). Pursuit involved considerable energy expenditure and capture was not necessarily easy. Gazelles that could elude baboons for any length of time were older than infants and almost certainly subadults in some cases.

Prey size is important because hunting entails some degree of control over animals that may struggle and/or flee. The study of Guinea baboons by Goffe and Fischer (2016) provides more specific information. The researchers estimated weights ranging from 10 kg to 14 kg for young antelopes killed by the baboons. Since male Guinea baboons weigh a maximum of about 26 kg, the best case for the hunters was that they were about 2.5 times the size of their prey. This seems to suggest some difficulty in wrestling the prey down. Such difficulties were explicit in the report on the olive baboons at Gashaka that struggled to control young antelopes (Sommer et al. 2016).

Persistence in hunting despite failure is another indicator of motivation. Where pursuit is necessary, unsuccessful chases take up additional time and energy. In the montane chacma baboon study (Allan et al. 2022), 4 of 12 hunts were unsuccessful, 2 of which targeted antelopes. Persistence has another dimension in the case of baboons at Gashaka (Sommer et al. 2016). The very low rate of meat eating may be due to the difficulty of finding prey in the wet forest. The fact that it occurs at all under these circumstances suggests that there is a predisposition to the behavior.

Persistence in the face of cost took a different form in at least one case arising from passive defense by prey rather than flight. Chacmas in the De Hoop Reserve fed on tortoises six times in 13 months, but failed in ten attempts to break the carapaces (Hill 1999).

Motivation for meat eating is further evidenced by baboon behavior in response to possession of a fresh carcass. The interest of others is sometimes expressed passively in the "vulture response" of sitting near the possessor and staring at him (Altmann and Altmann 1970; Sommer et al. 2016). These individuals might be rewarded by scraps left behind. In some of these situations, scraps are taken while the meat possessor is distracted (Allan et al. 2022).

Possessors of carcasses usually resist sharing. Females sometimes groom a male in order to get some meat or an entire carcass (Sommer et al. 2016). Voluntary sharing takes place between individuals with special connections, most often in male–female relationships (Goffe and Fischer 2016; Schrier et al. 2019; Strum 1981). In at least one case an adult female shared with her juvenile son (Sommer et al. 2016).

There are also attempted and successful appropriations. In one case there was "much fighting among the adult males" over an infant gazelle (Altmann and Altmann 1970: 154). In another case two coalition partners obtained a carcass from a third adult male (Sommer et al. 2016).

5.5 Summary and Discussion

Baboon and chimpanzee evidence leads to a strong hypothesis that hominins consumed invertebrate animals from the time of their origin from the LCA. Insects would have been foremost among terrestrial prey. Baboon analogies demonstrate that the earliest hominins could have obtained insects without using tools or otherwise penetrating nests or the ground. Swarming insects, such as grasshoppers and dragonflies, could be snatched from the air with high rates of success, or could be taken from trees. In at least one population, baboons turned over rocks to prey on invertebrates beneath them. These analogies suggest that terrestrial invertebrates could have provided early hominins with a small and probably irregular part of the faunivorous diet. However, this might have had disproportionate significance with regard to protein and/or other nutrients that plants did not provide with the same value, or at all.

Closely associated with lakes and other bodies of water from their beginnings, hominins would have had access to aquatic animals for food. Baboons (and a few other primates) take prey from both fresh and salt waters. Some, such as shellfish, were more difficult to process than insects, but baboons and other primates show that this was possible for early hominins. Baboons also show that fish could be obtained from ponds and the shallow edges of lakes. Some of the fish, dead or dying, are merely gathered by baboons. However, baboons can enter the water to hand-capture live fish.

Baboons also hand-capture terrestrial vertebrates, such as various species of hares and antelopes. Here again, combined with chimpanzee data, baboons provide evidence that the earliest hominins engaged in such behavior without leaving identifiable archeological remains. With regard to the further development of hominin faunivory, there is a key difference between chimpanzees and baboons. The main prey animals of chimpanzees are arboreal monkeys, while baboons find virtually all of their prey on the ground (and occasionally in adjacent waters). In this respect baboons are a better model for early hominin faunivory. Though still capable of some effective arboreal behavior, probably vertical climbing, the hominins would not have had the acrobatic agility of chimpanzees that allows them to capture monkeys in the trees or harass them to the point of falling to the ground. The decline in arboreal capability was probably related to increasing adaptation to terrestrial bipedalism.

Early hominins displayed a variety of bipedal adaptations, none of which were as efficient or speedy as the modern human gait. Even modern humans find it difficult to run down many species of animals. Early hominins could not have pursued prey in the same way as the typical baboon chase. However, baboons display several meat-getting behaviors that could have been used by early hominins. First, the bulk of the meat obtained by baboons comes from accidental encounters with small and helpless immatures, such as infant antelopes in grassy hiding places. Second, relay chases could have brought down fast but quickly tiring animals. Third, early hominins could have surrounded their prey. Finally, they might have learned to ambush their prey.

Potential for ambush is represented in baboons only by rare and rudimentary responses to potential prey. Surrounds are not reliably reported for hunting, but sometimes occur when baboons attack leopards. Relay hunts occur in baboons. It is not clear that the apparent cooperation is conscious, but the behavior pattern is there and available for elaboration, as it might have been for early hominins.

Some mammalian prey of baboons, such as subadult antelopes and sheep, approach 15 kg in weight. This is presumably the limit for baboons, since they struggle to bring these animals down. It is likely to be analogous to the last stage of hominin faunivory before weapons and social coordination made hunting a regular feature of hominin behavior. From that point on, increasingly large animals were taken and archeological evidence for hunting became clear-cut. By about 2 mya the marks of butchery show that hominins ate from the carcasses and the fact that they underlie the marks of carnivore teeth (if any) show that the hominins were there first. Scavenging might have become more important from that point on because weapons and social coordination facilitated competition with large carnivores.

Several lines of evidence indicate the motivational importance of meat to baboons (as well as chimpanzees): (1) They hunt even though it sometimes involves considerable expenditure of time and energy. (2) They persist in hunting despite failures. (3) Individuals show great interest in meat possessed by others, including the vulture response and attempts at appropriation. (4) Meat possessors resist sharing, sometimes taking evasive actions such as climbing trees. (5) Voluntary sharing only takes place within special relationships, such as a long-term affiliation between a particular male and female.

It seems likely that early hominin diets increasingly resembled those of savanna baboons as the hominins underwent two processes: expansion into more open habitats and adaptation to long-term aridification (Chaps. 1 and 2). Eventually the hominins built on that foundation with quantitative changes in their protein sources: more fish, more aquatic invertebrates, more and larger land mammals. The addition of big game hunting, probably by *Homo erectus*, was one of the markers of the end of the early hominin phase of human evolution.

A frequent topic of controversy is the role of females in the kind of hunting society that has just been described. Whenever hunting began to gain importance among early hominins, it was probably practiced largely by males as is the case among baboons. Some females probably engaged in some hunting, but most females would have been slowed by larger and more helpless offspring than those that characterized their ancestors. However, females could have obtained meat as part of their foraging activities, and bipedalism might have allowed them to accumulate and carry foods such as USOs in sufficient quantities to exchange with males for meat.

References

Ade B. Bloodthirsty baboons. Afr Wildlife. n.d.;39(1):25–28.

Allan A, LaBarge L, Anderson C, et al. Patterns of predation and meat-eating by chacma baboons in an Afromontane environment. Folia Primatol. 2022; https://doi.org/10.1163/14219980-bja10004.

Altmann SA, Altmann J. Baboon ecology: African field research. Chicago: University of Chicago Press; 1970.

Andrews P. Last common ancestor of apes and humans: morphology and environment. Folia Primatol. 2020;91:122–48. https://doi.org/10.1159/000501557.

Bugir C, Butynski TM, Hayward MW. Prey preferences of the chimpanzee (Pan troglodytes). Ecol Evol. 2021;11(3):1–9. https://doi.org/10.1002/ece3.7633

Bunn HT, Gurtov AN. Prey mortality profiles indicate that Early Pleistocene *Homo* at Olduvai was an ambush predator. Quater Internatl. 2014;322–323:44–53. https://doi.org/10.1016/j.quaint.2013.11.002.

Cheney DL, Seyfarth RM. Baboon metaphysics: the evolution of a social mind. Chicago: University of Chicago Press; 2007.

Davies JG, Cowlishaw G. Interspecific competition between baboons and black kites following baboon-ungulate predation in the Namib desert. J Arid Environ. 1996;34:247–9.

de Chevalier G, Bouret S, Bardo A, et al. Cost-benefit trade-offs of aquatic resource exploitation in the context of hominin evolution. Front Ecol Evol. 2022;10:812804. https://doi.org/10.3389/fevo.2022.812804.

DeVore I, Washburn SL. Baboon ecology and human evolution. In: Howell FC, Bourliere F, editors. African ecology and human evolution. New York: Aldine; 1963. p. 335–67.

Domínguez-Rodrigo M, Courtenay LA, Cobo-Sanchez L, et al. A case of hominin scavenging 1.84 million years ago from Olduvai Gorge (Tanzania). Ann N Y Acad Sci. 2021; https://doi.org/10.1111/nyas.14727.

Ferraro JV, Plummer TW, Pobiner B, et al. Earliest archaeological evidence of persistent hominin carnivory. PLoS One. 2013;8(4):1e62174–10.

Goffe A, Fischer J. Meat sharing between male and female Guinea baboons (*Papio papio*). Primate Biol. 2016;3(1):1–8. https://doi.org/10.5194/pb-3-1-2016.

Grine F, Lee-Thorp J, Blumenthal S, et al. Stable carbon isotope and molar microwear variability of South African australopiths in relation to paleohabitats and taxonomy. In: Schmidt CW, Watson JT, editors. Dental wear in evolutionary and biocultural contexts. New York: Academic Press; 2020. p. 187–223. https://doi.org/10.1016/B978-0-12-815599-8.00009-5.

Hamilton WJ III, Tilson RL. Fishing baboons at desert waterholes. Am J Primatol. 1985;8:255–7. https://doi.org/10.1002/ajp.1350080308.

Hamilton WJ III, Buskirk RE, Buskirk WH. Omnivory and utilization of food resources by chacma baboons, *Papio ursinus*. Am Nat. 1978;112:911–24. www.jstor.org/stable/2460166

Harding RSO. Predation by a troop of olive baboons (Papio anubis). Am J Phys Anthropol. 1973; 38(2):587–591. https://doi.org/10.1002/ajpa.1330380272

Hausfater G. Predatory behavior of yellow baboons. Behaviour. 1976;56:44–68.

Hill R. Size-dependent tortoise predation by baboons at De Hoop Nature Reserve, South Africa. S Afr J Sci. 1999;95(3):123–4.

Jackson HD. Chacma baboons preying on higher vertebrates. Arnoldia. 1978;8(25):1–8. (Miscellaneous publications, National Museum Rhodesia)

Lesnik J. Not just a fallback food: global patterns of insect consumption related to geography, not agriculture. Am J Hum Biol. 2017;29(4) https://doi.org/10.1002/ajhb.22976.

Lesnik J. Edible insects and human evolution. University of Florida; 2018.

Lewis MC, O'Riain MJ. The ecology of chacma baboon foraging in the marine intertidal zone of the Cape Peninsula, South Africa. In: Nowak K, Barnett AA, Matsuda I, editors. Primates in flooded habitats. Cambridge UK: Cambridge University Press; 2019. p. 148–51.

Lewis MC, West AG, O'Riain. Isotopic assessment of marine food consumption by natural-foraging chacma baboons on the Cape Peninsula, South Africa. Am J Phys Anthropol. 2017;165:77–93. https://doi.org/10.1002/ajpa.23332.

Mare C, Landman M, Kerley GIH. Rocking the landscape: Chacma baboons (*Papio ursinus*) as zoogeomorphic agents. Geomorphology. 2019;327:504–10. https://doi.org/10.1016/j.geomorph.2018.11.0280169-555X.

Martin JE, Tacail T, Braga J, et al. Calcium isotopic ecology of Turkana Basin hominins. Nat Comm. 2020;11:3587. https://doi.org/10.1038/s41467-020-17427-7.

Matsumoto-Oda A, Collins AD. Two newly observed cases of fish-eating in anubis baboons. Ltrs Evol Behav Sci. 2016;7(1):5–9. https://doi.org/10.5178/lebs.2016.41.

McKee JK. Observations on the carnivorous activities of chacma baboons at the Buxton Limeworks, Taung District, Bophuthatswana. S Afr J Sci. 1992;88:299–300.

Oliver JS, Plummer TW, Hertel F, Bishop LC. Bovid mortality patterns from Kanjera South, Homa Peninsula, Kenya and FLK-Zinj, Olduvai Gorge, Tanzania: evidence for habitat mediated variability in Oldowan hominin hunting and scavenging behavior. J Hum Evol. 2019;131:61–75. https://doi.org/10.1016/j.jhevol.2019.03.009.

Parkinson JA, Plummer TW, Oliver JS, Bishop LC. Meat on the menu: GIS spatial distribution analysis of bone surface damage indicates that Oldowan hominins at Kanjera South, Kenya had early access to carcasses. Quater Sci Rev. 2021;277(1):107314. https://doi.org/10.1016/j.quascirev.2021.107314.

Patterson DB, Braun DR, Allen K, et al. Comparative isotopic evidence from East Turkana supports a dietary shift within the genus *Homo*. Nat Ecol Evol. 2019;3(7) https://doi.org/10.1038/s41559-019-0916-0.

Rowell T. Forest living baboons in Uganda. J Zool Lond. 1966;149:344–364.

Schreier AL, Schlaht RM, Swedell L. Meat eating in wild hamadryas baboons: opportunistic trade-offs between insects and vertebrates. Am J Primatol. 2019;e23029 https://doi.org/10.1002/ajp.23029.

Sommer V, Lowe A, Jesus G, et al. Antelope predation by Nigerian forest baboons: ecological and behavioural correlates. Folia Primatol. 2016;87:67–90. https://doi.org/10.1159/00044583.

Stewart KM. The case for exploitation of wetlands environments and foods by pre-sapiens hominins. In: Cunnane SC, Stewart KM, editors. Human brain evolution: the influence of freshwater and marine resources. New York: Wiley; 2010. p. 137–71.

Strum SC. Processes and products of change: Baboon predatory behavior at Gilgil, Kenya. In: Harding RSO, Teleki G, editors. Omnivorous primates, gathering and hunting in human evolution. New York: Columbia University Press; 1981. p. 255–302.

Van Bocxlaer B. Paleoecological insights from fossil freshwater mollusks of the Kanapoi Formation (Omo-Turkana Basin, Kenya). J Hum Evol. 2020;140:102341. https://doi.org/10.1016/j.jhevol.2017.05.008.

Watts DP. Meat eating by nonhuman primates: A review and synthesis. J Hum Evol. 2020;149(Supplement 2):102882. https://doi.org/10.1016/j.jhevol.2020.102882.

Chapter 6
Predation on Hominins

6.1 Introduction

Early hominins lived with a greater number and a greater variety of predatory animals than exist in Africa today. Survival in these circumstances was a crucial aspect of early hominin evolution. Hominins were subject to ambush in relatively dense vegetation and pursuit in more open areas. Baboon analogies pertain to hominin susceptibility to various predators and potential responses to them. Carnivorans (mammals of the order Carnivora, as distinguished from carnivorous animals in general) must have presented the greatest danger to hominins. Particularly significant were leopards, their ancestors and relatives, and leopard-like saber-toothed felids. Baboon behavior suggests a variety of responses that would have been available to hominins, ranging from avoidance to counter-attacks.

Fossil evidence for predation on hominins is rare, which might seem to imply little danger, but observations of living baboons suggest that this is illusory. If the paleoanthropological record accurately reflects a low incidence of mortality from predation, it might be due to relatively effective countermeasures by the hominins. Baboon analogies suggest that this is the case.

6.2 Predators and Predation

Carnivorans are the greatest danger to extant baboons and this was probably true for early hominins as they evolved to an increasingly terrestrial way of life, and especially as they expanded into more open habitats (Willems and van Schaik 2017). The species that attack baboons all existed and/or had counterparts at various times during the era of early hominins (Lewis 2017).

G. E. King, *Baboon Perspectives on Early Human Ancestors*,
https://doi.org/10.1007/978-3-031-36769-4_6

Extant carnivorans display two basic modes of predatory attack (Hopley et al. 2023). One is ambush: the predator awaits its prey under some kind of cover. The other is pursuit (also called "coursing"). Pursuit implies a run that extends beyond the brief rush often made by an ambush hunter. A stalking phase may or may not be incorporated into either approach (Treves and Palmqvist 2007). Most carnivorans specialize, but some are skilled in both kinds of attack.

Inferences about the hunting behavior of prehistoric carnivorans can be made on the basis of morphology and comparative evidence. For instance, powerful forelegs suggest the grasping ability of an ambush hunter, while long legs indicate pursuit. Ambush hunters are most effective in woodland or in more open areas with large enough patches of cover (like modern leopards). Pursuit hunters are more effective in more open areas (like modern wild dogs). Jaws and teeth for crunching bone suggest scavenging rather than hunting; however, this signal is somewhat ambiguous because the extant spotted hyena is a specialized bone cruncher and also a formidable pack hunter (Kruuk 1972).

6.2.1 Potential Predators of Hominins

Early hominins lived among a variety of dangerous carnivorans (Table 6.1). Hominins seem to have originated during a time when modern carnivoran families, felids and hyaenids, became more common and replaced more archaic taxa (Lewis 2017). The earliest hominins (about 7 to 4 mya) shared their woodland habitats with saber-toothed cats (*Dinofelis* and *Machairodus*), both probably ambush hunters. The long-legged "hunting hyena" (*Chasmaporthetes*) might have been an ambush threat in woodland, but might also have attacked any hominin that wandered into the open. *Agriotherium*, a huge bear must have favored larger prey, but might have taken hominins in sudden woodland encounters and could have run them down in the open. With longer legs than extant bears, it was a faster runner. These and other ancient carnivorans are described in more detail in Table 6.2.

At neighboring sites in East Africa dating to more than 4 mya, *Australopithecus anamensis* thrived in mosaic environments; after 3.8 mya the hominins began to exploit grassland resources (Bobe et al. 2020). During this transitional period, danger from ambush predators presumably continued in woodland and arose more often near concentrations of trees and bushes in savannas. At the same time, hominins must have been exposed to more risk from pursuit hunters.

A host of carnivorans were added to the mix after 3.6 mya, including two more sabertooths, a giant hyena, and probable ancestors and/or close relatives of extant lions, leopards, cheetahs, and hyenas (Table 6.2). Many were "roughly the equivalent of modern species" (Lewis 2017: 35). The origin of taxa more like those in recent Africa, strengthens the argument for analogies with baboons. Meanwhile, the ancient saber-toothed genus *Dinofelis* and the "hunting hyena" *Chasmaporthetes* were still potential threats, as they persisted until about 1 mya.

Table 6.1 African large carnivores

Genus	Body mass in kg	Description	Habitat	Attack mode	Time Span
Machairodus	130–230	Saber-toothed cat	Open > woodland	Stalk/ambush	10 mya to 10 kya
Dinofelis	70–100 [150–230]	Saber-toothed cat	Woodland > open	Ambush/stalk	7 to 1 mya
Chasmaporthetes	50–70 [50–55]	Long-legged hyena	Open > woodland	Pursuit/pack	7 to 1 mya
Agriotherium	600–700	Fast-running bear	Woodland > open	Opportunist/ pursuit	6 to 3.6 mya
Megantereon	80–110 [40–65]	Saber-toothed cat	Woodland > open	Ambush/stalk	4(?) to 1mya
Homotherium	150–230 [115–345]	Saber-toothed cat	Open > woodland	Stalk/pursuit	4.2 to 1 mya
Pachycrocuta	100–130 [60, $n = 1$]	Giant hyena	All	Opportunist/ pursuit/pack	4(?) to 1 mya
Crocuta	~60 [40–65, *C. crocuta*]	Spotted hyena	Open > woodland	Pursuit/pack	4 mya to present
Hyaena	35–40 [35–45]	Striped hyena	All	Opportunist	4 mya to present
Parahyaena	35–40 [35–45]	Brown hyena	All	Opportunist	4.2 mya to present
Panthera leo	~170 [135–225]	Modern lion	Open > woodland	Stalking/brief pursuit	3.6(?) mya to present
Acinonyx	~50 [90, $n = 1$]	Fossil/ modern cheetah	Open > woodland	Stalking/long pursuit	3.6 mya to present
Panthera pardus	~55 [30–50]	Modern leopard	All	Stalking/ ambush	3.6(?) mya to present
Lycaon	25–40 [20–35]	Wild dog	Open > woodland	Pursuit/pack	2.4 mya to present

This table synthesizes information from Treves and Palmqvist (2007), Hopley et al. (2023), Bobe et al. (2020), Kuhn et al. (2016), and O'Regan and Reynolds (2009). Mass figures are from Treves and Palmqvist (2007); those in brackets are from South African species reconstructed by O'Regan and Reynolds (2009)

Paleoanthropological research can clarify some particular hominin–carnivore relationships, as shown by two examples from South Africa. Stable isotope research on fossils from one locality found that some potential predators had C_3 signatures similar to those of the hominins while others did not (Lee-Thorp et al. 2001). On this basis the strongest candidates for hominin predators were leopard, spotted hyena, and the saber-toothed *Megantereon.* The saber-toothed *Dinofelis* and the hyena *Chasmaporthetes*, having more mixed diets, were probably less likely to have attacked hominins. At another site, stratified remains of early *Homo* and leopards traced the relationship between the species across time in the context of cave

Table 6.2 Characteristics of some Plio-Pleistocene African carnivorans

Taxa	Time span	Limbs/body	Jaws/Teeth	Comments
Sabertooths				
Dinofelis	7–1 mya	Early and late species jaguar-sized; robust forelimbs for grappling (Hopley et al. 2023)	Moderate "sabers"—maybe slightly larger prey than jaguar (Anton n.d.)	Ambush
Megantereon	4–1 mya	Similar to leopard in body size but some closer to jaguar (Anton n.d.); short legs, very strong forelegs (Hopley et al. 2023)	Sabertooth features of neck and skull suggest killing larger prey than *Dinofelis* (Anton n.d.)	Ambush; less common than other sabertooths (Hopley et al. (2023)
Homotherium	4–1 mya	Short hind legs and back, poor for acceleration; moderate speed for longer distances; body size overlapped lions but lighter weight (Anton n.d.); long forelimbs with poor grasping ability (Werdelin and Lewis 2020)	Large incisors in arc like dogs and hyenas = greater role in grasping prey; possible pack hunter (Anton n.d.)	Short pursuit (maybe able to catch bipeds)
Other Felids				
Panthera cf. *pardus* (leopards)	3.6? mya to present	Fossil forms that were ancestors or close relatives of the extant leopard		Probably ambush
Panthera cf. *leo* (lions)	3.6? mya to present	Fossil forms that were ancestors or close relatives of the extant lion. Some larger		Probably emphasized prey larger than hominins
Acinonyx (cheetahs)		Fossil species larger than extant		Greater size might have allowed some predation on hominins
Hyaenids				
Chasmaporthetes	7–1 mya	"hunting hyena" with long legs		Pursuit
Parahyaena howelli	c. 4 mya		Generalized hyena without strong scavenging adaptations (Hopley et al. 2023; Werdelin and Lewis 2020)	Scavenging and pursuit (Hopley et al. 2023)
Pachycrocuta	4–1 mya	Giant hyena		

(continued)

Table 6.2 (continued)

Taxa	Time span	Limbs/body	Jaws/Teeth	Comments
Crocuta	2.5 mya–present		Earlier species not specialized for scavenging (Hopley et al. 2023)	Possible pursuit like extant species

References: Anton (n.d.), Hopley et al. (2023), Werdelin and Lewis (2020)

occupation (Pickering et al. 2008). The presence of leopards seemed to discourage hominin activity. It seems a reasonable speculation that the hominins still lacked the ability to drive the cats out of a desirable cave.

Some paleoanthropologists have argued that the relationship between hominins and their predators has been distorted by problems in excavation and interpretation. They maintained that the abundance and diversity of carnivores associated with hominins has been exaggerated because fossils from different time periods have been lumped together, forming a "palimpsest" (Hopley et al. 2023; O'Regan and Reynolds 2009). Others disagreed (e.g., Geraads and Daujeard 2011).

This is an important issue, but not a crucial one for the discussion presented here. Hominins during each time period lived with at least some of these carnivorans and were probably preyed upon by some of them. Furthermore, regardless of the number or taxonomic classification of coeval carnivoran taxa, hominins always faced the same two basic modes of attack: ambush and pursuit. The main shift in predator pressure probably came from increasing use of relatively open habitats by many hominin populations.

6.2.2 Baboon Predators

An abundance of large herbivores provided potential food for prehistoric carnivorans. This raises the question of whether or not predation on hominins was frequent enough to have a significant effect on hominin ecology and/or behavior (Treves and Palmqvist 2007). Contemporary Africa is broadly analogous to the Plio-Pleistocene situation in harboring a variety of large herbivores and large carnivorans, although the numbers and diversity of both are reduced (Willems and van Schaik 2017). The place of baboons in the modern biomes indicates that early hominins were subject to significant predation. All the large African carnivorans include baboons in their diets, even though they prefer other prey.

A survey of game reserves in southern and eastern Africa found that baboons had rarely been reported as prey, but the available data contained 97 instances of baboons killed by predators (Saayman 1971). Fieldwork by scientists began to add additional cases (Busse 1980; Stelzner and Strier 1981; Rhine et al. 1980). Synthesis of cases where a specific predator was identified or inferred provides the following distribution: leopard—78; lion—14; hyena—8. Cowlishaw (1994) obtained a comparable result in a survey of reports from across sub-Saharan Africa: leopards preyed on baboons at every site and lions in fewer localities, while hyenas of all species provided only a few cases.

Reports of baboons killed by leopards have continued to accumulate and provide more details (e.g., Bidner et al. 2018; Allan et al. 2021). Ordinarily, baboons are a relatively small part of the leopard diet (no more than 5% by biomass), but the rate increases under some circumstances. During a 5-month study of three female leopards in a montane region of South Africa, baboons equaled impala as the most frequent prey, each with 18 kills amounting to around 20% of the total biomass (Jooste et al. 2012).

Even where the percentages are low, baboons sometimes rank among the most frequent prey of leopards. Research in South Africa placed chacma baboons among the three most common animals in the leopard diet (along with bushbuck and duiker) (Williams et al. 2018). Another South African study also found baboons to be among the "main prey species" of leopards (Greyling et al. 2023).

Hyaenid predation on early hominins may have been somewhat higher than it is for extant baboons, because the hyenas were more diverse and numerous. However, the baboon evidence suggests that felids were always the most important in this regard. All early hominins lived with leopards, leopard ancestors or relatives, or leopard-like sabertooths such as *Megantereon* (Table 6.2).

Rowell (1966) suggested that baboon victims are "occasional young stragglers," which seems to imply that predation has no significant behavioral or demographic effect on baboon groups. However, data from studies cited above tell a different story. Of 18 cases for which age/sex categories were reported, the baboons killed were 9 adult males, 5 adult females, and 4 immatures (Busse 1980; Stelzner and Strier 1981; Rhine et al. 1980; Saayman 1971). The survey by Cowlishaw (1994) also indicated that adult males were taken more frequently than females or young. Only in the unusual situation described by Jooste et al. (2012), where baboons were killed with the same frequency as impala, was there no sex difference among the victims.

It is possible that leopards prefer adult male baboons because they provide more meat than any other age-sex class. Other hypotheses were summed up by Cowlishaw (1994): (1) many adult males are peripheral and distant from nearest neighbors; (2) maturing and adult males go through solitary periods when transferring between groups; (3) leopards would prefer smaller baboons but usually find them in the center of a group.

Records of predation on baboons have been affected by two developments. First, hunting has greatly diminished carnivore populations in most parts of Africa. Second, probably because of hunting, the presence of human observers tends to result in avoidance by carnivores (Washburn et al. 1965). Consequently, data collected over the last 70 years likely represent the minimum of predation on baboons. Baboons probably faced greater danger from predators before that, and early hominins probably faced even greater danger.

6.2.3 Body Mass of Predators and Prey

Body mass is an important element in predator–prey relationships. One basic factor is the degree to which meat was essential to the predators. Studies of extant carnivores have established that species in which individual weight exceeds 21.5 kg are obligate meat eaters (Carbone et al. 1999; O'Regan and Reynolds 2009). That is, they must have a large quantity of meat on a regular basis to survive. All of the carnivores in Table 6.1 exceeded that threshold (Lewis 2017; O'Regan and Reynolds 2009).

This implies a readiness to exploit a variety of sources for meat. Some meat can come from scavenging, but this may set up intense and potentially dangerous competition among obligate carnivores. Even extant spotted hyenas, highly adapted for scavenging, are also very effective hunters (Kruuk 1972). Ungulates were undoubtedly the mainstay of the ancient carnivoran diet, but the need for meat might have led many to include early hominins (Treves and Palmqvist 2007).

The size of a predator relative to potential prey is important for two reasons. First, the predator must be large enough to take down the prey. Second, the predator must not be so large that the prey is not worth the effort (caloric expenditure) under ordinary circumstances. Application of this principle to hominin-baboon analogy requires an adjustment because early hominins were significantly larger than extant baboons (Chap. 2).

Median size of adult baboons, across all species, for adult males and females, is roughly 20 kg (range of about 10 to 30 kg, Fischer et al. 2019). Leopards and hyenas tend to be almost twice the size of baboons, and lions are about 7 to 11 times larger (cf. Table 6.1). These ratios provide the adjusted analogy for early hominins. According to recently applied data and techniques, the median body mass of early hominins, including earliest *Homo*, was about 42.5 kg (range of 25–60 kg, average ~43 kg, Ruff and Wood 2023). The relative sizes suggest that hominins might have been a small but regular part of the diet of many carnivores in the range of roughly 80–400 kg. This applies with varying degrees of probability to five of the extinct genera in Table 6.1

Several qualifications are in order. First, a powerful animal like a leopard can kill prey of its own weight, such as an extant human (Camaros et al. 2015), or greater, such as a gorilla (Fay et al. 1995). Therefore, even large adult male hominins were potential prey for early leopards or their close relatives, or sabertooths with similar capabilities. Extant leopards prefer prey in the range of 10–40 kg with the strongest preference at about 25 kg (Hayward et al. 2006), which suggests that female and young hominins may have been particularly at risk.

Second, even the smaller predators in Table 6.1 could have preyed on young hominins and might also have taken adult females. This is especially so if the females were much smaller than males as some interpretations of the fossils propose (Chap. 2). Extant cheetahs, for example, sometimes expand their prey base by killing juveniles of larger species (Annear et al. 2023). Extant cheetahs are not known to prey on baboons, but some of their prehistoric ancestors and relatives were larger.

Third, carnivores hunting in packs can bring down animals much larger than themselves. This is true of extant spotted hyenas and wild dogs, and may apply to fossil taxa such as the hyenas *Pachycrocuta* and *Chasmaporthetes*. Based on the behavior of extant hyenas, the hunting group might be a large pack or just two or three individuals. Smaller groups might profitably hunt smaller prey, such as hominins.

6.3 Defenses Against Carnivorans

The evidence for carnivorans that were contemporary with early hominins raises the question of how hominins survived this danger. Baboons demonstrate a number of different ways in which hominins might have coped.

6.3.1 Avoidance and Flight

Baboons can avoid predators by tactical movement during travel, such as staying away from dense vegetation that would provide cover for leopards (Altmann and Altmann 1970). Another avoidance tactic is to locate activities according to levels of danger (Hill and Weingrill 2007). A study of chacma baboons found that they spent more time feeding in low-risk relatively food-poor habitat than in a high-risk food-rich habitat; they also preferred the low-risk areas for other activities, including resting and grooming (Cowlishaw 1997a, b).

Baboons also keep predators at a distance by associating with certain other species. Buffalo and elephant intimidate predators (Fig. 6.1). Antelopes such as impala and bushbuck add to vigilance (Fig. 6.2), complementing baboon eyesight with their sense of smell; baboons respond to their warning barks (Washburn and DeVore 1961). Badenhorst (2018) inferred such behavior for early hominins, including *Australopithecus africanus*. He noted that the extinct *Gazella helmoedi* is thought to be a South African representative of the impala lineage that gave rise to one of the important "partners" of extant baboons.

Given warning, baboons are likely to flee. However, flight is complicated when the threat is "interior" rather than "exterior" (Itani 1967). That is, the predator seems to appear suddenly in the midst of a group. One reaction is rapid random motion that probably serves to confuse the predator. When a leopard jumped from bushes into the midst of a baboon group, the baboons responded with "a flurry of animals running in several directions" while uttering "a sudden barrage of loud, plosive vocalizations" (Altmann and Altmann 1970).

When baboons flee from danger, they seek shelter in trees or on cliffs. A study of wild baboons demonstrated the tactical relationship between refuges and behavioral

Fig. 6.1 Cape buffalo. Baboons associate with animals, such as buffalo and elephants, that intimidate predators. (Photo by Glenn King. Manyara, Tanzania)

Fig. 6.2 Baboons with an impala. The vision of the baboons and the olfactory sense of the antelopes complement each other in predator defense. (Photo by Glenn King. Tarangiri, Tanzania)

ecology. If refuges were readily available, a high-risk area was used more intensively. If refuges were scarce, the baboons adopted a time-minimizing strategy and left the area as soon as possible (Cowlishaw 1997a, b). As long as early hominins retained a significant degree of climbing ability (Chap. 2), they probably responded in similar ways.

However, reaching a refuge, even if close by, must have been more of a problem for hominins than for baboons. As bipeds, the hominins would not have the same speed as baboons. This might not have been a great issue in the woodland habitats of the earliest hominins, but that would have changed with increasing use of open areas (Willems and van Schaik 2017). Probably another factor was the declining arboreal abilities of hominins as they committed to a terrestrial way of life (Chap. 2). Here the defensive capabilities of baboons become an important analogy for early hominins.

6.3.2 Active Defense

Willems and van Schaik (2017) performed a literature survey and comparative analyses of primate "counter-attacks" against predators, covering 16 species of 13 genera. Of the 40 reports that they found, 60% pertained to either baboons or chimpanzees. The authors drew analogies to early *Homo*, but populations of later *Australopithecus* were also comparable in terms of increasingly diverse habitats, including savannas (Chap. 2). The general conclusions about primate counterattacks were:

1. They were more common than expected among terrestrial taxa in open habitats.
2. They were often unprovoked and proactive.
3. Males were four times more likely to be involved than females.
4. Males were 6.4 times more likely to jointly counterattack carnivorans compared with all other predators combined.
5. In 25% of the relevant reports (8 of 36), the predator was killed.

The contribution of baboons to these results is clear. Male baboons engage in active defense against predators (Fig. 6.3). Several adult male baboons can injure or kill a leopard and these are the usual odds in a daytime confrontation (Cowlishaw 1994; Cheney and Seyfarth 2007; Devore and Hall 1965). A single male can intimidate smaller carnivores, such as cheetahs (Baenninger et al. 1977; DeVore and Washburn 1963) and jackals (Altmann and Altmann 1970).

Counteraggression against leopards, the most frequent attackers, involves some risk. In one fight, for example, three males were injured (an adult, a subadult, and a juvenile). With regard to adaptive significance, males may be defending their offspring and/or actual or potential mating partners as well as themselves (Willems and van Schaik 2017). Because male baboons are individually dangerous and because they often cooperate (or at least synchronize their behavior), benefits of such behavior tend to outweigh the costs.

Fig. 6.3 A male chacma baboon displays the long canine teeth that may be used to defend himself and other members of his troop. (Photo by Curt Busse. Okavango, Botswana)

The survey by Cowlishaw (1994), which focused entirely on baboons, found that "retaliation occurred against a wide variety of predators and was almost always successful" (p. 300). This was based on a total of 15 encounters, 6 of them with leopards. In the course of these encounters four leopards and one defender were killed. More than one male was involved in at least five of nine events. In a single instance that involved a female defender, the predator was undeterred. In general leopards avoid baboons during the day (Hayward et al. 2006), although they attack refuges at night (see Sect. 6.5 below). This is particularly supported by "numerous studies" in South Africa (Jooste et al. 2012).

Baboons sometimes engage in "proactive" aggression against leopards. Kiffner et al. (2013) reported a female leopard with two cubs that were "mobbed" by a troop of forty baboons. During the interaction under discussion, "three adult male baboons persistently attacked the leopards" (Kiffner et al. 2013: 169). It was inferred that they killed a cub: during 30 min of observation after the interaction ended, only one cub was seen. This apparent result parallels the killing of a leopard cub by chimpanzees (Hiraiwa-Hasegawa et al. 1986), suggesting that early hominins might have had a similarly aggressive relationship with leopards and perhaps some of their felid predecessors.

Baboons are more intimidated by lions (Cowlishaw 1994), but sometimes respond aggressively (Altmann and Altmann 1970). In one incident a large chacma baboon troop used sustained and strenuous threats to coerce a lioness and cubs into vacating the baboons' travel path (Saayman 1971). Busse (1980) reported similar cases of "harassment." Baboons sometimes followed lions as far as several hundred meters until the carnivores moved away from the troop. Adult and subadult males performed this behavior more often than females.

6.3.3 The "Male Rampart"

Washburn and DeVore (1961) presented a model of baboon defense that some later authors referred to as the "male rampart." This melodramatic phrase obscures the nature and variation of the pattern, which simply involves one or more adult males taking a position between the group and potential or actual danger. Sometimes the "rampart" forms because males advance toward the threat. In other cases, it is the result of males retreating more slowly than other group members and then turning toward the predator (Altmann and Altmann 1970). Willems and van Schaik (2017), based on their comparative study, suggested that the so-called "rampart" hypothesis should be revived.

Rowell (1966) described such a deployment among olive baboons in a forest habitat. However, she emphasized that it occurred only when a minor cause of alarm elicited no response from adult males while others ran away. "A stronger stimulus produced precipitate flight, with the big males well to the front and the last animals usually the females carrying heavier babies" (p. 362). This vision of male terror titillated a number of writers, who neglected to ask what stimulus produced "precipitate flight." Although Rowell heard alarm calls associated with leopard and lion, "no other interactions with these predators were seen or deduced" (p.359). The cause of observed baboon flight was the approach of humans. Other baboon researchers have noted that baboons quickly learn to avoid humans, especially if any baboons are killed by them (Washburn and Hamburg 1965; Stolz and Saayman 1970).

Anderson (1986) included four baboon species in a general survey of primate responses to domestic dogs. Every baboon species displayed instances of aggression as well as flight. Males interposed themselves between their groups and the threat. Three species of baboons chased dogs and chacma baboons sometimes killed them. More recently, Zinner et al. (2000) saw comparable behavior in hamadryas baboons. Because they were crop raiders, the baboons risked being killed by humans or their dogs. Adult and subadult males effectively "shielded" other group members against dogs. They also preceded others into the dangerous areas of agricultural fields. The term "shield" is probably more appropriate than "rampart" for this behavior. A rampart is monolithic and static. Shields are portable and flexible in their use.

6.3.4 Use of Weapons

Active predator defense is one of those issues where different lines of evidence come together: in this case from baboons, chimpanzees, and paleoanthropology. Baboon and chimpanzee analogies indicate that early hominin males would have engaged in joint defense (Willems and van Schaik 2017). However, baboons and chimpanzees have long, sharp canine teeth; early hominin canines were reduced in size and more like incisors (Chap. 2). The hominins would probably have needed other weapons to defend themselves.

Chimpanzee evidence suggests that early hominins could have developed effective clubs and thrusting spears. In a field experiment, chimpanzees picked up branches that had been provided by researchers and used them to attack a moving mechanical leopard (Kortlandt 1980). In recent field research chimpanzees were observed killing galagos in tree holes with pointed branches, reminiscent of thrusting spears (Pruetz et al. 2015). Given upper body anatomy like chimpanzees, early hominins could have wielded such weapons to defend against predators. The uniquely hominin trait of bipedalism would have made it possible to carry such weapons continuously and use them at a moment's notice.

Stones are another possible weapon. For years there were sporadic and questionable accounts from diverse sources of chacma baboons defending themselves by throwing stones. Then primatologists reported that three troops in southern Africa dropped and threw stones at them from the tops of cliffs on numerous occasions (Hamilton et al. 1975). Escape movements and vocalizations indicated that the baboons regarded the humans as a threat. This report from southern Africa elicited accounts from other scientists about similar incidents in Kenya (Pickford 1975) and Sudan (Pettet 1975).

Treves and Palmqvist (2007) mentioned the use of stones, but expressed doubt that such "simple projectiles" could deter predators that are persistent in pursuit or "emboldened" by the numbers of a pack. This conclusion is implicitly connected with the authors' hypothesis that early hominins foraged in small groups of mutually trusting individuals that countered predation by maintaining a high level of vigilance. An alternative hypothesis emerges from the baboon model that they reject. A large group like a baboon troop might produce a barrage of stones sufficient to discourage many carnivores. Bipedal hominins could carry stones during group movement, especially adults who were not carrying young.

Again, chimpanzee analogy also comes into play. Chimpanzees demonstrate learned accuracy in throwing that might represent the potential of the LCA. Following on reports of individual differences in aimed throwing in wild and captive apes, Hopkins et al. (2012) studied chimpanzees to seek the neuroanatomical correlates of the behavior. They found that, in the brains of reliable throwers, the proportion of white matter to gray matter was significantly higher in the homologue of Broca's area and in the motor-hand area of the precentral gyrus. In addition, asymmetries in white matter within both brain regions were larger in the hemisphere contralateral to the chimpanzee's preferred throwing hand. The researchers suggested that a neural pattern like this could have provided the foundation for intense selection on increased motor skills associated with throwing in hominins.

To summarize the preceding argument: primate analogies lend themselves to a scenario in which early hominins combined (1) throwing ability evolved from a chimpanzee-like ancestor with (2) large baboon-like groups containing numerous males with baboon-like tendencies toward active defense against predators. Hopkins and colleagues also suggested a connection with the origins of speech and language, because of the involvement of a homologue for Broca's area. Speculation that the hominins in question had the rudiments of speech suggests that such communication might have helped to coordinate barrages of rocks.

6.4 Other Predators

Baboon analogies support sparse paleoanthropological evidence that early hominins faced dangers from air and water as well as land. Predation by raptors is almost certain; crocodiles and snakes are possibilities. The frequency of these events probably fell far short of attacks by leopards and other carnivorans.

6.4.1 Raptors

The skull of a juvenile *Australopithecus africanus* from Taung, South Africa, displays damage that is nearly identical to that which is inflicted on contemporary monkey skulls by the talons of eagles (Berger 2006). The fossil was the lone hominin found among the remains of mostly small animals like the assemblages formed by modern raptors (Berger and Clarke 1995). Though no additional cases have been discovered, there is no reason to think that the fate of the Taung child was unique. Damage to prehistoric monkey skulls also indicates attack by crowned eagles or similar raptors (Gilbert et al. 2009).

The Taung child is estimated to have weighed 10–12 kg. The African crowned eagle (*Stephanoaetus coronatus*) of today preys on mammals weighing up to 20 kg and prefers primates (specifically, mangabey monkeys). However, neither the crowned eagle nor any other raptor is known to have killed any primate weighing more than 13 kg (McGraw and Berger 2013). These data suggest immunity to attack by such raptors for most adult baboons and all adult hominins. However, juveniles like the Taung child, who move independently of their mothers, are and were vulnerable.

Baboon behavior toward raptors is comparable to their reactions to carnivorans. In two encounters with crowned eagles, olive baboons gave alarm calls that caused infants and juveniles to rush from trees and seek shelter under bushes or in close proximity to adults (Paciência et al. 2017). In both encounters adult male baboons took the lead in climbing trees and threatening the eagle, which withdrew. Similar behavior took place in encounters between hamadryas baboons and a different species of large eagle. In response to alarm calls, immatures rushed to adults while adult males threatened the eagle (Zinner et al. 2000; Zinner and Pelaez 1999).

6.4.2 Crocodiles

The lakes where hominins may have caught fish and gathered invertebrates were inhabited by crocodiles, enemies for which hominins would have had no active defense. *Australopithecus anamensis* at Kanapoi in Kenya lived near a lake containing at least three species of crocodiles. One of them, comparable to extant Nile crocodiles, was described as "gigantic" (Brochu 2020). Bones of early *Homo* from Olduvai Gorge seemed to display tooth marks characteristic of crocodiles and suggested that one or two individuals lost a leg in the encounters (Brochu et al. 2010). The researchers noted that the reptiles were common in lake and river basins throughout much of Africa and speculated that they were common predators on hominins.

Doubts have been raised about the crocodile interpretation of the skeletal evidence. Applying an alternative statistical method, McPherron et al. (2022) concluded that the original results are indistinguishable from a null model based on random data. On the other hand, baboon analogy supports the plausibility of crocodile predation on hominins. Interactions between the species were seen in the Okavango Delta of Botswana (Cheney and Seyfarth 2007). On three occasions a crocodile jumped out of shallow water to attack adult male baboons. Two of the baboons escaped injury, but the third was bitten on the face, arms, and a leg.

Baboons in this area become extremely vigilant when crossing water (Fig. 6.4) and anyone watching them "would have no doubt that they fear and despise these reptiles" (Cheney and Seyfarth 2007).These experienced observers had the impression that the baboons understood slightly different alarm calls that distinguished crocodiles from lions when they were near water. The bark for a mammalian predator set them running for trees while the (apparent) crocodile bark resulted in the baboons running a short distance from the water and then stopping to watch. It seems likely that early hominins behaved in similar ways in similar situations.

6.4.3 Snakes

Isbell (2009) argued that danger from snakes was an important factor in primate evolution. Headland and Greene (2011) presented evidence that this has been a continuing danger for hominins down to the present day. The former author emphasized vipers and the latter constrictors. There is a small amount of evidence for both from baboons. Isemonger (cited by Altmann and Altmann 1970) reported two

Fig. 6.4 A group of chacma baboons begin a water crossing in the Okavango swamp, where crocodiles are a danger. (Photo by Curt Busse. Okavango, Botswana)

python attacks on baboons (one successful) and an instance of baboons fighting a python at night. He also reported one case of a young baboon bitten by a puff-adder.

6.5 Sleeping Sites

Choice of sleeping sites is a prominent feature of primate adaptations. Multiple factors are involved (Markham et al. 2016), but safety from predators is probably the most frequently discussed in the literature. The probability of predation described above indicates that this was a major issue for hominins.

6.5.1 Sleeping Above the Ground

Early hominins, diurnal and using sight as their primary sense, were more vulnerable at night. As long as they retained some of their ancestral climbing ability, they probably used night refuges like those of baboons, i.e., trees or cliffs. An early study of Guinea baboons suggested that a regularly used sleeping tree was "a safe refuge from leopards" (Anderson and McGrew 1984). However, the data on actual predation might have been limited by the fact that the study concentrated on the transition from darkness before sunrise until the baboons had left the tree.

Fig. 6.5 A vervet monkey. This species and its close relatives are widely distributed in Africa, but are not able to use the full range of habitats exploited by baboons. (Photo by Curt Busse. Okavango, Botswana)

Isbell et al. (2018) used GPS to study collared leopards, baboons, and vervet monkeys (Fig. 6.5). They found that the two monkey species were equally vulnerable to leopards, but in very different ways. Vervets were vulnerable on the ground, but safe in trees at night because they are small enough to roost at the ends of flexible branches that leopards cannot negotiate. Baboons were vulnerable at night because they are too large to occupy terminal branches and/or hide in dense vegetation. During the day baboons were not attacked by leopards, presumably because of their active defense and propensity for counter-attacks. These results confirmed earlier reports of baboon defense during the day (e.g., DeVore and Washburn 1963) and vulnerability at night (e.g., Busse 1980).

Body size figures prominently in this comparison of baboons and hominins. Isbell et al. (2018) hypothesized that the size of early hominins, as with baboons discussed above, would have limited their ability to take refuge in trees at night but would have facilitated defense on the ground during the day. An additional factor that they mention is that baboons are within the range of prey size preferred by leopards (10–40 kg; cf. Hayward et al. 2006), as was true of early hominins. Vervets are much smaller at about 3–6 kg.

GPS data demonstrated the interest of leopards in baboons. Collared leopards remained near baboon-occupied sleeping sites longer than vacant ones (Bidner et al. 2018). Baboons were most often killed by leopards at or near their sleeping sites. This may be relevant to the observation on Guinea baboons that the first individual to leave the tree in the morning was usually an adult male (Anderson and McGrew 1984).

Bidner and colleagues found that leopards visited riverine sites more frequently than cliffside sites. They interpreted this to mean that leopard visits were often due to factors other than baboon presence; however, it seems possible that the leopards were aware that chances of success were greater where baboons slept in trees rather than on cliffs. Baboons tend to prefer cliffs to trees when both are available (Bidner et al. 2018; Hamilton 1982). Leopards are able to kill baboons on cliffs as well as in trees (Busse 1980); however, it seems possible that primates can more readily escape by using their grasping hands and feet to gain sections of the rock face that even the agile leopard cannot reach.

A recent study of climbing ability in humans may be relevant. Because our species is able to execute so many discrete activities, researchers often disagree on which were the movements most essential to the evolution of our species. Carroll (2021) elaborated on a recently introduced premise for analysis of this problem: the performance gap between female and male athletes narrows in sports, which most reflect movements with evolutionary significance. He investigated rock climbing and found that female climbers are some of the best in the world, a trend that is not found in any other major sport. Carroll's conclusion is that the exceptional ability of female rock climbers is evidence for the existence of sex-blind musculoskeletal adaptations to facilitate essential movements. Baboon evidence suggests that this was the use of cliffs to reduce the risk of predation.

Regulation of sleep patterns may be one more defense against predation. A study of olive baboons found that they lost sleep time when in less familiar locations, regardless of how long they had slept the prior night or how much they had physically exerted themselves the preceding day (Loftus et al. 2022). They also exhibited synchronized patterns of waking throughout the night, which suggests that maintaining sleep homeostasis is secondary to remaining vigilant.

6.5.2 Sleeping on the Ground

Starting with the premise of predation danger from leopards, indicated by their baboon studies, Isbell et al. (2018) raised the question of how hominins protected themselves at night when they eventually had to sleep on the ground. They made the novel suggestion that the hominins surrounded themselves with *bomas* constructed from thorny branches, like those that are still used in Africa today.

To the extent that a primate analogy might shed light on this hypothesis, it seems likely that chimpanzee nesting techniques would be more relevant than anything that might be seen in baboons. However, a complementary hypothesis can be constructed with the addition of baboon evidence. As described above, some baboons have been reported to throw and roll rocks down hills as a defensive measure. If early hominins chose hills or other high points for ground sleeping, they might have deterred predators with rocks. Chimpanzee-like upper body anatomy would have made accurate throwing possible and bipedalism would have made it possible to carry rocks to the sleeping ground to accumulate for the night. To extend the speculation one step further, the rocks might also have been used to reinforce or anchor the bomas.

Sleeping on the ground would presumably have entailed a need for a higher level of vigilance. Hominins might have been prepared for this by a baboon-like pattern of sleep. As noted above, baboons exhibited synchronized patterns of waking throughout the night, which suggests that maintaining sleep homeostasis is secondary to remaining vigilant when sleeping in risky habitats (Loftus et al. 2022).

6.6 Summary and Discussion

Early hominins lived among a greater number and variety of predatory animals than exist in Africa today. The greatest danger came from carnivorans (i.e., species of the order Carnivora, as opposed to meat-eating animals in general). These included modern species such as leopards and lions and, before them, their ancestors and close (extinct) relatives. Hominin environments also included species that have no close analogs today. One was a long-legged hyena that probably hunted rather than scavenged. Another was a giant bear with long legs, faster than any extant bear.

These predators were not all equally dangerous to early hominins. Distinguishing among them depends in large part on relative body mass. The predator must be large enough to subdue the prey, and the prey must be large enough to be worth the predator's expenditure of energy. A second consideration is the mode of attack, either ambush or pursuit. All things considered, including baboon analogy, the greatest enemies of early hominins were probably leopards, leopard ancestors and relatives, and those saber-toothed felids that were similar to leopards.

Baboon behaviors for dealing with carnivorans could have been used by early hominins. Tactical movement includes avoidance of possible ambush situations and adjustment of time spent in foraging areas to the degree of risk. Predators can be kept at a distance by associating with intimidating large herbivores, such as elephants and buffalo. Associations with certain antelopes multiplies vigilance.

When baboons flee from carnivorans, they seek refuge in trees or on cliff faces. Fossil evidence for retention of ancestral climbing ability suggests that this was an option for early hominins as well. However, bipedal hominins could not have reached such refuges as quickly as quadrupedal baboons do. This lends added importance to analogies from baboons regarding active defense ("counter-attack"). Early hominins, larger than baboons, might have confronted some of their predators. As in baboons and for the same reasons, males would have taken the primary role. First, they were larger than females and not primarily responsible for immatures. Second, they benefited from the behavior because they were defending actual or potential mates and actual or potential offspring. Third, the cost was relatively low because the behavior entailed joint action against a lone predator such as a leopard or less formidable predators such as cheetahs. Finally, the hominins might have begun using weapons at a very early point in their evolution. The main evidence for this comes from chimpanzees, but there are plausible reports of baboons using stones against perceived predators.

Evidence for prehistoric raptor attacks consists of talon punctures in the skull of one juvenile *Australopithecus* and in the skulls of a number of monkeys. Raptors sometimes attack baboons and other primates in the present, but apparently with little success against baboons. Baboons use some of the same tactics against raptors that they employ against mammalian carnivores.

No such tactics are effective against extant or prehistoric crocodiles. Early hominins were probably like baboons in using avoidance or extreme vigilance when near larger bodies of water. However, the danger was not ever-present, as it was in the case of carnivorans. Snakes occasionally attack primates for food in the case of constrictors and defense in the case of venomous snakes. It has been argued that snake attacks were an important factor in primate evolution. However, there are very few reports of baboons killed by snakes.

Predation is the main reason that baboons and other primates sleep in trees or on cliff faces. As noted above, retention of arboreal capabilities probably allowed early hominins to do the same thing. Primates prefer cliffs to trees where both are available. This may be due primarily to leopards, which are agile tree climbers and hunt baboons at night. Early hominins may have responded to the same danger in the same way. Even modern humans are quite capable of climbing trees, and proficiency in rock climbing is demonstrated by modern proponents of the sport with a minimal gap between men and women.

Eventually hominins began to spend nights on the ground, but perhaps later in their evolution than some researchers think. When they did, two baboon-like patterns may have helped to prepare them: group vigilance through sporadic waking of individuals, and the use of rocks for defense from heights.

References

Allan ATL, Bailey AL, Hill RA. Consistency in the flight and visual orientation distances of habituated chacma baboons after an observed leopard predation. Do flight initiation distance methods always measure perceived predation risk? Ecol Evol. 2021; https://doi.org/10.1002/ece3.8237.

Altmann SA, Altmann J. Baboon ecology: African field research. Chicago: University of Chicago Press; 1970.

Anderson JR. Encounters between domestic dogs and free-ranging non-human primates. App Anim Behav Sci. 1986;15(1):71–86. https://doi.org/10.1016/0168-1591(86)90024-9.

Anderson JR, McGrew WC. Guinea baboons *(Papio papio)* at a sleeping site. Am J Primatol. 1984;6:1–14.

Annear E, Minnie L, Andrew K, Kerley GIH. Can smaller predators expand their prey base through killing juveniles? The influence of prey demography and season on prey selection for cheetahs and lions. Oecologia. 2023;201(3):1–12. https://doi.org/10.1007/s00442-023-05335-8

Anton M. Behavior of *Homotherium* in the light of modern big cats. Die *Homotherium*-Funde von Schonigen. n.d.;9-23

Badenhorst S. Possible predator avoidance behaviour of hominins in South Africa. S Afr J Sci. 2018;114(7/8):Art. #a0274. https://doi.org/10.17159/sajs.2018/a0274.

Baenninger R, Estes RD, Baldwin S. Anti-predator behaviour of baboons and impalas toward a cheetah. Afr J Ecol. 1977;15(4):327–9. https://doi.org/10.1111/j.1365-2028.1977.tb00414.x.

Berger LR. Brief communication: predatory bird damage to the Taung type-skull of *Australopithecus africanus* Dart 1925. Am J Phys Anthropol. 2006;131(2):166–8. https://doi.org/10.1002/ajpa.20415.

Berger LR, Clarke R. Eagle involvement in accumulation of the Taung child fauna. J Hum Evol. 1995;29(3):275–99. https://doi.org/10.1006/jhev.1995.1060.

Bidner LR, Matsumoto-Oda A, Isbell LA. The role of sleeping sites in the predator-prey dynamics of leopards and olive baboons. Am J Primatol. 2018;80:e22932. https://doi.org/10.1002/ajp.22932.

Bobe R, Manthi FK, Ward CV, et al. The ecology of *Australopithecus anamensis* in the early Pliocene of Kanapoi, Kenya. J Hum Evol. 2020;140 https://doi.org/10.1016/j.jhevol.2019.102717.

Brochu CA. Pliocene crocodiles from Kanapoi, Turkana Basin, Kenya. J Hum Evol. 2020;140:102410. https://doi.org/10.1016/j.jhevol.2017.10.003.

Brochu CA, Njau J, Blumenschine RJ, Densmore LD. A new horned crocodile from the Plio-Pleistocene hominid Sites at Olduvai Gorge, Tanzania. PLoS One. 2010;5(2):e9333. https://doi.org/10.1371/journal.pone.0009333.

Busse C. Leopard and lion predation upon chacma baboons living in the Moremi Wildlife Reserve. Botswana Notes Records. 1980;12:15–21. www.botsoc.org.bw

Camarós E, Cueto M, Lorenzo C, Valentín, et al. Large carnivore attacks on hominins during the Pleistocene: a forensic approach with a Neanderthal example. Archeol Anthropol Sci. 2015; https://doi.org/10.1007/s12520-015-0248-1.

Carbone C, Mace GM, Roberts SC, Macdonald DW. Energetic constraints on the diet of terrestrial carnivores. Nature. 1999;402:286–8. https://doi.org/10.1038/46266.

Carroll C. Female excellence in rock climbing likely has an evolutionary origin. Curr Res Physiol (Journal Preproof). 2021; https://doi.org/10.1016/j.crphys.2021.01.004.

Cheney DL, Seyfarth RM. Baboon metaphysics: the evolution of a social mind. Chicago: University of Chicago Press; 2007.

Cowlishaw G. Vulnerability to predation in baboon populations. Behaviour. 1994;131:293–304. https://doi.org/10.1163/156853994X00488.

Cowlishaw G. Refuge use and predation risk in a desert baboon population. Anim Behav. 1997a;54:241–53.

Cowlishaw G. Trade-offs between foraging and predation risk determine habitat use in a desert baboon population. Anim Behav. 1997b;53:667–86.

DeVore I, Washburn SL. Baboon ecology and human evolution. In: Howell FC, Bourliere F (eds) African ecology and human evolution. Aldine, Chicago; 1963. p. 335–367.

DeVore I, Hall KRL. Baboon social behavior. In: DeVore I, editor. Primate behavior, field studies of monkeys and apes. New York: Holt Rinehart Winston; 1965. p. 53–110.

Fay JM, Carroll R, KerbisPeterhans JC, Harris D. Leopard attack on and consumption of gorillas in the Central African Republic. J Hum Evol. 1995;29:93–9. https://doi.org/10.1006/jhev.1995.1048.

Fischer J, Higham JP, Alberts S, et al. The Natural History of Model Organisms: Insights into the evolution of social systems and species from baboon studies. eLife 2019;8:e50989

Geraads D, Daujeard C. Carnivore diversity in the African Plio-Pleistocene: a reply to O'Regan and Reynolds (2009). J Hum Evol. 2011;60:813–5. https://doi.org/10.1016/j.jhevol.2011.01.012.

Gilbert CC, McGraw WS, Delson E. Brief communication: Plio-Pleistocene eagle predation on fossil cercopithecids from the Humpata Plateau, southern Angola. Am J Phys Anthropol. 2009;139:421–9. https://doi.org/10.1002/ajpa.21004.

Greyling E, Comley J, Cherry MI, et al. Facilitation of a free-roaming apex predator in working lands: evaluating factors that influence leopard spatial dynamics and prey availability in a South African biodiversity hotspot. PeerJ. 2023;11:e14575. https://doi.org/10.7717/peerj.14575.

Hamilton WJ III. Baboon sleeping site preferences and relationships to primate grouping patterns. Am J Primatol. 1982;3(1–4):41–53. https://doi.org/10.1002/ajp.1350030104.

Hamilton WJ III, Buskirk RE, Buskirk WH. Defensive stoning by baboons. Nature. 1975;256:488–9. https://doi.org/10.1038/256488a0.

Hayward MW, Henschel P, O'Brien J, et al. Prey preferences of the leopard (*Panthera pardus*). J Zool. 2006;270(2):298–313. https://doi.org/10.1111/j.1469-7998.2006.00139.x.

Headland TN, Greene HW. Hunter-gatherers and other primates as prey, predators, and competitors of snakes. Proc Natl Acad Sci USA. 2011;108:20865-20866, E1470–4.

Hill RA, Weingrill T. Predation risk and habitat use in chacma baboons (*Papio hamadryas ursinus*). In: Gursky S, Nekaris KAI, editors. Primates and their predators. New York: Kluwer Academic; 2007. p. 339–54.

Hiraiwa-Hasegawa M, Byrne R, Takasaki H, Byrne JM. Aggression toward large carnivores by wild chimpanzees of Mahale Mountains National Park, Tanzania. Folia Primatol. 1986;47(1):8–13. https://doi.org/10.1159/000156259.

Hopkins WD, Russell JL, Schaeffer JA. The neural and cognitive correlates of aimed throwing in chimpanzees: a magnetic resonance image and behavioural study on a unique form of social tool use. Philos Trans R Soc B. 2012;367:37–47. https://doi.org/10.1098/rstb.2011.0195.

Hopley PJ, Cerling TE, Crete L, et al. Stable isotope analysis of carnivores from the Turkana Basin, Kenya: evidence for temporally-mixed fossil assemblages. Quatern Int. 2023;650:12–27. https://doi.org/10.1016/j.quaint.2022.04.004.

Isbell LA. The fruit, the tree, and the serpent: why we see so well. Cambridge MA: Harvard University Press; 2009.

Isbell LA, Bidner LR, Van Cleave EK, Matsumoto-Oda A. GPS-identified vulnerabilities of savannah-woodland primates to leopard predation and their implications for early hominins. J Hum Evol. 2018;118:1–13. https://doi.org/10.1016/j.jhevol.2018.02.003.

Itani J. Postscript by the editor. Primates. 1967;8:295–6. (cited by Altmann and Altmann 1970)

Jooste E, Pitman RT, van Hoven W, Swanepoel LH. Unusually high predation on chacma baboons *(Papio ursinus)* by female leopards *(Panthera pardus)* in the Waterberg Mountains, South Africa. Folia Primatol. 2012;83:353–60. https://doi.org/10.1159/000339644.

Kiffner C, Ndibalema V, Kioko J. Leopard (*Panthera pardus*) aggregation and interactions with Olive baboons (*Papio anubis*) in Serengeti National Park, Tanzania. Afr J Ecol. 2013;51:168–71.

Kortlandt A. How might early hominids have defended themselves against large predators and food competitors. J Hum Evol. 1980;9(2):79–112. https://doi.org/10.1016/0047-2484(80)90066-4.

Kruuk H. The spotted hyena, a study of predation and social behavior. Chicago: University of Chicago Press; 1972.

Kuhn BF, Hartstone-Rose A, Lacruz AIR, et al. The carnivore guild circa 1.98 million years: biodiversity and implications for the palaeoenvironment at Malapa, South Africa. Palaeobiol Palaeoenv. 2016; https://doi.org/10.1007/s12549-016-0245-0.

Lee-Thorp J, Thackeray F, van der Merwe J. The hunters and the hunted revisited. J Hum Evol. 2001;39(6):565–576. https://doi.org/10.1006/jhev.2000.0436

Lewis ME. Carnivore guilds and the impact of hominin dispersals. IN Boivin N, Crassard R, Petraglia M, editors. Human Dispersal and Species Movement From Prehistory to the Present. Cambridge University Press; 2017. p. 29–61. https://doi.org/10.1017/9781316686942.003.

Loftus JC, Harel R, Nunez CL, Crofoot MC. Ecological and social pressures interfere with homeostatic sleep regulation in the wild. eLife. 2022;11:e73695. https://doi.org/10.7554/eLife.73695.

Markham CA, Alberts SC, Altmann J. Haven for the night: sleeping site selection in a wild primate. Behav Ecol. 2016;27(1):29–35. https://doi.org/10.1093/beheco/arv118.

McGraw WS, Berger, LR. Raptors and primate evolution. Evol Anthropol. 2013;22(6):280–293. https://doi.org/10.1002/evan.21378.

McPherron SP, Archer W, Otarola-Castillo ER, et al. Machine learning, bootstrapping, null models, and why we are still not 100% sure which bone surface modifications were made by crocodiles. J Hum Evol. 2022;164:103071. https://doi.org/10.1016/j.jhevol.2021.103071.

O'Regan HR, Reynolds SC. An ecological assessment of the southern African carnivore guild; a case study from Member 4, Sterkfontein, South Africa. J Hum Evol. 2009;57:212–22. https://doi.org/10.1016/j.jhevol.2009.04.002.

Paciência FMD, Baluya D, Mbaryo P, et al. Olive baboons' (*Papio anubis*) response towards crowned eagles (*Stephanoaetus coronatus*) at Lake Manyara National Park. Primate Biol. 2017;4:101–6. www.primate-biol.net/4/101/2017/

Pettet A. Defensive stoning by baboons. Nature. 1975;256:549.

Pickering T, Egeland C, Dominguez-Rodrigo M, et al. Testing the "shift in the balance of power" hypothesis at Swartkrans, South Africa: Hominid cave use and subsistence behavior in the Early Pleistocene. J Anthropol Archaeol. 2008;27:30–45. https://doi.org/10.1016/j.jaa.2007.07.002.

Pickford M. Defensive stoning by baboons. Nature. 1975;256:549–50.

Pruetz JD, Bertolani P, Boyer Onto K, et al. New evidence on the tool-assisted hunting exhibited by chimpanzees (*Pan troglodytes verus*) in a savannah habitat at Fongoli, Sénégal. R Soc Open Sci. 2015;2 https://doi.org/10.1098/rsos.140507.

Rhine RJ, Norton GW, Roertgen WJ, Klein HD. The brief survival of free-ranging baboon infants (*Papio cynocephalus*) after separation from their mothers. Int J Primatol. 1980;1:401–409. https://doi.org/10.1007/BF02692282

Rowell T. Forest living baboons in Uganda. J Zool Lond. 1966;149:344–64.

Ruff CB, Wood BA. The estimation and evolution of hominin body mass. Evol Anthropol. 2023;32(4):223–237. https://doi.org/10.1002/evan.21988

Saayman GS. Baboon's responses to predators. Afr Wldlf. 1971;25:46-49.

Stelzner J, Strier K. Hyena predation on an adult male baboon. Mammalia. 1981;45:259–260.

Stolz LP, Saayman GS. Ecology and behavior of baboons in the northern Transvaal. Annals Transvaal Mus. 1970;26:99–143.

Treves A, Palmqvist P. Reconstructing Hominin interactions with mammalian carnivores (6.0–1.8 Ma). In: Gursky SL, Nekaris KAI, editors. Primate anti-predator strategies. Developments in primatology: progress and prospects. Boston: Springer; 2007. https://doi.org/10.1007/978-0-387-34810-0_17.

Washburn SL, DeVore I. The social life of baboons. Sci Am. 1961;204:62–71.

Washburn SL, Hamburg DA. The implications of primate research. In DeVore I (ed) Primate behavior: New York: Holt Rinehart Winston; 1965 p607-622.

Washburn SL, Jay PC, Lancaster JB. Field studies of Old World monkeys and apes. Science. 1965;150:1541–1547. https://doi.org/10.1126/science.150.3703.1541.

Werdelin L, Lewis ME. A contextual review of the Carnivora of Kanapoi. J Hum Evol. 2020;140:102334. https://doi.org/10.1016/j.jhevol.2017.05.001.

Willems EP, van Schaik CP. The social organization of *Homo ergaster*: inferences from anti-predator responses in extant primates. J Hum Evol. 2017;109:11–21. https://doi.org/10.1016/j.jhevol.2017.05.003.

Williams KS, Williams ST, Fitzgerald LE. Brown hyaena and leopard diets on private land in the Soutpansberg Mountains, South Africa. Afr J Ecol. 2018;2018:1–7. https://doi.org/10.1111/aje.12539.

Zinner D, Pelaez F, Berhane D. Anti-predator behavior of male hamadryas baboons *Papio hamadryas* in Eritrea. Afr Primates. 2000;4(1&2):54–8.

Zinner D, Pelaez F. Verreaux's eagles (*Aquila verreauxi*) as potential predators of Hamadryas baboons (*Papio hamadryas hamadryas*) in Eritrea. Am J Primatol. 1999;47:61–66. https://doi.org/10.1002/(SICI)1098-2345(1999)47:13.0.CO;2-K.

Chapter 7
Social Organization and Male–Male Relationships

7.1 Introduction

Baboons provide diverse options for hypotheses about early hominin social organization. They pertain to both synchronic variation and long-term changes. This chapter begins with the ideas of some archeologists about prehistoric human social ecology, and goes on to explore ways in which baboon social patterns articulate with the archeological model. As in traditional ideas about baboon and human societies, males play important roles. These are considered here and used to evaluate a longstanding idea about male–male relationships in human evolution that has been called "men in groups." The role of females in early hominin evolution will become increasingly prominent in Chaps. 8 and 9.

7.2 Archeological Perspectives on Prehistoric Hominin Society

Archeologists and paleoanthropologists have often turned to extant and recent human hunter-gatherers for models of prehistoric hominin behavior. As biologically modern humans, these peoples can only represent the later stages of human evolution. Primates like baboons and chimpanzees are more appropriate for the earlier stages. However, human and primate sources can be articulated with one another to develop a more dynamic view of hominin social evolution.

Archeologists at a recent conference largely agreed on several inferences about prehistoric human society (Goren-Inbar and Belfer-Cohen 2020), using concepts that are readily applied to early hominin societies as modeled by baboons. First, groupings of 20–50 individuals were common. Second, these groups were not rigidly separated from one another. Third, group size was dynamic, varying from small

G. E. King, *Baboon Perspectives on Early Human Ancestors*,
https://doi.org/10.1007/978-3-031-36769-4_7

bands to large aggregations. Fourth, the various groupings served a variety of functions and purposes, including subsistence and predator defense. This "elasticity" probably derived from both social behavior and ecological constraints.

There was general agreement among the archeologists that larger groups facilitated increased vigilance and stronger cooperation. These features would be adaptive in defense against predators and rival hominin groups, as well as in competition with predators and other hominins for hunted or scavenged prey. Presumably, these social features would have been especially important in earlier hominin evolution when safety in numbers compensated for limited technology. Further developments, perhaps including cooperative hunting, would have stimulated major changes in brain, behavior, technology, and life history (Stout et al. 2018). The period around 2-1.7 mya may have been pivotal for the developments that transformed the genus *Homo* (Chaps. 4 and 5).

Some archeological research provides direct evidence for a relatively late shift in social organization, well after the first appearance of *Homo*. Comparison of two Early Pleistocene archeological sites revealed the same spatial pattern in the accumulation of bones and stone tools. Statistical tests showed intense interdependence between both types of materials in both spatial distribution and intensity (Dominguez-Rodrigo and Cobo-Sanchez 2017). The researchers inferred that the high-density single clusters characterizing these sites represent communal use of the same spot for processing and consuming animal carcasses. They argued that this does not support the presence of individual nuclear families, which would indicate human-like social organization. Multiple clusters of remains characterize the assemblages left by the camps of recent hunter-gatherers. Baboon behavior articulates with hunter-gatherer models and archeological evidence in a multidimensional reconstruction of early hominin social organization and its transformations.

7.3 The Troop: A Baboon Model for Early Hominin Society

Baboons fall into two broad categories as far as basic social structure is concerned (Fischer et al. 2019). The simpler form occurs in the "COKY" species—chacma, olive, Kinda, and yellow baboons (Jolly 2020). This is the troop, typically composed of multiple males and females of all age categories (Fig. 2.2). The troop usually travels as a unit and, if it spreads out for feeding, contact is maintained with vocalizations. Troop size varies from roughly 10–200, but many troops fall into the 20–50 range postulated by the archeologists.

Dunbar et al. (2018) studied recurring troop sizes in three baboon species and hypothesized an underlying pattern that makes certain values particularly common. The values were approximations to 20, 40, 80, and 160. The researchers posited that these tendencies are responses to varying demographic factors. The lower end of the scale is determined by predation risk, because larger groups are needed for increased vigilance and defense. The upper end is determined by the maximum size that can be attained before limits on resource availability require that the troop split permanently, usually into two daughter groups. Baboons observed in a habitat with high predator density fissioned at a mean size of about 65 while the mean size in a low-predator habitat was about 30. Analysis of the data suggested an optimal size of about 40, which is close to the mean group size for the whole genus. In the terms discussed in Chap. 3, application of these data to early hominins entails a genus-level referential model, strengthened by a conceptual context based on ecological and demographic principles.

The term *troop* has been used in some cases to designate groupings of primates (including baboons) that are larger than the COKY troops and organized differently (e.g., Kummer 1968). To avoid confusion, I will use the term *maximum social group* (MSG) to denote the largest group to which individuals belong in any primate social system. More specifically, this is the largest group that involves its members in any kind of affiliative or tolerant social interactions with any regularity. The troop is the maximum social group of COKY baboons in every habitat that they have occupied. Given such adaptability, a similar society could have persisted in early hominins as they expanded from woodlands into a wide variety of mosaic habitats and eventually survived in relatively open savannas.

7.3.1 Relations Between Troops

The baboon perspective suggests that intergroup relationships of early hominins were complex and variable. Intertroop relationships of baboons vary from hostility to avoidance to tolerance to occasional brief associations that give rise to supra-troop groupings. Research on wild baboons has never discovered anything like the highly publicized "wars" between chimpanzee communities. Some early accounts described aggression between troops, especially among chacma baboons (reputed to be the most aggressive species). However, there were virtually no fatalities in these reports. For example, Saayman (1971) observed 58 encounters among four chacma troops in South Africa. Seventy-six percent of these meetings were nonantagonistic and occasionally two troops intermingled and traveled together for a while. There was "overt fighting" in just 14 of the 58 encounters, and two "spectacular and tumultuous fights" took place. Yet Saayman did not mention any serious injuries, much less deaths. Chacma baboons, with a reputation for aggression, provide evidence suggesting that hominin competition for resources entailed only limited fighting and encounters that more often involved noncontact behaviors.

Baboon troops commonly avoid each other rather than be drawn into conflict. This was often true of chacma baboons as well as other species (DeVore and Hall 1965). Continuing research with more sophisticated methods has confirmed the pattern. Markham et al. (2013) used GPS to investigate temporal landscape partitioning in yellow baboons. They monitored collared subjects to synchronously record the hourly locations of five troops for about 900 days. They used behavioral, demographic, and life history data to measure factors affecting the use of overlap areas. On an annual scale, home ranges of neighboring groups overlapped substantially.

However, home ranges overlapped less when space use was assessed over shorter time scales, indicating that the troops were seldom in the same area at the same time. Neighboring groups were in close spatial proximity to one another on fewer days than expected from random movement. Nevertheless, yellow baboons engaged in some significant intergroup competition (Markham et al. 2013). When direct conflict occurred, losing groups used the area surrounding the interaction less than previously. Visual contact usually resulted in displacement of one group by the other.

In circumstances of scarcity, baboon groups might fight each other over sleeping places (cliffs or trees) or water sources (Altmann and Altmann 1970; Stolz and Saayman 1970). Similar conflicts probably took place among early hominins. They needed water from scarce sources in dry savannas and they needed arboreal refuges from predators until they found ways to sleep safely on the ground. Competition could arise if tall trees or cliffs were scarce. Competition for females may be another factor in hostility between troops. In some baboon species, troop females are herded away from other groups during conflicts (Kitchen et al. 2004; Saayman 1971).

Baboons may defend space and boundaries, but these patterns are ultimately related to resources (Hamilton et al. 1976). Two chacma troops in the Namib Desert defended a boundary near a waterhole. In the Okavango Swamp in Botswana, large troops defended well-defined boundaries of small home ranges that were relatively dense in plant species and energy sources. The size and configuration of troop space and the arrangement of resources within each space influence the likelihood of intertroop encounters and the expression of spatial defense.

The diverse interrelationships of baboon troops and resources suggest that early hominins engaged in a variety of intergroup encounters, ranging from hostility to avoidance to temporary affiliation. These diverse options contradict some popular ideas about intergroup violence and warfare being fundamental features of human evolution. If anything, the expansion into more open habitats almost certainly resulted in larger home ranges with reduced contact between troops, except possibly in the case of limited crucial resources such as water.

7.3.2 *Fission–Fusion*

Troop organization is often somewhat flexible and allows for groupings of various sizes (as envisioned by the archeologists for prehistoric hominins). A troop usually moves as a unit but in some circumstances, especially food scarcity, subgroups forage independently and then reunite. In this respect baboons sometimes resemble the chimpanzee system of fission-fusion. Baboons differ from chimpanzees in that the parties come together again as a single troop. This temporary fission should not be confused with the kind of fission in which a baboon troop permanently divides into two or more "daughter" troops.

Anderson (1981) described "subtrooping" in two troops of chacma baboons in South Africa. It varied with the season, being more frequent when temperatures and food availability decreased. Some subtroops frequented only certain parts of the larger troop range and remained separate from other groups for up to several days. Anderson speculated that lack of predation, coupled with genetic isolation for several generations, might have facilitated the pattern that she observed. From a functional viewpoint, it probably increased the efficiency of resource use during less favorable seasons, especially for less dominant individuals.

Comparing her observations with those of others, Anderson found that her population was "relatively unique" with regard to seasonality of the phenomenon, consistency of membership in subgroups, and the frequent occurrence of one-male subtroops. Nevertheless, she suggested that subtrooping is a facultative potential of all baboon species that requires little genetic change to develop. Her hypothesis is supported by later research on olive baboons living in a different kind of environment in a different part of Africa—the Gashaka forest in Nigeria. The larger troop of the two that were studied displayed "regular subgrouping" into parties averaging 15 individuals (Kunz and Linsenmair 2008). Aldrich-Blake et al. (1971) reported subtrooping in arid conditions in Ethiopia.

There may be an ecological convergence with chimpanzees here. The olive baboons used the forests more often than expected by chance, had large home ranges, and were highly frugivorous. They spent about 50% of their feeding time on the fruits and seeds of at least 79 woody plant species. In contrast to Anderson's chacma baboons, fission–fusion at Gashaka was a response to abundance (of fruit) rather than scarcity.

Since early hominins probably shared a strong preference for fruit with *Pan* (and other apes), fission–fusion may have been more prominent in the woodland hominin populations. Later on, during the expansion into arid habitats, the same capability would have been available to deal with scarcity as in the South African baboons. The fission–fusion capability is consistent with the social flexibility of human hunter-gatherers envisioned by the archeological conference (Goren-Inbar and Belfer-Cohen 2020).

In the context of fission–fusion, one-male subgroups are prominent in the reports of chacma baboons in South Africa (Anderson 1981) and olive baboons at Gashaka, Nigeria (Kunz and Linsenmair 2008). In the latter case, for example, the proportion of one-male groups varied from 50% to 63%. Differentiation of these one-male groups within troops may be a model for the origin of multilevel societies in baboons and early hominins.

7.4 Multilevel Societies

Hamadryas and Guinea baboons display the multilevel society, in which smaller social groups are nested within larger ones, creating three or four levels of social organization. In both species the foundation of the system is a tier composed of one-male groups, each containing one adult male with one or more adult females. Affiliation at the different levels serves varied (but overlapping) biological, ecological, and social functions, including foraging, predator defense, and shared sleeping locations (Fischer et al. 2019; Kummer 1968; cf. Goren-Inbar and Belfer-Cohen 2020). In contrast to these complex societies, the COKY troop is essentially an independent one-level system. Departures from that status, such as subtrooping or affiliation between two troops are occasional and temporary.

The two main categories of baboon social organization—the troop and the multilevel society—can be linked to each other and to early hominins in two alternative ways. One model places them in a sequence, with multilevel societies emerging from troops (e.g., Swedell and Plummer 2019). The alternative is to place one or the other at the base of hominin evolution and derive species-typical forms of human social organization directly from it.

Multilevel organization, whether viewed as the original form of hominin society or an intermediate stage, fits well with the archeological model described at the beginning of this chapter (Goren-Inbar and Belfer-Cohen 2020). Large groups regularly separate into smaller groups and then come together again. Some groups within the structure typically fall into the 20–50 range. These intermediate social units contain semi-independent subgroups and belong to a larger community that may number several hundred. The groupings at different levels provide somewhat different contributions to basic functions.

The basic social unit in both multilevel baboon species is a group that revolves around one male who is affiliated with one or more females as social and reproductive partners. The hamadryas male in this position has been termed the "leader." The comparable figure in Guinea baboons is called the "primary male." The term *central male* will be used here to encompass both. It distinguishes the adult male with a pivotal role from other members of the group. The others may include one or more additional adult or nearly adult males that are subordinate in social and reproductive terms. This basic unit has been given different names by students of hamadryas baboons and Guinea baboons. Here, the term *unimale group* (cf. Hex et al. 2021) will be applied to both.

7.4.1 Multilevel Society in Hamadryas Baboons

The following description of hamadryas society is based on two recent distillations (Evans et al. 2022; Swedell and Plummer 2019) of a long history of research by many primatologists, especially the seminal work of Hans Kummer (1968). The unimale group of hamadryas baboons is commonly called the "one-male unit," abbreviated OMU. The core of each OMU is a "leader" male (central male) who has a close social relationship with one or more females that may last for years. These females are also his sexual partners when they are available, but the social relationship continues through anestrus periods, including infant care. The OMU may also contain one or more "follower" males that have social but not sexual access to the females. The OMU is the most stable group in hamadryas social organization and is the minimal foraging group.

A distinctive feature of the hamadryas OMU is that it is formed by successive takeovers of females by various means: (1) a subadult male "recruits" a juvenile female from her natal group; (2) a follower "inherits" one or more females from his leader; (3) a male challenges a leader for his females; (4) a male opportunistically obtains a female when her leader can no longer defend her.

The next tier in hamadryas multilevel organization is the *clan*. This is an association based on kinship among males (Städele et al. 2015). A clan may be as small as 2 OMUs or as large as 13 or more (Schreier and Swedell 2009). In addition to OMUs a clan contains "bachelor" males that are not affiliated with any OMU. Though capable of dispersing into its constituent OMUs, the clan tends to be spatially cohesive within larger groups and as a separate foraging unit. It also provides for defense against predators. Clan males have been observed forming a shield against hyenas while other members retreated. Clan males also groom each other more often than they do nonmembers. The clan is fully documented in just one hamadryas population, located at the Filoha area in Ethiopia.

The *band* has been equated with the COKY troop because of its spatial and functional unity, composition, and size. It is a multi-male/multi-female group that combines several OMUs along with bachelor males, and several clans if they are present. Most bands probably have 50–100 members, though some contain 200 or more. The largest bands likely result from unusual resources in certain regions. Each band has its own area on the sleeping cliff. In the morning friendly bands often travel for a short distance together, though they always split apart and go in different directions for the rest of the day. Eventually, as the day goes on, each band fragments into its OMU components, though it often reassembles at a waterhole around midday. One study found that a band was more likely to break up into OMUs where general food availability was low and into clans when preferred resources were not available (Schreier and Swedell 2012).

The largest grouping of hamadryas baboons has been called the "troop." This is confusing because it is nothing like the troop in COKY baboons. The term maximum social group or MSG (coined above) will be used here. The hamadryas MSG may consist of hundreds of individuals, most of which have very limited social relationships with each other. It is an aggregation of bands that tolerate each other at the same sleeping cliffs at night. This allows for the use of a vital and limited resource in a desert region where trees are scarce. During daytime activity the hamadryas MSG has no significance.

7.4.2 Multilevel Society in Guinea Baboons

The social life of Guinea baboons is so far known from just one locality in a fairly rich environment (Zinner pers. comm.). The basic multilevel structure of this society is similar to that of hamadryas baboons, but there are major differences in behavior and social dynamics. These differences, especially those that characterize male–female relationships, make Guinea baboons an alternative model for the social behavior of early hominins.

Unimale groups of Guinea baboons, the fundamental units within the multilevel society, have been called "reproductive units" or just "units" (Dal Pesco et al. 2022). Each is composed of one "primary" male and 1–6 females. A primary (central) male has largely exclusive affiliative and sexual relationships with the females in his group. Females maintain exclusive social and mating relationships with one male at a time.

Unimale groups of Guinea baboons differ sharply from hamadryas OMUs with respect to the status and role of females. First, in contrast to accumulation of females by male hamadryas, female Guinea baboons play an active role in forming and maintaining relationships with males (Goffe et al. 2016). Second, a female Guinea baboon, though usually close to her central male, spatially and socially, has freedom of movement. Third, a female is free to leave the relationship at any time and affiliate with another male. *Male-focal Unit* (MFU) might be an appropriate term for these flexible groups.

Guinea baboons also differ from hamadryas with regard to secondary males or followers. These males can be identified in both species as associating with unimale groups and having social but not sexual relationships with the females (Goffe et al. 2016). However, in Guinea baboons a single bachelor is often associated with several unimale groups (in roughly 67% of cases) (Dal Pesco et al. 2021).

Two higher levels of social organization can be compared to those of hamadryas baboons in structural terms, but functional equivalence is dubious. A Guinea baboon "party," which is composed of several unimale groups and their secondary males, is the equivalent of the hamadryas clan. Two or three of these Guinea baboon clans regularly aggregate into "gangs" that can be compared with hamadryas bands. However, it seems that there is little to identify them other than overlapping home ranges (Patzelt et al. 2014). Guinea baboon clans vary in size from about 10–40 members and the Guinea baboon band averages about 70 (Fischer et al. 2017). The maximum social groups of the species are barely social, being identified by a common home range.

7.4.3 *Possible Societies of Early Hominins*

The troop and the multilevel society in baboons can be considered two alternative sources for reconstructing the social organization of early hominins. They can also be viewed as representing two stages in the social evolution of hominins. Chapais (2008) postulated two such stages at the beginning of hominin social evolution. The first of these stages was a "promiscuous" multimale-multifemale group. Chapais compared this postulated group to the chimpanzee community, but his description also applies to the baboon troop. The next stage, inferred from hamadryas baboons and other primate data, was a "multiharem" group. Chapais saw this as the foundation for the multifamily community that became the modal pattern for humans. Others have also inferred such a sequence in hominin evolution (e.g., Swedell and Plummer 2012, 2019).

7.5 Male Philopatry

An important feature of human and animal societies is the movement of maturing individuals from one group to another. Primate group size and composition are maintained in large part by emigration and immigration. The typical situation is that one sex usually or almost always leaves and the other stays in the birth group. The tendency of one sex to stay is called *philopatry* and departure from the natal group is often termed *dispersal*. Comparison of baboon societies suggests that philopatry shifted from females to males in early hominin evolution.

7.5.1 Male Philopatry in Humans and Baboons

In the majority of human societies, recent and past, male philopatry was the predominant pattern of residence. That is, most men lived their lives in the communities to which they were born while women left when they married. This is opposite to the pattern of troop-living baboons, in which males leave their natal groups. However, both of the multilevel *Papio* species resemble humans in that females leave their natal groups to enter a sociosexual relationship with a male in another group. For example, adult females in one hamadryas baboon study transferred between clans nineteen times while only one adult male transfer was seen. This is broadly analogous to exogamy in most traditional human societies and may represent an early hominin origin of the practice.

7.5.2 The Frontier Hypothesis

Jolly (2020) described and elaborated a hypothesis about baboon expansion and social organization that he and others suggested as a model for hominins (e.g. Fischer et al. 2017; Swedell and Plummer 2012). Male philopatry is a crucial factor in this "Frontier Hypothesis," which suggests that a population norm of male philopatry evolved in the common ancestor of hamadryas and Guinea baboons. It was a response to a demographic context peculiar to the frontier of a rapidly expanding population with repeated group fission (fission in this context refers to permanent separation between a group and one or more daughter groups). Jolly's emphasis on demography distinguishes this scenario from other hypotheses, most of which revolve around ecological factors.

According to the Frontier Hypothesis, the common ancestor of the two multilevel baboon species was probably similar to extant Guinea baboons, including male philopatry. Other social features evolved to accommodate male philopatry: one-male units, multilevel society, male–male tolerance, and some degree of female dispersal. The larger context for these events goes back to a southern origin for the genus *Papio*. The genetic population structure of living baboons preserves evidence of the initial expansion of *Papio* from the south. Immediately after the expansion, male-philopatric, multilevel populations with a general physical and behavioral resemblance to Guinea baboons occupied the whole northern range of the genus. Hamadryas baboons presumably changed further in response to less productive habitats in the Horn of Africa. Subsequently, olive and yellow baboons shifted to female-philopatric systems and replaced most of the northern populations as males moved from one troop to another.

The Frontier Hypothesis is based on established patterns of papionin population structure and demography, as well as more general evolutionary theory and models related to dispersal, invasive expansion, and frontier effects on population structure and evolution (Jolly 2020). It seems to be an example of a referential model embedded in the framework of a conceptual model (Chap. 3).

According to Jolly, the Frontier Hypothesis has far-reaching implications for the reconstruction of early hominin evolution. First, it brings a baboon model more in line with chimpanzee evidence. The behavior of extant chimpanzees and humans suggests that their Last Common Ancestor was male-philopatric, and the Frontier Hypothesis posits a similar origin for *Papio*. Second, the sequence of evolutionary events in the Frontier Hypothesis contradicts the scenario of baboon troops giving rise to multilevel societies through increasing fission–fusion and strengthened male–female bonds (see Chap. 8). The result is two mutually exclusive theories explaining major changes in hominin social organization: troop to multilevel society or the reverse. Finally, the Frontier Hypothesis seems to render the behavior of olive and yellow baboons as having little relevance to reconstructing early hominin behavior. These implications are all highly debatable, but they must be considered in future work. For the present, I will continue with more conventional comparisons that encompass hominins and all the baboon species.

7.6 Male–Male Relationships in Troops

Relationships between adult males are one important component of social organization in humans and other primates. The behavior of baboons in troops provides several important analogies for male–male relationships in early hominins. These involve aggression, dominance, cooperation, and tolerance. More complicated relationships in multilevel societies will be discussed in the next section.

7.6.1 Aggression and Dominance Among Males

The evolution of male–male aggression in hominins and its implications for the behavioral tendencies of modern humans have been the subject of bitter conflict in scientific and popular publications. The focus often has been on the most extreme forms of aggression: injurious and lethal violence. As a result, reconstructions of early hominin behavior have been plagued by controversies about male aggressiveness and levels of violence.

Some early descriptions of baboon behavior (in some cases captives in poor conditions) emphasized male dominance, achieved by aggression. This was one stimulus for the backlash in which many writers rejected any baboon model for early hominin behavior (Chap. 3). However, extensive research on wild baboons has developed a more complex and varied picture of conflict and conflict resolution among males. Thus, the baboon perspective can address questions about both causes and limits on aggression in the diverse habitats occupied by early hominins. How much violence was likely to occur in large troops and how much could be sustained in the face of external dangers? To what extent might it have been mitigated by stable dominance relationships? Which resources were involved in the competition?

Fighting between male baboons quickly draws attention from observers because it is noisy, may involve vigorous chases (Fig. 7.1), and occasionally results in serious injuries inflicted by large canine teeth. However, baboons have less spectacular ways to regulate male interactions. Threats are usually sufficient to maintain status differences and simple displacements (forcing another individual to move away

Fig. 7.1 An adult male olive baboon chases another male after a fight. In many such chases the fleeing individual is never caught. In some, the roles are reversed at some point. (Photo by Glenn King, Tarangiri, Tanzania)

Fig. 7.2 A male chacma baboon with a typical fight wound. (Photo by Curt Busse. Okavango, Botswana)

from a spot) are daily reminders. This is the case because aggression is not an "instinct," but a set of tactics in pursuit of social advantage.

This explains the pattern reported by Kitchen et al. (2005) in which most chases and fights, including those that led to injuries, were between males of similar rank, that is, those that were involved in serious competition for dominance. Opponents of disparate ranks clashed most frequently in contests that involved resources of high fitness value: meat, estrous females, and endangered infants. Results of the study suggested that competitive encounters among male baboons follow patterns predicted by evolutionary game theory (Kitchen et al. 2005). Given similar social structure and environmental circumstances, there is no reason to think that hominin aggression was any more intense or common (Fig. 7.2).

7.6.2 Coalitions

A male coalition, consisting of two and occasionally three individuals, can play a major role in mating effort in a troop. Male baboons strive to establish *consort relationships* in which they more or less monopolize mating opportunities with females when they are fertile. In a study of yellow baboons, coalitions succeeded in causing a consort turnover in 35 of 55 attempts (Alberts et al. 2003). This was the main cause of deviations from the system of access to females based on individual dominance. Selection of a coalition partner seems to be based mainly on fighting ability (Bercovitch 1988). Having an affiliative relationship may be a factor in forming a coalition, but it can also be explained as cementing an existing partnership (Noë and Sluijter 1995). Recurrent partners rarely spend extra time in affiliative behavior and their spatial proximity can be explained by the common focus of the coalition. It may be that coalitionary behavior is each partner's opportunistic response to a situation, and only appears to be cooperative. This interpretation avoids anthropomorphism, but seems to be undercut by the occurrence of preliminary recruiting behaviors. Signals exchanged by prospective partners include head flagging, staccato grunting, and mounting. The baboons seem to anticipate the cooperative relationship.

Low cost is an important factor in coalitionary behavior. Comparative evidence indicates that the behavior requires little time (usually less than 30 min) and energy (mostly threatening with occasional short bursts of running and fighting) and that wounds are rare and usually slight. In short, joining a coalition is a low-cost behavior with the prospect of a substantial benefit. This suggests the possibility that the behavior was favored by natural selection in a way that could also apply to early hominins.

7.6.3 The Kinda Baboon Alternative (?)

The Kinda species provides an alternative model for male–male relationships in baboon troops and possibly for early hominins (Petersdorf et al. 2019). Living in relatively large groups and displaying less sexual dimorphism than the other species, Kinda male dominance is organized by *queuing* rather than contest, that is, waiting in a metaphorical line for an opportunity to reproduce. New males entering a troop accept a place at the end of the line rather than fighting with resident males. Alpha males have long tenures.

An important factor in the Kinda system is *sperm competition*, which means that reproductive advantage over other males comes from fertilizing as many females as possible with large quantities of high-quality sperm. The hallmark of this adaptation is large testis in proportion to body mass. Kinda baboons outdo all other baboon species in this regard. This feature tends to negate the relevance of Kinda baboon behavior to early hominins. Lacking evidence of testis size in early hominins, we have to look to the evolutionary end result in humans. Humans have not evolved a comparable adaptation. Relative testis size in proportion to body mass is average for primates (Dixson 2012).

7.7 Male–Male Affiliation in Multilevel Species

In troops, coalitions are often situational and temporary (although partners in some cases may display a degree of affiliative behavior). Baboons in multilevel societies have more organized and longer-term alliances, especially between leader and follower. Males develop enduring relationships with other males that are characterized by general tolerance, friendly interaction, and occasional coalitionary behavior.

These male–male associations link the central males of unimale groups with each other and with their followers. Kinship accounts for some but perhaps not all of these relationships. Within this common framework, hamadryas and Guinea baboons differ in some ways that may offer alternative or complementary analogies for early hominins.

7.7.1 Male–Male Relationships in Hamadryas Baboons

Hamadryas baboons have a complicated system of ongoing male–male relationships (Evans et al. 2022). One aspect is the takeover of females by one central male from another. In some cases, a younger and/or healthier male takes females from another who can no longer defend them. This may be a cost to the losing male, but there is little or no aggression. In other takeover attempts, one fit male challenges another and serious fights may result.

Takeovers are sporadic events and most of the time central males seem to "respect" one another's possession of females (Evans et al. 2022). This state of affairs seems to depend on recurring communication between males in the form of *notifications*, that is, a set of ritualized signals that seem to convey mutual trust (see Chap. 11 for details). Another possible factor is that leader males who limit aggression toward other males may be more successful in attracting followers, who contribute to maintaining females. In sum, the hamadryas system entails occasional aggression between males to obtain females and restraint of aggression between males at most times (mediated by notifications) (Fig. 7.3). Evans et al. (2022) argue that this combination of tactics was favored by natural selection and provides an analogy for social relationships in early hominins.

In hamadryas baboons, each follower is affiliated with only one OMU, and not every OMU has a follower. The functions of hamadryas followers seem to be focused on relations within the OMU. In a system where females are mainly acquired by coercion, hamadryas leaders with followers had longer tenures as leaders, acquired more females, and sired more infants than males without followers (Choudhury et al. 2015). Follower relationships with OMU females may encourage the females to adhere to the leader as long as possible. A central male may be aware of this function, but a follower male is readily tolerated because he belongs to the same clan as the leader and is likely to be a relative. Leaders and followers are maternal relatives more often than expected by chance (Städele et al. 2016).

Kinship is also a factor in the hamadryas clan, a group in which males prefer to associate with relatives (Städele et al. 2015). Two anecdotes suggest the strength of clan ties (Colmenares et al. 2006). First, a juvenile male stayed with his father when his mother was abducted into another clan. Second, an infant of an abducted female returned to his natal clan on his own when he reached the age of 3 years. Closer kinship is illustrated by the case of a captive colony in which maternal brothers had the most cohesive relationship among males.

Fig. 7.3 Tolerance among three male hamadryas baboons in the Alexandria Zoo. (Photo by Hatem Moushir. Resized for publication. Wikimedia license: https://creativecommons.org/licenses/by-sa/4.0/)

7.7.2 Male–Male Relationships in Guinea Baboons

Kalbitzer et al. (2015) compared male chacma and Guinea baboons with regard to competition, aggression, and dominance. Chacmas were more frequently involved in agonistic interactions and displayed consistent dominance relationships. Although theory predicts that the intensity of male competition is higher if many males compete for access to few females, differences in the ratio of males and cycling females could not explain the species difference in agonistic behavior. In a study of Guinea baboons, "agonistic interactions were generally rare and largely restricted to a few dyads" (Patzelt et al. 2014).

Male chacmas of high rank and those engaged in sexual activity showed elevated levels of stress hormones (glucocorticoids) and also tended to show elevated testosterone levels. There were no such effects in Guinea baboons. In related contrasts with chacma baboons, male Guinea baboons in a group do not form linear dominance hierarchies and dominance relationships between individuals are less consistent than in chacmas. An attempt to quantify dominance relations found that a clear male dominance hierarchy could not be established due to the high degree of uncertainty in individual rank scores. The only clear tendency was that bachelor males were more likely to be found at the low end of the scores (Dal Pesco et al. 2021).

Male Guinea baboons in the same "party" (the equivalent of the hamadryas clan) are tolerant and friendly toward one another. In one study, 80% of dyads displayed greeting interactions (comparable to hamadryas notifications) and a third of them engaged in affiliative behaviors such as sitting in close proximity (Patzelt et al.

2014). However, most males have a small number of "preferred partners," usually two or three, said to have a "strong bond" (Dal Pesco et al. 2022). These are friendly and enduring relationships characterized by high spatial tolerance, support in agonistic interactions, and occasional grooming sessions (Fischer et al. 2017; Fig. 7.4). Continuing research showed that these bonds could be stable for at least 4 years (the full length of the study) and were not affected by the males' relationships with females (Dal Pesco et al. 2021).

Dal Pesco et al. (2021) pursued the issue of relatedness between males with strong bonds. They confirmed that these relationships are differentiated, equitable, and stable over time, and that there are no clear dominance patterns between partners. Relatedness was assessed with genetic material obtained from fecal samples. The evidence demonstrated that average relatedness was significantly higher between strongly bonded males, suggesting that kin bias contributes to the social preferences of males

Dal Pesco et al. (2022) tested the evolutionary significance of male bonding in Guinea baboons with behavioral and paternity data collected over 45 months. Strong bonds did not lead to reproductive success. Instead, males that spent less time socializing with other males were associated with a higher number of females and sired more offspring. Reproductively active males still maintained bonds with other males, but adjusted their time budgets. They may have maintained strong bonds with the most intense ritualized greetings, such as mounting and genital manipulation

It is not clear why males maintained these relationships at some cost or risk, since they did not contribute to reproductive success. Two possible factors are the effect of close kinship (Dal Pesco et al. 2021) and the benefit of occasional

Fig. 7.4 A male Guinea baboon grooms another male. (Photo by Dietmar Zinner. Niokolo Koba National Park, Senegal)

coalitionary support (Patzelt et al. 2014). The absence of reproductive success means that any analogy with hominins would not be based on natural selection. Perhaps similar social contexts resulted in parallel development of strong bonds with male kin.

Bachelor males among Guinea baboons play a different role than hamadryas followers (Dal Pesco et al. 2021). Most central males among Guinea baboons have at least one associated bachelor, and a majority have more than one. A bachelor is likely to associate with several unimale groups and he interacts with the central males as well as their female associates. Average relatedness between a central male and associated bachelors is higher than in any other relationship except for central males who are closely bonded to each other.

The kinship bond between primary and secondary male parallels the hamadryas baboon. However, the social functions seem to differ. While the hamadryas follower focused on his leader, the multiple ties of Guinea baboon bachelors seem to contribute more to social solidarity within and across unimale groups.

7.7.3 Men in Groups

Tiger (1969) surveyed all-male groups across diverse cultures. He characterized this behavioral tendency as male bonding and hypothesized that it had an evolutionary basis, which he attributed to the need for cooperation among men in hunter-gatherer societies. Since Tiger's work, much more has been learned about male–male association in primates, including baboons. This work supports and modifies Tiger's ideas. Developments include information about dominance and aggression, and about coalitions and bonds. Dominance, aggression, and coalitions are important in male–male relations in most baboon troops. These features are modified in a human-like direction in the multilevel societies, suggesting what the transition in early hominins might have been like.

Rodseth (2012) set forth similar arguments. Humans live in modular societies with a minimum of two levels of organization, the conjugal family and the local community. Yet any human community is likely to contain at least one other social unit, a same-sex association such as a men's "club" or "brotherhood." What has been called "bachelor threat" in other mammalian species is also a problem in human societies, but tensions between married men and bachelors are often eclipsed by the need for warriors to defend the local community. The ethnographic record includes many cases in which fraternal security takes precedence over conjugal bonds, resulting in the physical segregation of the sexes, including husbands and wives. At the extreme, a husband usually sleeps at a men's house while making regular visits to his conjugal family. Though this pattern is classically associated with tribal Amazonia and Melanesia, Rodseth sees it as part of a continuum of variation in small-scale societies worldwide.

The two multilevel baboon species suggest an early hominin foundation for these associations, long before they became culturally elaborated and differentiated. In both baboon species, male–male bonds pervade social entities beyond the unimale groups. These relationships include kinship among many of the males, friendships where there is no kinship, and dominance-subordination between primary and secondary males. Control of mating is often considered a motivation and/or function of these male associations in humans and this seems to be anticipated by the exclusion of secondary males in both baboon species.

Male–male relationships in Guinea baboons form a network that seems to be a possible model for the beginning of the human pattern of all-male groups, as expounded by Tiger and Rodseth. However, further research that focused on the adaptive benefits of these relationships cast doubt on the hypothesis that they were favored by natural selection. If this result applies to hominins, "men in groups"was either selected for in later hominins (hunting parties?) or originated as a cultural phenomenon for reasons educed by cultural anthropologists.

7.8 Summary and Discussion

Baboons provide diverse options for hypotheses about early hominin social organization, pertaining to variation and change. Based on recent human hunter-gatherers, members of an archeological conference came to some agreement about probable early human societies. Groupings of 20–50 were common, but not rigidly separated from one another. Size changed with various functions and purposes. Larger groups were especially important for defense and cooperation when technology was limited. These inferences from human hunter-gatherers articulate well with baboon evidence, suggesting deep evolutionary roots for basic features of hominin social organization.

The original hominin society might have been much like the baboon troop. This is a multi-male/multi-female social entity found in four of the six baboon species. Medium-sized troops fit into the 20–50 range. Larger troops may be important for defense against predators or other troops. Competition for resources may lead to fighting, but displays and avoidance are more common. Occasionally tolerance allows the brief formation of a super-troop. When a troop is too large for its resources, it has two ways to compensate. It can fission temporarily into smaller foraging subgroups, or it can fission permanently into two groups. The flexibility of the troop suggests that it could have been the social unit of early hominins as they expanded from their woodland base to encompass increasingly diverse habitats. Some archeological evidence suggests that the hominin troop could have persisted for millions of years, until major changes in the hominin way of life took place around 2 mya.

Variation among baboon species provides a series of potential models or sources of analogy for early hominins. Chacma baboons in southern Africa may represent the original troop organization of *Papio*. Olive and yellow baboons to the north display some more complex forms of social organization, such as the formation of coalitions to achieve mating success. Kinda baboon troops are unusually large, often including more than 200 members. Kinda baboons also differ from the other species in that reproductive competition is carried out through sperm competition rather than aggression.

Hamadryas and Guinea baboons are organized into multilevel societies. The basic unit in each of these systems is a unimale group that is composed of a single central adult male affiliated with one or more adult females. This group can forage separately, but it is integrated into larger social groups. The higher levels of social organization manifest numbers consistent with the archeological model from hunter-gatherers and include units for travel, foraging, and predator defense.

Common features of the multilevel baboons can be summarized as follows: (1) Unimale groups, each composed of one central male, one or more affiliated females who are social and reproductive partners, and one or more followers. (2) Often one or more secondary males with social but not sexual access to the females. (3) One or two higher levels of social organization, identifiable by groups with consistent membership. (4) An ephemeral maximum social group, defined by mutual tolerance among constituent groups with regard to a sleeping places and/or home range overlap. (5) Male philopatry: males tend to stay in one group while females leave (the circumstances of female departure are very different in the two species).

Baboon troops and multilevel societies can be considered alternative models for early hominin social organization. Alternatively, they can be linked as two stages in hominin evolution. The latter view has been advanced by some social anthropologists. Male philopatry and female dispersal have been predominant characteristics of human society and similar patterns distinguish the multilevel baboon societies from the troops.

An influential scenario called the "Frontier Hypothesis" suggests that male philopatry was a key development in the expansion of both baboons and hominins to occupy a wide range of habitats. The hypothesis further suggests that male philopatry was a causal factor in the origin of unimale groups and multilevel society. The postulated *Papio* phylogeny casts doubt on the relevance of olive and yellow baboons for analogies to early hominins. While accepting this as a subject for future debate, this work continues with the application of all baboon species to the goal of reconstructing early hominin behavior patterns.

One such topic is the evolution of male–male relationships, which have been the subject of bitter debate about the role of aggression in the human evolutionary heritage. Early descriptions of baboons in troops, especially popular accounts, tended to focus on injurious violence. However, even among chacma baboons, reputed to be the most aggressive baboon species, disputes are more often settled by displays or by established dominance relations. Much the same is true of olive and yellow baboons, with the added feature of coalitions for cooperation in aggressive confrontations.

Aggression is a tool to obtain status and resources, including mating opportunities. Kinda baboons suggest an alternative in which sperm competition largely substitutes for direct competition and aggression. However, the relevance of this pattern to hominins is questionable because of the difference in the main physical correlate.

Sperm competition is represented by relatively large testis in proportion to body mass, a trait found in Kinda baboons but not in humans.

Male philopatry could have set the stage for new long-term relationships among male hominins. While cooperation in troops takes the form of temporary coalitions, multilevel societies display long-term relationships among central males and between them and secondary males. These affiliations are often based on kinship. Peace and solidarity are maintained in part through the exchange of ritualized communication. Secondary males may have social but not sexual relationships with females in unimale groups.

Within this general framework, there are important differences between hamadryas and Guinea baboons that suggest alternative models for early hominins. Hamadryas males control females and sometimes acquire them in aggressive confrontations with other males; Guinea baboon females choose their affiliations with males and are free to change them without interference. Central males among hamadryas "respect" each other's possession of females; Guinea baboon males have strong bonds with a few others and are "tolerant" of most other males. Hamadryas followers each associate with one unimale group; Guinea baboon bachelors have multiple connections. Hamadryas followers seem to contribute to the tenure and female holdings of the central male; Guinea baboon bachelors seem to function as a network that contributes to the solidarity of the larger social unit.

Stronger bonds among males, in association with more complex and ongoing forms of cooperation, suggest an early evolutionary foundation for patterns of male bonding and group formation in human societies. Culturally elaborated forms in recent humans have been characterized as "men in groups"and attributed to cooperative hunting. While hunting and other recently developed human patterns may well have contributed to the phenomenon, baboons show us that it could have originated in early hominins.

References

Alberts SC, Watts HE, Altmann J. Queuing and queue-jumping: long term patterns of reproductive skew among male savannah baboons. Anim Behav. 2003;65:821–40.

Aldrich-Blake FPG, Bunn TK, Dunbar RIM, Headley PM. Observations on baboons, *Papio anubis*, in an arid region in Ethiopia. Folia Primatol. 1971;15:1–35. https://doi.org/10.1159/000155365.

Anderson CM. Subtrooping in a chacma baboon (*Papio ursinus*) population. Primates. 1981;22:445–58. https://doi.org/10.1007/BF02381237.

Altmann SA, Altmann J. Baboon ecology: African field research. University of Chicago Press, Chicago; 1970.

Bercovitch F. Coalitions, cooperation and reproductive tactics among adult male baboons. Anim Behav. 1988;36(4):1198–209. https://doi.org/10.1016/S0003-3472(88)80079-4.

Chapais B. Primeval kinship, how pair-bonding gave birth to human society. Cambridge, MA: Harvard University Press; 2008.

Choudhury S, Pines M, Saunders J, Swedell L. The adaptive value of secondary males in the polygynous multi-level society of hamadryas baboons. Am J Phys Anthropol. 2015;158:501–13. https://doi.org/10.1002/ajpa.22804.

Colmenares F, Esteban MM, Zaragoza F. One-male units and clans in a colony of hamadryas baboons (*Papio hamadryas hamadryas*): effect of male number and clan cohesion on feeding success. Am J Primatol. 2006;68(1):21–37. https://doi.org/10.1002/ajp.20204.

Dal Pesco F, Trede F, Zinner D, Fischer J. Kin bias and male pair-bond status shape male-male relationships in a multi-level society. Behav Ecol Sociobiol. 2021;75:article 24. https://doi.org/10.1007/s00265-020-02960-8.

Dal Pesco F, Trede F, Zinner D, Fischer J. Male–male social bonding, coalitionary support and reproductive success in wild Guinea baboons. Proc R Soc B. 2022;289:20220347. https://doi.org/10.1098/rspb.2022.0347.

DeVore I, Hall KRL. Baboon social behavior. In: DeVore I, editor. Primate behavior, field studies of monkeys and apes. New York: Holt Rinehart Winston; 1965. p. 53–110.

Dixson AF. Primate sexuality, comparative studies of the prosimians, monkeys, apes, and humans. New York: Oxford University Press; 2012.

Dominguez-Rodrigo M, Cobo-Sanchez L. The spatial patterning of the social organization of modern foraging *Homo sapiens*: a methodological approach for understanding social organization in prehistoric foragers. Palaeogeogr Palaeoclimatol Palaeoecol. 2017;488(S7) https://doi.org/10.1016/j.palaeo.2017.06.008.

Dunbar RIM, MacCarron P, Robertson C. Trade-off between fertility and predation risk drives a geometric sequence in the pattern of group sizes in baboons. Biol Lett. 2018;14:20170700. https://doi.org/10.1098/rsbl.2017.0.

Evans KD, Swedell L, Chowdhury S. Respect for males amid suppression of females: selective use of aggression and fitness correlates in the male-dominated society of hamadryas baboons. J Hum Evol. 2022;165 https://doi.org/10.1016/j.jhevol.2022.103151.

Fischer J, Kopp GH, Dal Pesco F. Charting the neglected West: the social system of Guinea baboons. Am J Phys Anthropol. 2017;162:15–31. https://doi.org/10.1002/ajpa.2314.

Fischer J, Higham JP, Alberts S, et al. The natural history of model organisms: insights into the evolution of social systems and species from baboon studies. elife. 2019;8:e50989. https://doi.org/10.7554/eLife.50989.

Goffe AS, Zinner D, Fischer J. Sex and friendship in a multilevel society: behavioural patterns and associations between female and male Guinea baboons. Behav Ecol Sociobiol. 2016;70(3):323–36. https://doi.org/10.1007/s00265-015-2050-6.

Goren-Inbar N, Belfer-Cohen A. Reappraisal of hominin group size in the Lower Paleolithic: an introduction to the special issue. J Hum Evol. 2020;144:102821. https://doi.org/10.1016/j.jhevol.2020.102821.

Hamilton WJ III, Buskirk R, Buskirk WH. Defense of space and resources by chacma (*Papio ursinus*) baboon troops in an African desert and swamp. Ecology. 1976;57:1264–72.

Hex SBSW, Tombak K, Rubenstein DI. A new classification of mammalian uni-male multi-female groups based on the fundamental principles governing inter- and intrasexual relationships. Behav Ecol Sociobiol. 2021;75(11):1–31. https://doi.org/10.1007/s00265-021-03046-9.

Jolly CJ. Philopatry at the frontier: a demographically driven scenario for the evolution of multilevel societies in baboons (*Papio*). J Hum Evol. 2020;146:102819. https://doi.org/10.1016/j.jhevol.2020.102819.

Kalbitzer U, Heistermann M, Cheney DL, et al. Social behavior and patterns of testosterone and glucocorticoid levels differ between male chacma and Guinea baboons. Horm Behav. 2015;75:100–10. https://doi.org/10.1016/j.yhbeh.2015.08.013.

Kitchen DM, Cheney DL, Seyfarth RM. Factors mediating inter-group encounters in savannah baboons (*Papio cynocephalus ursinus*). Behaviour. 2004;141:197–218. https://doi.org/10.1163/156853904322890816.

Kitchen DM, Cheney DL, Seyfarth RM. Contextual factors mediating contests between male chacma baboons in Botswana: effects of food, friends and females. Int J Primatol. 2005;26(1):105–25. https://doi.org/10.1007/s10764-005-0725-y.

Kummer H. Social organization of hamadryas baboons, a field study. Chicago: The University of Chicago; 1968.

Kunz BK, Linsenmair KE. The disregarded West: diet and behavioural ecology of olive baboons in the Ivory Coast. Folia Primatol. 2008;79(1):31–51. https://doi.org/10.1159/000108384.

Markham AC, Guttal V, Alberts S, Altmann J. When good neighbors don't need fences: temporal landscape partitioning among baboon social groups. Behav Ecol Sociobiol. 2013;67:875–84. https://doi.org/10.1007/s00265-013-1510-0.

Noë R, Sluijter AA. Which adult male savanna baboons form coalitions? Int J Primatol. 1995;16(2):77–105. https://doi.org/10.1007/BF02700154.

Patzelt A, Kopp GH, Ndao I, et al. Male tolerance and male–male bonds in a multilevel primate society. Proc Natl Acad Sci USA. 2014;111(41):14740–5. https://doi.org/10.1073/pnas.1405811111.

Petersdorf M, Weyher A, Kamilar JM, et al. Sexual selection in the Kinda baboon. J Hum Evol. 2019;135(S37) https://doi.org/10.1017/j.jhevol.2019.07.007.

Rodseth L. From bachelor threat to fraternal security: male associations and modular organization in human societies. Int J Primatol. 2012;33(5):1194–214. https://doi.org/10.1007/s10764-012-9593-4.

Saayman GS. Aggressive behaviour in free-ranging chacma baboons (*Papio ursinus*). J Behav Sci. 1971;1(3):77–83.

Schreier AL, Swedell L. The fourth level of social structure in a multi-level society: ecological and social functions of clans in hamadryas baboons. Am J Primatol. 2009;71:948–55. https://doi.org/10.1002/ajp.20736.

Schreier AL, Swedell L. Ecology and sociality in a multi-level society: ecological determinants of spatial cohesion in hamadryas baboons. Am J Phys Anthropol. 2012;148(4):580–8. https://doi.org/10.1002/ajpa.22076.

Städele V, Van Doren V, Pines M, et al. Fine-scale genetic assessment of sex-specific dispersal patterns in a multi-level primate society. J Hum Evol. 2015;78:103–13. https://doi.org/10.1016/j.jhevol.2014.10.019.

Städele V, Pines M, Swedell L, Vigilant L. The ties that bind: maternal kin bias in a multi-level primate society despite natal dispersal by both sexes. Am J Primatol. 2016;78(7):731–44. https://doi.org/10.1002/ajp.22537.

Stolz G, Saayman LP. Ecology and behaviour of baboons in the Northern Transvaal. Ann Transv Mus. 1970;26:99–143.

Stout D, Rogers MJ, Jaeggi AV, Semaw S. Archaeology and the origins of human cumulative culture: a case study from the earliest Oldowan at Gona, Ethiopia. Curr Anthropol. 2018;60(3):283–448. https://doi.org/10.1086/703173.

Swedell L, Plummer T. A papionin multi-level society as a model for hominin social evolution. Int J Primatol. 2012;33:1175–93.

Swedell L, Plummer T. Social evolution in Plio-Pleistocene hominins: insights from hamadryas baboons and paleoecology. J Hum Evol. 2019;137:1–15. https://doi.org/10.1017/j.jhevol.2019.102777.

Tiger L. Men in groups. New York: Random House; 1969.

Chapter 8
Male–Female Relationships

8.1 Introduction

Baboon social diversity offers a number of models and analogies for male–female relationships in early hominins. Several of them may represent the earliest beginnings of human-like family units within the hominin lineage. *Family* is defined here as a long-term social unit consisting of one male affiliated with one or a few females. This describes social units found in the great majority of human societies known to anthropology.

In both humans and baboons, these relationships involve more than sex. However, *sexual strategies* are an important factor. The term refers to patterns of behavior that are likely to produce an adaptive result for the behaving animal. In both humans and baboons, sexual strategies vary from mutual bonding to aggression and resistance. In both humans and baboons, males provide widely varying amounts of parental care to immature individuals.

More than in some other areas of behavior, comparison of reproduction in baboons and hominins requires consideration of differences between them, because these place more stringent limits on analogies. Accordingly, the chapter begins with a brief discussion of baboon sexuality. The basic features are shared with many other primates, so the concepts will be familiar to anyone who has studied or read widely about primate behavior. These features will be briefly summarized for readers without such background.

8.2 Some Basic Features of Baboon Sexuality

The topic of sexuality is an example of the need to focus on analogies rather than comprehensive models in comparing baboons with hominins. Differences between baboons and humans show that baboon sexuality cannot be taken as an integrated

G. E. King, *Baboon Perspectives on Early Human Ancestors*,
https://doi.org/10.1007/978-3-031-36769-4_8

"model"for early hominin behavior. Nevertheless, there are specific aspects of baboon sexuality that can provide informative analogies (vide Dixson 2012).

As in other Old World monkeys, a typical adult female baboon who is not pregnant or lactating goes through an ovarian cycle (also called a menstrual cycle). It lasts for roughly a month (there are individual, populational, and species variations). Hormonal and behavioral changes take place around the time of ovulation. This period is often called estrus, a term that is also applied to other mammals. A number of researchers reject this usage because it implies a rigidity that does not characterize monkeys and apes and because it seems to create an artificial distinction between humans and other primates (Dixson 2012). However, the term is still widely used by primatologists and it conveniently denotes a segment of the female cycle during which there are significant hormonal and behavioral changes that are connected with ovulation and thus with the highest probability of conception.

This idea of estrus is useful as long as flexibility in the rest of the cycle is kept in mind. In contrast to most other mammals, copulation may occur at any point in the cycle. It can be said that the female may be sexually receptive at any time, that is, willing or at least able to accept copulation with a male. However, two behavioral changes occur around the time of ovulation (Beach 1976; Dixson 2012) and can be considered features of estrus. One is proceptivity: the female actively seeks copulation. The other is attractivity: the female stimulates greater sexual interest in males.

Female baboons are like females in many other primate species in that a sexual swelling of the perineal area is associated with estrus. The pink swelling is an attractive signal to males. This is one of the most striking differences between baboons and extant humans. Since females of both *Pan* species display comparable swellings, it is not clear what the situation was in early hominins. Whatever the answer to that question might be, not all baboon analogies for male–female relations are affected by the problem.

Whether as a sign of receptivity or proceptivity (depending on the time in the cycle), the female baboon presents her hindquarters to a male. Copulations may take place in the context of consortship (or mate guarding), in which a male tries to stay close to a female and copulate with her multiple times while preventing other males from doing so. The female may be an active participant in the relationship, rejecting other males who try to replace her consort.

Potential baboon-hominin analogies include sexual conflicts, that is, ways in which the reproductive interests of the male and female are somehow contradictory. The most drastic of these is infanticide by males. In baboons, as in other primate species (and other mammals as well), infanticide has been interpreted as a reproductive tactic (Hausfater and Blaffer-Hrdy 1984; van Schaik and Janson 2000). Killing an infant is adaptive for a male who has replaced another male in a relationship with the mother, because it ends the nursing period with the result that the female becomes sexually available. Though the reproductive hypothesis of infanticide is widely accepted, it should be noted that it still has opponents (Rees 2009). Perhaps the most common argument is that there is insufficient evidence for its adaptive significance (Dixson 2012).

An alternative explanation is that the behavior eliminates a rival male's genes. This interpretation is countered by the fact that juveniles are rarely killed (van Schaik and Janson 2000; Muller and Wrangham 2009; Zipple et al. 2017). Sparing juveniles fits the sexual availability hypothesis because, according to the hypothesis, the killing of an infant ends the nursing period and results in a female's return to the estrus cycle. No such advantage comes from killing a juvenile. Rather than being a potential contribution to reproductive success, trying to kill a juvenile may entail costs in terms of energy expenditure, loss of feeding time, neglect of social relationships, and some danger from defense of the juvenile by other adults.

8.3 Human Families and Primate Analogies

"One key question in social evolution is the identification of factors that promote the formation and maintenance of stable bonds between females and males beyond the mating context" (Goffe et al. 2016). A useful starting point is the general features of family in extant humans (including recent and historically known), viewed in cross-cultural perspective. Family is loosely defined here as a long-term relationship between one male and one or more females that usually involves reproductive effort (and economic effort in humans, but this does not enter into hominin-baboon comparisons). An operational definition for "long-term" could be persistence from establishment of a bond through the childhood of one offspring (Fisher 1992). However, significantly longer male–female relationships exist in both humans and baboons.

Cross-cultural and historical surveys show that such relationships have been part of human social organization across virtually all cultures and all known time periods. Demography dictated that monogamy was most common, but polygyny was allowed in the majority of societies and desired by many men if not most (Gregersen 1983). The typical human mating relationship has been a relatively long-term association between one male and one to three females (King 2003). Polyandry has occurred in just a few societies under unusual conditions (certain rules of land tenure, for example). A strict rule of monogamy (often violated) became common in the world with the recent spread of Christianity.

Extended families and other larger kinship groups became common with the advent of complex societies, mostly based on farming. The family in the narrow sense became submerged in these groupings, but persisted in virtually every known culture and has become prominent again in modern societies where mobility is greater. The cross-cultural importance of the family suggests an evolutionary origin (Chapais 2008; Swedell and Plummer 2019). This is not to say there is a simple genetic basis for the phenomenon. Rather, the hypothesis is that a complex genetic basis underlies the motivations and behavioral tendencies of human males and females that result in the formation of families.

Baboons provide several perspectives on this process, including the form, social context, and proximal mechanisms that might have been involved in the antecedent social patterns in early hominins (Chapais 2008; Petersdorf et al. 2019). A crucial factor in analogies between male–female bonds in baboons and early hominins is that this kind of relationship emerged and persisted within communities, whether in troops or multilevel societies.

An alternative evolutionary approach to the human family is comparison with "pair-bonded" species, such as gibbons and titi monkeys. There is an extensive controversy about the definition and nature of pair-bonding itself, which is beyond the scope of this book (see, for example, Fernandez-Duque et al. 2020). The key point here is that, in most of the species originally cited as pair-bonded (e.g., gibbons), each male–female pair is relatively isolated from others. Interactions are largely limited to territorial conflicts and to the transfer of maturing individuals from one group to another.

Humans and baboons differ from such species in several important respects. First, a male may be involved in a long-term affiliation with more than one female. Second, it is rarely (or never) the case that two adult males are in the same relationship with a female. Third, such associations are combined into larger communities (Chapais 2008; Städele et al. 2021).

8.4 Male–Female Associations in Troops

Human families in most or all cultures entail several characteristics that can be compared to social patterns in baboon troops: (1) long-term affiliation (generally measured in years) between a particular male and female(s); (2) a sexual relationship between such partners; (3) production of shared offspring; (4) some degree of paternal care or support for the offspring. COKY baboons all display at least some relevant traits. Chacma baboons, yellow baboons, and olive baboons are broadly similar to one another in male–female relationships that are potential analogies for early hominins (Hawley et al. 2023). Kinda baboons seem to differ in several respects, according to initial reports. They may offer an alternative analogy for male–female relationships in early hominins.

8.4.1 Special Friends (Primary Associations)

Adult males and females in many troops form relationships that have been called "special friendships" (Smuts 1985). These can be defined as "long-term, nonsexual social bonds between unrelated (to the females) males and anestrous females" (McLester 2020). Sexual behavior is not excluded from these relationships; the point of the definition is that the connection involves more than mating (Fig. 8.1). These relationships can lead to any or all of the following social patterns, which are possible analogies for early hominin male–female interactions: copulation, production of offspring, male care for immatures.

Fig. 8.1 A male olive baboon grooms an anestrous (currently infertile) female, an indicator of Special Friendship. (Photo by Glenn King. Manyara, Tanzania)

Some researchers now use the term "primary associations" instead of special friendships (Hawley et al. 2023; Städele et al. 2021). This seems overly broad to me, since it can literally refer to the mother–offspring relationship as well as others. Long usage has given "special friend" the meaning intended here (McLester 2020). In any event, whatever the terminology, relationships like these could have been the basis for differentiation of families within troops of early hominins (Fogel et al. 2021; Städele et al. 2021; Fig. 8.2).

8.4.2 Benefits of Special Friendships

The following discussion draws on several recent reviews, where numerous references to earlier literature can be found: Hawley et al. (2023), Städele et al. (2019, 2021). It pertains to chacma baboons, olive baboons, and yellow baboons—abbreviated here as COY (COKY without the Kinda baboons, which differ in some important ways and will be discussed separately). In all of the COY species, some males form close associations with pregnant and lactating females. These associations can be identified by rates of proximity and grooming, and reduced levels of aggression.

Evidence for the significance of these relationships has been gathered from long-term studies of the COY species. Most of it points to the function of the male behavior as being parental rather than mating effort (Fig. 8.3). The males are in many cases the sires of infants born to associated females, to which they devote varying degrees of paternal care. The argument against mating effort comes from genetic analyses that show that a male is not especially likely to father a special friend's next infant.

Fig. 8.2 A male olive baboon, with an infant huddled against him, is groomed by a female. This social configuration is typical of Special Friendships in the species. (Photo by Glenn King. Manyara, Tanzania)

Fig. 8.3 A male olive baboon grooms an infant with a female next to them. (Photo by Glenn King. Manyara, Tanzania)

Fig. 8.4 A male chacma baboon with an infant, presumably the offspring of a Special Friend. (Photo by Curt Busse. Okavango, Botswana)

In one or more COY species, males provide one or more kinds of care for offspring (Fig. 8.4). These include the following: (1) selectively supporting infants and juveniles in social conflicts; (2) protecting them from predation, infanticide, and/or harassment by other troop members (e.g., female rivals of the mother); (3) providing access to valuable resources, such as high-quality foods, and facilitating efficient feeding. By performing such behaviors, a male may contribute to the perpetuation of his own genes as well as benefiting the fitness of the mother.

The benefits of male–female bonds in humans and baboons extend beyond reproductive success. Noting research findings that humans of both sexes who are more socially integrated have longer lives, Campos et al. (2020) asked whether the same thing applied to yellow baboons for which they had 35 years of life history data. They modeled the factors in age-specific adult mortality in both sexes and compared survival trajectories with social bonds over lifetimes. Both males and females with strong bonds to the opposite sex lived significantly longer. Mortality risk was lowered by about 28% in males and more than 30% in females. Campos and colleagues also found a sex difference in that males with higher dominance rank for their age appeared to have shorter lifespans. This makes clear that longer life in males is actually due to social bonding itself rather than a dominance status that facilitates bonding.

Persistence of the special friend bond varies among *Papio* species. In chacma baboons it ends when the female becomes sexually receptive again (Baniel et al. 2016). In olive baboons it can last for years, entailing intimate behaviors spanning multiple periods when the females are not sexually available (anestrus). Male olive baboons, for example, spend a great deal of time with anestrus females, groom them, and huddle with them (Smuts 1985; Fig. 8.2).

Olive baboons are distinctive in other ways (Städele et al. 2019). Dominance status is less sharply defined, which suggests that reproductive skew is less than in the other species. Relatively larger testes indicate that sperm competition is a factor. Coalitions diminish the significance of individual dominance (which is also true in yellow baboons). Male olive baboons relate to the infants of special friends in a variety of ways: they greet, inspect, hold, and carry those infants more than any others (Hawley et al. 2023).

Though the emphasis in research findings has shifted to parenting, there is evidence that the special friend relationship may function in pursuit of mating as well as parenting. Hormonal study of yellow baboons determined that increased testosterone in males is positively correlated with both mating effort and paternal behavior (Onyango et al. 2013). Male olive baboons also seem to combine a degree of mating effort with the parenting pattern. A study that combined behavioral and genetic data showed that sires have stronger bonds with their infants' mothers than do other males and that these bonds sometimes persist past weaning age (Städele et al. 2019). In many cases the sire of the previous infant is still a close partner of the female when she nurses her subsequent offspring. Males who have the strongest bonds with females that have resumed cycling, but are not currently sexually receptive, may be likely to sire the female's next offspring. In more than one-third of the cases, a female's successive infants were sired by the same male. The researchers concluded that development of stable breeding bonds and paternal investment seem to be grounded in the formation of close ties between males and anestrus females. All of these features suggest that olive baboons may be the best single-species model for the foundation of the hominin family in a troop context.

Hawley et al. (2023) considered the fact that male care for immatures is rare in mammalian societies with multiple males and females, as opposed to those that live in well-defined pairs. Why are baboon troops an exception to this rule? According to Hawley, this may be the result of unusual characteristics such as higher paternal certainty, high risk of infanticide, and a long juvenile developmental period. The long juvenile period would certainly have applied to early hominins, based on common origin with the *Pan* species. High risk of infanticide is a doubtful explanation since known variation among *Papio* species runs from high (chacma baboons) to relatively low (olive and yellow baboons to virtually zero (Kinda and Guinea baboons). Higher paternal certainty is a questionable attribution for troops, since females may benefit from copulations with multiple males (Nunn 1999). For example, confusion of paternity may counter infanticide. Another possibility is that a female might choose a special friend from among her sexual partners. High paternal certainty may have become a factor in the unimale groups of multilevel societies.

Relationships like that of special friends seem to be rare in chimpanzees, despite occasional signs of male–female affection and a few cases of recurring sexual connections. Städele et al. (2021) noted that comparable associations do take place in one chimpanzee population and that these differ from baboons in that mating effort seems to be the predominant function. This may be regarded as an alternative analogy for female–male relationships in early hominins. On the other hand, it is plausible that such a pattern existed in the LCA, but later shifted to a system more like that of baboons. Such a shift might have been a response to changing environmental factors, such as increased food competition and predator pressure.

8.4.3 From Special Friends to Families?

In terms of social structure, the transition from special friendships to human-like families in early hominins need not have been very complicated. A male baboon may have several female special friends. Females may also have more than one, but some evidence (a small amount at this point) suggests that two friends is typical (Städele et al. 2021). Starting with special friends in early hominins, the transition to a family-like structure would have required each female to focus on just one male. Guinea baboons may eventually provide a clue as to why this happened, since females in that species voluntarily associate with one male at a time. As far as hominin males are concerned, concentration on one or two females might have increased the probability of successful reproduction. The male could copulate with the females throughout their cycles, obviating the need for a special signal of ovulation. The male could also provide more concentrated parental effort to his offspring. Groups like these are characteristic of multilevel societies (see below).

8.4.4 Conflicts and Reconciliations

Conflicts can occur in any relationship (preferred foods are often a source of friction in primates). In many primate species, such conflicts may be followed by reconciliation, that is, positive behaviors (such as certain facial expressions and grooming) that restore the relationship. In chacma baboons, heterosexual opponents exhibit friendly post-conflict reunions, almost exclusively between males and pregnant/lactating females who have formed tight social bonds. This is in accord with the valuable relationship hypothesis, which predicts rates of reconciliation to increase with the fitness consequences of the bond between the erstwhile opponents. In chacma baboons, males are as likely as females to initiate reconciliation, suggesting that males in this species play a role in maintaining heterosexual friendships that is greater than previously appreciated (Webb et al. 2019), bringing them more in line with the other baboon species.

8.5 Male–Female Interaction in Kinda Baboons

Kinda baboons seem to have an equivalent of special friendship, although there are distinctive features. Based on measurements of grooming and proximity, Schneider-Crease et al. (2022) determined that each Kinda individual, regardless of sex, had a top male and a top female among their affiliates (Fig. 8.5). Females were likely to have a single top male partner, while being affiliated with more than one female.

Weyher et al. (2014) provided details of grooming patterns that indicated strong male motivation to maintain these relationships. Males initiated 25 of 27 interactions (93%). Females terminated 16 of these interactions, 3 were mutual, and 6 were terminated by males.

Both types of relationship (top male and top female) were correlated with rapid maturation of infant behavior (female rank was also important in this regard). Possible adaptive functions of rapid maturation include lower infant mortality and shorter interbirth intervals for the mother. One possible proximate cause of this phenomenon is that the mother's bonds are used to recruit helpers with infant care. Another is that a more secure mother may invest less in proximity to her mobile infant, facilitating independence.

Petersdorf et al. (2019) argued that Kinda baboons provide a distinctive and perhaps superior model for the hominization of female–male relationships. Compared to other COKY species, Kinda baboons are postulated to be somewhat more like

Fig. 8.5 A Kinda baboon pair with an infant. Sexual dimorphism in this species is the least pronounced in the genus *Papio*. (Photo by Kenneth Chiou. Kafue National Park, Zambia. Resized for publication. Wikimedia license https://creativecommons.org/licenses/by-sa/4.0/)

Table 8.1 Key features of Kinda baboons compared with other baboon Species

Trait	Kinda baboons	Other COKY baboons
Troop size	200+	Typically <200
Environmental seasonality	Highly seasonal	Varied
Degree of estrous synchrony	High	Low to intermediate
Size of sexual swellings	Small	Medium to large
Relative testis size, mm^3/kg	2.28	1.61, 1.62, 2.16
Sexual dimorphism in body mass, M:F	1.55	1.71 to 1.83
Sexual dimorphism in canine basal area, M:F	1.96	2.75, 2.89, 3.14
Male immigration	Nonaggressive	Often aggressive
Male–female grooming	Often male-driven	Usually female-driven

The table illustrates several distinctive features of Kinda baboons that are the basis for analogies between Kinda baboon and hominins, as explicated by Petersdorf et al. (2019). It is a simplified version of the table in the Petersdorf paper with additional data from the text

later hominins and modern humans in several respects (Table 8.1). The reasoning is as follows: A large troop in a highly seasonal environment means that many females will be in estrus during a relatively short period of time. Consequently, it would be difficult for one male to monopolize all or even a large percentage of the estrus females. This situation minimizes the value of aggressive competition in males, which leads to a reduction in sexual dimorphism.

The alternative to aggressive competition is sperm competition, that is, a male competes with others by fertilizing females with as much sperm as possible. The genital proportions of Kinda baboons are consistent with this interpretation. Males have large testes that produce competitive quantities of sperm (Petersdorf et al. 2019). Since mating success is nonaggressive, immigrating males are nonaggressive. In addition to sperm production, a viable male strategy is to increase mating opportunities by providing benefits to females—at a minimum, grooming.

According to Petersdorf and colleagues, the key to the evolution of Kinda baboons is their distinctive ecological niche. The Kinda case shows how selective pressures can affect mating systems by shifting them toward traits like those that emerged during hominin evolution. The postulated scenario combines a Kinda model with paleoanthropological information that places early hominins in increasingly seasonal and heterogeneous environments. This factor and/or unusual group size affected male mating opportunities by reducing the potential for monopolization of females. As a consequence, sexual dimorphism declined and female mate choice became more important.

There are problems with this interesting scenario and its use as a model for early hominins. First, Kinda baboons live in a limited geographic area; other baboons have expanded to occupy much the same range of habitats as early hominins. Second, the relative seasonality of the Kinda habitat is in some doubt (Zinner pers. comm.). Third, it is debatable as to whether aggressive competition was diminished in hominins. Reduction in sexual dimorphism, for example, is viewed by some as evidence of reduced aggression; however, it can also be explained by more use of extrasomatic weapons. Fourth, sperm competition is not predominant in humans, as shown by anatomy and physiology of human male genitalia (Dixson 2012). Evidence from olive baboons suggests that both sperm competition and aggressive competition could have been important in early hominins. Finally, more generally, the small sample size for the species increases the chances of errors in the research results (Fuchs et al. 2018).

8.6 Male–Female Associations in Multilevel Societies

In common with special friendships, the unimale groups of multilevel baboon societies entail mating and offspring. However, they differ in having a higher level of paternal certainty. Compared to the special friend relationships in troops, the unimale groups of hamadryas and Guinea baboons are more like human families in two important respects. First, they are associations that combine one male with one or a limited number of females. Second, these associations are embedded in multilevel communities. However, within this common framework, there are crucial differences between the species regarding sociosexual dynamics within unimale groups. These present quite different analogies for the reconstruction of early hominin evolution.

8.6.1 Male-Focal Groups in Guinea Baboons

In Guinea baboons the male-focal unit (Chap. 7) is based on female choice rather than male coercion. Goffe et al. (2016) found that each female was mainly found in close proximity to one specific male (within 2 m), which they called the "primary male." The term used here, for comparative purposes is "central male" (Chap. 7). Female–male grooming was mostly confined to the central male. So were ritualized greetings between male and female, which involved hand touches, embraces, hip touches, and mounting. Copulations were almost all confined to the central males, but male–female social interactions were not strongly affected by female reproductive state. Central males handled infants more frequently than other males did.

Each female maintained exclusive social and mating relationships with one male at a time, while males might be affiliated with as many as six females. The stability of these male–female relationships varied considerably because females were free to shift from one male to another (Fischer et al. 2017; Goffe et al. 2016). During a 17-month study period, half of the females transferred between different males one or multiple times. Other than the focus on a single male at a time, female–male relationships in Guinea baboons seem similar to the special friendships in troops with regard to affiliative behavior in both sexes. This suggests one pathway for hominin male–female relations to have transitioned from troop to multilevel society.

8.6.2 One-Male Units in Hamadryas Baboons

Hamadryas baboons provide a very different model for the transition of male–female relationships into the multilevel context. The basic unit of everyday life in hamadryas baboons is an OMU that typically contains one or two females, although a few units contain many more (Fig. 8.6). Hamadryas females have been accumulated by the male through several different means (Chap.7) and the male restricts their movement.

Females are strongly oriented to the "leader" and their most serious fights with each other concern grooming access to the male. Females do not leave the central male unless appropriated by another male. Benefits received from the leader include protection from predators and from infanticide by outsiders and he shields them during storms. While the group travels, a male will wait for a female who is slowed by age, injury, or carrying a dead infant. The OMU is the reproductive unit. Leaders are not known to copulate with outside females. Females limit overt copulations to the central male. However, they may engage in surreptitious copulations with other males. This could create paternity confusion that would lessen the chance of infanticide by males who could potentially take over the OMU in the future.

Amann et al. (2017) investigated the determinants of takeovers in a population of wild hamadryas with a dataset of 172 events gathered over 20 years. In contrast to other baboon species, hamadryas males show no bias with regard to female reproductive state at the time of the takeover. However, they prefer to assimilate females that have never given birth (as evidenced by their physical features). In other words, the males target females with long-term reproductive potential, a strategy that is consistent with the high degree of stability in OMUs. This preference is paralleled

Fig. 8.6 A hamadryas male with two females. (Photo by Steven Gotz. Oakland Zoo)

Fig. 8.7 A hamadryas male with a female and infant. (Photo by Steven Gotz. Oakland Zoo)

by studies of human mating that demonstrate a male preference for a nulliparous female as a potential partner in a long-term relationship.

Hamadryas baboons may be considered an alternative to Guinea baboons as a model for early hominin male–female relationships. An alternative is to view the two species in sequence. Jolly (2020) proposed that hamadryas social organization evolved from a system like that of Guinea baboons. He attributed this development to the distinct environment of hamadryas, but similar pressures may have existed elsewhere. Baboons, like hominins, adapted to arid habitats throughout Africa. If Jolly's proposed evolutionary sequence for baboons is correct, hominin evolution may have stabilized at a stage comparable to Guinea baboons or it may have gone on to a greater resemblance to hamadryas baboons. This is speculation and the particulars of such a transition in hominins would be the subject of future research and discussion. A related issue of great importance is the role of sexual aggression in hominin evolution (Figs. 8.6 and 8.7).

8.7 Sexual Aggression

Sexual threat and violence in humans are of great concern and evolutionary explanations are particularly controversial (Baniel et al. 2017). Different baboon analogies suggest alternative possibilities for early hominins. The interactions in question vary from trivial spats to injurious attacks. There is speculation that sexual violence in hominin evolution is related to the desire for sexual privacy.

8.7.1 Forms of Sexual Aggression

Several forms of sexual aggression by males against females occur in mammals. Primates are among the orders in which such behavior is most common (Cassini 2021). One of these is infanticide, which is known to occur in four of the six baboon species. The frequency and context vary significantly across the four species..

Forced copulation ("rape") is rare in primates and is not known to occur in baboons. A questionable explanation is that female resistance is a sufficient deterrent, despite the degree of sexual dimorphism in all baboons. Male baboons seem to have few inhibitions about attacking females. An alternative is that the cries of a female might bring other males to intervene. This might be true in some cases, but not in those where an alpha male is completely dominant. Instead, and contrary to some ideas about rape in humans, male baboons seem not to have any strong motivation to achieve copulations by direct force, much less a specific innate system for such behavior.

The reference to "direct" force is necessary because baboons do provide evidence of more generalized aggression, called *coercion*, that indirectly leads to sexual success. This is characterized as coercion or intimidation. Two other types of sexual aggression will also be considered here. *Harassment* is interference in a copulatory relationship by one or more others. *Punishment* is aggression received by a female from a male in response to actual or perceived copulation with another male.

8.7.2 Sexual Aggression in Troops

Baniel et al. (2017) began a study of chacma baboons with the problem of human sexual violence in mind. They noted that such behavior is "widespread across human populations" and postulated that their work would have implications for its evolutionary origins. Their specific goal was to test the sexual coercion hypothesis, which interprets male aggression against females as intimidation directed toward dominating a female when she becomes fertile. In other words, the function of repeated male aggression is delayed mating benefits.

Baniel and colleagues observed two troops of chacma baboons in Namibia. Results of the study supported the sexual coercion hypothesis and cast doubt on alternatives. First, male aggression was related to the reproductive state of the targeted females: cycling females were attacked much more often than those in other reproductive stages. Second, the attacks caused injuries, that is, they were costly to the females and could not be ignored. Third, a female who received more aggression from a male during her cycle was more likely to be in a consort relationship with him when ovulating at the end of the cycle. These factors add up to intimidation as a mating strategy for males.

The data ruled out two alternative explanations for the aggression. It was not punishment because aggression rarely followed when females mated with rivals. It was not a signal of prowess to potential mates because there was no correlation between the consort result and the male's general aggression in the troop. Clutton-Brock (2017) commented that, "It would not be surprising if male coercion of females was common in early hominins."

The results of the chacma study were consistent with research on 12 troops of olive baboons in Gombe National Park, Tanzania. A more general study of wounding patterns included the result that cycling females (adult and adolescent) were much more likely to be wounded than females in other reproductive stages (MacCormick et al. 2012). These researchers, like Baniel and colleagues, concluded that sexual coercion was the function of the behavior. Similarly, female yellow baboons in Kenya were at the greatest risk of injury on days when they were most likely to be ovulating (Archie et al. 2014).

Kinda baboons appear to differ from the other COKY species with regard to sexual aggression. This is in accord with the pattern of sperm competition in place of confrontation between males (see above). Sexual dimorphism is less than in other baboons and the thin canines of the males are less suited to attack. Petersdorf et al. (2019) suggested that this is analogous to evolution in the human direction that took place "after *Australopithecus*"(i.e., early *Homo*).

Kinda males make no attempt to monopolize females. Therefore, there is no function for coercion, punishment, or infanticide. As far as forced copulation is concerned, none has been observed. Perhaps female resistance is most likely to be effective in the baboon species with the lowest level of sexual dimorphism.

As was noted previously, the distinctive ecology and demography of the Kinda baboons casts some doubt on their relevance to early hominin behavior. At the least, though, it can be said on the basis of Kinda behavior that troop organization does not entail sexual violence under all circumstances.

8.7.3 Sexual Aggression in Multilevel Societies

Infanticide and coercion occur in hamadryas baboons, but the pattern differs from COKY species. Infanticide takes place when a male takes over a group of females from another male. The infant mortality rate for these events can be as high as the 67% recorded at Filoha. Under ordinary circumstances, the rate of infant deaths from all causes is 13% (Swedell et al. 2014). If a takeover occurs, pregnant females may respond physiologically with pregnancy termination. This curtails their investment in offspring that are likely to be killed anyway, mitigating the effect of infanticide on their own fitness. As far as humans are concerned, although earlier conclusions may have been exaggerated, recent research confirms that step-fathers are significantly more likely to kill infants than are biological fathers (Nobes et al. 2019). Furthermore, as in baboons and other primates, the behavior is largely limited to infants as opposed to juveniles, with a borderline in humans at about 4 years of age. This focus on infants is analogous to the pattern in baboons and other primates that functions to return a lactating mother to sexual availability.

In hamadryas baboons there is a high level of aggression directly against the female(s) during any takeover. This is apparently necessary to break the bond with the former leader and coerce the females into following the new leader. Once the new unimale group is consolidated, the rate of aggression declines. In contrast to COKY baboons, there is no need for coercion to promote sexual consorts in a stable unimale group. Another difference between hamadryas and COKY baboons is that punishment, rather than coercion, is an important feature of a stable OMU. A female who goes too far from the group is likely to be chased and attacked with a stereotyped neck bite. Analogous aggressive behavior may have developed in early hominins when the social context shifted to unimale groups in multilevel societies.

While the hamadryas model was the only one available, it seemed that sexual aggression might have been an integral part of the earliest hominin multilevel societies. We now know of a different pattern of behavior in a Guinea baboon population. Little or no aggression between the sexes has been observed. Females are not taken over, herded, or punished (Fischer et al. 2017). Male-driven grooming may facilitate affiliation and sexual interaction.

This possible analogy for early hominins tells us that the origin of the family in a multilevel society does not require rigidly male-dominated OMUs like those in hamadryas baboons. If early hominins were like Guinea baboons, the patterns of sexual violence seen in extant *Homo sapiens* might have come into existence during a later period. Though possibly an evolutionary change under new circumstances, this might also be a case of cultural developments (e.g., patriarchal ideals about lineage) overriding biology.

8.7.4 Concealment of Mating and Sexual Privacy

Concealment of mating is widespread among humans and in a few other species. Ben Mocha (2020) postulated that this maintains two important relationships for the male: (1) mating control over a partner and (2) continued cooperation with other males that might be jealous or try to intervene if they perceived the sexual behavior.

Baboon evidence suggests the female viewpoint: chacma females initiate fewer copulations in the presence of adult male bystanders. This might be to avoid punishment and/or aggressive interference from rival males (Baniel et al. 2019). A related finding is that female baboons are able to suppress copulation calls that are ordinarily associated with male ejaculation (Vaglio et al. 2020), which could summon other males to the scene and generate competition.

8.8 Summary and Discussion

Male–female relationships in baboons offer several perspectives on early hominin behavior and the possible origins of the human family. Salient features of family in this context are (1) a long-term relationship between one male and one or several females (usually no more than two or three); (2) a sexual relationship between the partners with the consequent possibility of offspring; and (3) some degree of male care for the offspring.

Relevant relationships between particular males and females exist in baboon troops. These are called special friendships or primary associations. Either a male or a female can have several special friends. For males these relationships may lead to mating opportunities and the opportunity to care for their own offspring. For females the benefit is protection for themselves as well as their offspring. Protection may be needed against predators, infanticidal males, or harassment by competitive troop members. The minimum duration of a special friendship extends to the time that the offspring is weaned, but is longer in some species. In olive baboons, such relationships may last for years and entail affiliative (affectionate) behaviors such as proximity and huddling. Kinda baboons suggest a modification of this pattern in which there is less competition among males to obtain female associates and more male investment in those associates, manifested in male-driven grooming.

To go from the special friend pattern to something more like the human family would require one or more females to maintain a long-term relationship with just one male. Two possible pathways to this outcome are represented by the one-male units in hamadryas baboons and the male-focal associations in Guinea baboons. Hamadryas one-male units involve coercion of females as they are added to the group. Coercion continues in the form of threats and neck-biting, although females display affiliative behavior toward the male after being integrated into the group. In Guinea baboons, females choose to associate with a particular male and are free to switch from one male to another. Even so, some of these associations last for years. These may be alternatives, but it is also possible that the hamadryas system evolved from ancestors similar to Guinea baboons and that a parallel sequence took place in early hominins.

Sexual aggression is a troublesome issue in human behavior and baboons offer diverse perspectives on possible early hominin origins. Sexual coercion involves persistent attacks on a particular female in order to intimidate her into sexual compliance. This behavior occurs in three of the troop-living species. In hamadryas baboons, intimidation takes place at the time that a female is incorporated into the OMU. Infanticide is another form of sexual aggression that occurs in these baboon species, probably as a way to make newly acquired females sexually available. A tendency toward such behavior in humans may be represented by murder rates that are significantly higher for step-fathers than biological fathers.

Sexual aggression, whether against females or infants, seems not to occur in two *Papio* species: the troop-living Kinda baboons and the multilevel Guinea baboons. Both are less studied than the other species at this point, so sexual aggression may be seen in future research. However, it seems highly unlikely at this point that sexual aggression occurs at the same level as in other baboons, if at all. If these species represent the origins of human families and sexuality, the sexual violence that occurs in humans might have evolved in a later phase or might be a product of culture in more recent times.

Sexual privacy in humans may be related to sexual aggression in one or more ways. It has been suggested that it allows a male to monopolize a female without direct knowledge of other males. This may preserve male alliances and/or avert danger to both the male and the female. Baboon evidence suggests the female viewpoint: proceptive females initiate fewer copulations in the presence of adult male bystanders. This might be to avoid punishment and/or aggressive interference from rival males. A related finding is that female baboons are able to suppress copulation calls that are ordinarily associated with male ejaculation, which could summon other males to the scene and generate competition.

References

Amann AL, Pines M, Swedell L. Contexts and consequences of takeovers in hamadryas baboons: female parity, reproductive state, and observational evidence of pregnancy loss. Am J Primatol. 2017; https://doi.org/10.1002/ajp.22749.

Archie EA, Tung J, Clark M, et al. Social affiliation matters: both same-sex and opposite-sex relationships predict survival in wild female baboons. Proc R Soc B. 2014;281:20141271.

Baniel A, Cowlishaw G, Huchard E. Stability and strength of male-female associations in a promiscuous primate society. Behav Ecol Sociobiol. 2016;70:771–5. https://doi.org/10.1007/s00275-017-2100-8.

Baniel A, Cowlishaw G, Huchard E. Male violence and sexual intimidation in a wild primate society. Curr Biol. 2017;27(14) https://doi.org/10.1017/j.cub.2017.07.013.

Baniel A, Delaunay A, Cowlishaw G, Huchard E. Oestrous females avoid mating in front of adult male bystanders in wild chacma baboons. R Soc Open Sci. 2019;6(1):181009. https://doi.org/10.1098/rsos.181009.

Beach FA. Sexual attractivity, proceptivity and receptivity in female mammals. Horm Behav. 1976;7:105–38. https://doi.org/10.1016/0018-506X(76)90008-8.

Ben Mocha Y. Why do human and non-human species conceal mating? The cooperation maintenance hypothesis. Proc R Soc B. 2020;287:20201330. https://doi.org/10.1098/rspb.2020.1330.

Campos FA, Villavicencio F, Archie EA, et al. Social bonds, social status and survival in wild baboons: a tale of two sexes. Philos Trans R Soc B. 2020;375:20190721.

Chapais B. Primeval kinship, how pair-bonding gave birth to human society. Cambridge MA: Harvard University Press; 2008.

Clutton-Brock T. Behavioural ecology: sexual conflict in baboons. Curr Biol. 2017;27(18):R1008–10. https://doi.org/10.1017/j.cub.2017.07.027.

Dixson AF. Primate sexuality, comparative studies of the prosimians, monkeys, apes, and humans. 2nd ed. New York: Oxford University Press; 2012.

Fernandez-Duque E, Huck M, Van Belle S, Di Fiore A. The evolution of pair-living, sexual monogamy, and cooperative infant care: insights from research on wild owl monkeys, titis, sakis, and tamarins. Am J Phys Anthropol. 2020;171(S70):1–25. https://doi.org/10.1002/ajpa.24017.

Fischer J, Kopp GH, Dal Pesco F, et al. Charting the neglected West: the social system of Guinea baboons. Am J Phys Anthropol. 2017;172(S73):15–31. https://doi.org/10.1002/ajpa.23144.

Fisher HE. Anatomy of love, the natural history of monogamy, adultery, and divorce. New York: WW Norton; 1992.

Fogel AS, McLean EM, Gordon JB, et al. Genetic ancestry predicts male-female affiliation in a natural baboon hybrid zone. Anim Behav. 2021;180(1):249–78. https://doi.org/10.1017/j.anbehav.2021.07.009.

Goffe AS, Zinner D, Fischer J. Sex and friendship in a multi-level society: behavioural patterns and associations between female and male Guinea baboons. Behav Ecol Sociobiol. 2016;70(3) https://doi.org/10.1007/s00275-015-2050-7.

Gregersen E. Sexual practices: the story of human sexuality. New York: F. Watts; 1983.

Hausfater G, Blaffer-Hrdy S, editors. Infanticide, comparative and evolutionary perspectives. Hawthorne: Aldine; 1984.

Hawley CR, Patterson SK, Silk JB. Tradeoffs between mating effort and parenting effort in a polygynandrous mammal. iScience. 2023;26(7):106991. https://doi.org/10.1016/j.isci.2023.10699191.

Jolly CJ. Philopatry at the frontier: a demographically driven scenario for the evolution of multilevel societies in baboons (*Papio*). J Hum Evol. 2020;146:102819. https://doi.org/10.1016/j.jhevol.2020.102819.

King G. Traditional cultures, a survey of nonwestern experience and achievement. Long Grove: Waveland Press; 2003.

MacCormick HA, MacNulty DR, Bosacker AL, et al. Male and female aggression: lessons from sex, rank, age, and injury in olive baboons. Behav Ecol. 2012;23(3):684–91. https://doi.org/10.1093/beheco/ars021.

McLester E. Special friendships among baboons. In: Shackelford TK, Weekes-Shackelford VA, editors. Encyclopedia of evolutionary psychological science. Springer International Publishing; 2020. https://doi.org/10.1007/978-3-319-16999-6_1242-1.

Muller MN, Wrangham RW (eds) Sexual coercion in primates and humans, an evolutionary perspective on male aggression against females. Cambridge MA; Harvard University Press.; 2009. https://doi.org/10.4159/9780674054349.

Nobes G, Panagiotaki G, Russell Jonsson K. Child homicides by stepfathers: a replication and reassessment of the British evidence. J Exp Psychol Gen. 2019;148(6):1091–102. https://doi.org/10.1037/xge0000492.

Nunn C. The evolution of exaggerated sexual swellings in primates and the graded-signal hypothesis. Anim Behav. 1999;58:229–46.

Onyango PO, Gesquiere LR, Altmann J, Alberts SC. Testosterone positively associated with mating effort and paternal investment in savanna baboons (Papio cynocephalus). Horm Behav. 2013;73:430–7.

Petersdorf M, Weyher A, Kamilar JM, et al. Sexual selection in the Kinda baboon. J Hum Evol. 2019;135(S37) https://doi.org/10.1017/j.jhevol.2019.07.007.

Rees A. The infanticide controversy, primatology and field science. Chicago: University of Chicago Press; 2009.

Schneider-Crease I, Weyher A, Mubemba B, et al. Stronger maternal social bonds and higher rank are associated with accelerated infant maturation in Kinda baboons. Anim Behav. 2022;189(S63):47–57. https://doi.org/10.1016/j.anbehav.2022.04.011.

Smuts BB. Sex and friendship in baboons. Hawthorne: Aldine; 1985.

Städele V, Roberts ER, Barett BJ, et al. Male-female relationships in olive baboons (*Papio anubis*): parenting or mating effort? J Hum Evol. 2019; https://doi.org/10.1017/j.jhevol.2018.09.003.

Städele V, Vigilant L, Strum SC, Silk J. Extended male–female bonds and potential for prolonged paternal investment in a polygynandrous primate (*Papio anubis*). Anim Behav. 2021;174(1):31–40. https://doi.org/10.1017/j.anbehav.2021.01.017.

Swedell L, Plummer T. Social evolution in Plio-Pleistocene hominins: insights from hamadryas baboons and paleoecology. J Hum Evol. 2019;137:1–15. https://doi.org/10.1017/j.jhevol.2019.102777.

Swedell L, Leedom L, Saunders J, Pines M. Sexual conflict in a polygynous primate: costs and benefits of a male-imposed mating system. Behav Ecol Sociobiol. 2014;78:273. https://doi.org/10.1007/s00275-013-1741-3.

Vaglio S, Ducroix L, Villanueva MR, et al. Female copulation calls vary with male ejaculation in captive olive baboons. Behaviour. 2020;157(8–9):1–17. https://doi.org/10.1173/1578539X-bja10024.

Van Schaik CP, Janson CH, editors. Infanticide by males and its implications. Cambridge UK: Cambridge University Press; 2000.

Webb C, Baniel A, Cowlishaw G, Huchard E. Friend or foe: reconciliation between males and females in wild chacma baboons. Anim Behav. 2019;151:145–55. https://doi.org/10.1017/j.anbehav.2019.03.017.

Weyher AH, Phillips-Conroy JE, Fourrier MS, Jolly CJ. Male-driven grooming bouts in mixed-sex dyads of Kinda baboons (*Papio kindae*). Folia Primatol. 2014;85:178–91. https://doi.org/10.1159/000362544.

Zipple MN, Grady JH, Gordon JB, et al. Conditional fetal and infant killing by male baboons. Proc R Soc B. 2017;284:20162561. https://doi.org/10.1098/rspb.2016.2561.

Chapter 9
Female–Female Relationships

9.1 Introduction

Baboon analogies suggest various ways in which early hominin females may have related to each other. Enduring relationships between females occur in troops and in multilevel societies. Females maintain dyadic relationships with each other as well as network connections. Female–female interactions include competition as well as affiliation. Food, unless abundant, is a primary source of conflict. In some species, females may compete for preferred males.

Competitive behaviors among females resemble those of males: displacement, threat, and physical aggression. Physical attacks between females seem to be generally of low intensity (though escalated under certain circumstances). Females never engage in aggressive displays such as the wahoo contests of male chacma baboons. Baboon evidence is consistent with many recent findings about female–female relationships in humans regarding both affiliation and competition, which suggests that they originated in (or were perpetuated in) early hominins,

9.2 Female–Female Affiliation

Some contributors to a recent symposium on relationships between women demonstrated interest in possible evolutionary origins for these relationships. Fox et al. (2022) allowed that the capacity of women to form bonds with each other might come from "a long evolutionary history." Chimpanzees, and especially human hunter-gatherers, were prominent in some of these discussions, but many of the points made are amenable to comparison with baboon behavior. Differentiated affiliative relationships between pairs of female baboons are common in troops. These are often, but not always, based on kinship. Kramer (2022) endorsed the

G. E. King, *Baboon Perspectives on Early Human Ancestors*,
https://doi.org/10.1007/978-3-031-36769-4_9

evolutionary perspective, noting that "an ancient feature of human sociality is living in multilevel societies." This applies to early hominins as well as humans, and leads to analogies in hamadryas and Guinea baboons.

9.2.1 Female Affiliation in Humans

Patrilocality is a human social pattern in which the marriage of a woman entails taking up residence with her husband and his family, often in another community. It is the human equivalent of male philopatry. Patrilocality has been depicted as a fundamental feature of human evolution (Reynolds 2022), but this view has been strongly challenged. Based on considerable cross-cultural evidence, including hunter-gatherers, Rucas and Alami (2022) argue that patrilocality was far from universal in the human evolutionary past. Perhaps more important, they note that patrilocality rarely resulted in complete isolation of female blood relatives. In recent forager cultures (most of them with flexible residence patterns), a woman was often related to at least a few others in her husband's community. In societies with polygyny, a woman had about a 10% chance of being in a household with a sister or other close relative. In addition, women were in many cases free to visit relatives in other groups.

Reynolds (2022) maintained that female friends were important to women because "many female ancestors were surrounded by unrelated individuals upon marriage." Although this may be somewhat exaggerated (Rucas and Alami 2022), it must have been true that many women were in such situations during some significant part of their lives. Research shows that women and girls are more strongly attached to friends than are males. They place great value on these relationships and report higher levels of trust and communication than do males.

Female kin and friends provided cooperation in a variety of endeavors, such as food processing and child care. Beyond immediate cooperation for various purposes, "a large body of evidence" shows that such social support is strongly linked to better health and increased longevity. Social ties are more critical predictors of disease risk than body weight or physical exercise (Reynolds 2022). Friends as well as kin can be allies in conflicts and both can help a woman with child care and other needs. Studies of various human societies have amassed evidence that access to kin and other close affiliates has a positive effect on fertility. In Gambian women, for example, the presence of mother and sisters is associated with enhanced fertility (Kramer 2022).

9.2.2 *Female–Female Dyads in Troops*

The trend toward downgrading patrilocality in human evolution fits comfortably with the hypothesis that very early hominin societies were like the female-philopatric troops of baboons. More specifically, baboon troops display female–female relationships that are comparable to those described above for humans. The evidence is especially strong for kinship, but also suggests a role for friendship (Fig. 9.1). Primate societies like baboon troops have been characterized as "female-bonded" (Wrangham 1980).

Two studies of Amboseli yellow baboons (Silk et al. 2006a, b) and a study of Okavango chacma baboons (Silk et al. 2010) investigated the strength and durability of female–female bonds in troops. The results for both species showed that adult females formed close, equitable, supportive, and enduring social relationships. Based on measures of proximity and grooming, the strongest bonds connected a female with her mother and sisters; beyond that, bond strength decreased as relatedness decreased. The details of relationships in the two studies were somewhat different, but the theme of strong female–female bonds emerged from both. The durability of social bonds in yellow baboons followed the same patterns as bond strength. Fourteen dyads maintained close relationships for at least 5 years: mother/daughter (five pairs), maternal sisters (five), paternal kin (three).

These studies also indicated that if close maternal kin are not available, females seek alternatives. Bonds formed with paternal sisters are ordinarily intermediate in strength between ties to maternal sisters and to nonkin. A strong attraction to similar-aged females may be explained by the fact that such peers are often paternal

Fig. 9.1 A young female olive baboon grooms an older female. They may be mother and daughter. (Photo by Glenn King. Manyara, Tanzania)

half-sisters or cousins. However, some of these peers might actually be unrelated, and, even if related, the baboons might not be aware of the relationship. In short, the motivation to create these bonds might be a need for friendship such as was described above for human females.

The baboons also paralleled humans with regard to the physiological effects of social ties. Other research concluded that females' bonds with other females enhanced reproductive success, as in humans. This effect was independent of infant attractiveness or long-term bonds with specific males, that is, special friends (Silk et al. 2009). Females who formed the strongest social bonds with other adult females had the highest survivorship among their daughters and sons (at least after the age of high infanticidal risk). These benefits were unrelated to female dominance rank and increased quantitatively with the strength of a female's social bonds. Although females formed the strongest bonds with their closest kin, the mere presence of kin seemed to be less important to offspring survival than the quality of those relationships.

Several proximate factors might contribute to the connection between a female's social bonds and her offspring's survival. The first is related to predation, which is the primary cause of juvenile and adult mortality in the Okavango baboons (Cheney et al. 2004). Females with stronger and more secure social relationships may be less spatially peripheral to the troop while they are feeding during the day and while perching in sleeping trees at night. These positions could make them and their offspring less vulnerable to predators. Second, females with strong social bonds may be better shielded from social conflict within the troop and therefore able to feed more efficiently (Silk et al. 2003).

9.2.3 Female Groups and Networks in Humans and Baboons

More extensive connections than dyads have linked human females together in the historical and evolutionary past (Kramer 2022). These ranged from small groups to extensive networks. Some groups of women are informal and temporary (though often recurrent), forming for specific purposes such as collecting or processing food. Less frequently, female coalitions countered male aggression. Women's social networks coordinated a variety of activities such as visiting, helping, or exchange. For example, a gift-giving network among Kalahari foragers covered extensive areas and linked female partners who could provide each other with refuge during local environmental crises (Kramer 2022).

More formal networks called sodalities organize women (and men) to carry out a variety of functions (Kramer 2022). Some support individuals, such as the sick, old, and widowed. Others have societal functions, such as coordination of ceremonies. Despite the varied cultural definitions of networks and sodalities, the cross-cultural ubiquity of these social structures suggests an underlying evolutionary basis. All of these relationships have been culturally elaborated, but the tendency to form similar social structures can be seen in baboons, which suggests that there are roots in human evolution. Baboon analogies show how the foundations for these entities might have evolved and developed in early hominins.

Cooperative food collection does not occur in baboons, but females do group together in the course of foraging. King et al. (2011) investigated the factors involved in co-feeding in desert-living chacma baboons. They found that feeding groups were significantly correlated with social bonds measured by grooming rather than kinship. Females followed male grooming partners to favorable food patches and were joined by other females with whom they had social bonds. Similar grouping tendencies in early hominins might have provided the basis for more cooperative food collection and sharing as savanna expansion made good food patches harder to find. If the primary basis was social bonding rather than kinship, as in these baboons, this would have added flexibility to the formation of the hominin groups.

In addition to organizing ordinary activities, grooming connections can alleviate stress during difficult times. Wittig et al. (2008) tracked stress in female chacmas, measured by glucocorticoid levels, during a period of social instability caused by dominance competition between males. Stress increased in the week after the upheaval and females narrowed their grooming activities to a few preferred partners, mostly close kin. Changes in GC levels were not related to giving or receiving grooming. Rather, stress was relieved by participating in small grooming networks with reliable preferred partners. Wittig and colleagues noted research with humans that showed women to be more likely than men to seek social contact in times of stress and to turn to smaller support groups (see also Crockford et al. 2008).

Female baboons do not group together to counter aggression from the much larger males, but they do occasionally display the capacity to form coalitions in competition with each other. In a study of females in several yellow baboon troops, a little more than 2% of all dyadic disputes led to intervention by a third female (Silk et al. 2004). In most cases, high-ranking females supported others of high rank who were maternal relatives. That is, they supported the existing hierarchy. However, coalitions played an important role in reorganization of the female hierarchy in the troop that was ancestral to the three troops in the study.

Silk et al. (2004) cited other reports of female coalitions, some of which might represent greater frequency of the behavior. Yellow baboon females at another site intervened "regularly" and attacked other females "jointly." Female olive baboons intervened in disputes once every 14 h. Female chacma baboons intervened in 32 disputes during 15 months of observation. A similar capability in early hominin females might have been increasingly devoted to mutual defense against male aggression during the evolutionary reduction of sexual dimorphism in size.

A social network analysis of chacma baboons evaluated the comparative benefits of dyadic bonds against several network measures (Cheney et al. 2016). Results suggested that females with strong dyadic bonds also showed high *eigenvector centrality*, a measure of the extent to which an individual's partners are connected to others in the network. Eigenvector centrality was a better predictor of offspring survival than dyadic bond strength. Having preferred partners who are themselves the preferred partners of several others may facilitate social interactions by making encounters more predictable and less stressful. Similarly, such connections may facilitate the formation of new bonds if a female's primary partner dies. The offspring of individuals with high eigenvector centrality may also be less peripheral and at lower risk of predation—the primary cause of mortality for juveniles and adults in this population.

Parallels with modern humans suggest that the baboon patterns are analogous with adaptations of early hominins. Baboon researchers compared their findings to epidemiological and clinical studies of humans that indicate that social support has beneficial effects on health and well-being across the life span, including direct effects on reproduction such as the birth of heavier infants (Archie et al. 2014; Silk et al. 2003). Experimental studies in humans suggest that social integration is the cause not the consequence of improved health outcomes. Social support is thought to moderate the deleterious effects of chronic stress and thereby enhance well-being. Human studies have varied considerably in their measures of social support, ranging from detailed measures of the strength and quality of individual social bonds to composite measures of individual social integration and experience. All point to reduced mortality risk.

In some baboon field studies female relationships with other adults of both sexes are correlated with offspring survivorship, an effect that was unrelated to dominance rank or ecological conditions (Silk et al. 2003). Complementary work focused on longevity (Archie et al. 2014). Females with social connections to either males or females lived longer than isolated individuals. Females with strong connectedness to individuals of both sexes lived longest.

9.3 Female–Female Relationships in Multilevel Societies

In one view of hominin evolution presented earlier, troops were replaced by multilevel societies. Baboon multilevel societies, in comparison to troops, are characterized by shifts from female to male philopatry and from multi-male organization to unimale groups within larger communities. Nevertheless, some of the same female–female relationships and interactions can be seen in both kinds of society. Assuming a transition from troops to multilevel societies in early hominins, analogy indicates that such connections would have been preserved, though presumably in modified form. Two *Papio* species suggest the possibilities.

9.3.1 Female–Female Relationships in Guinea Baboons

Relatively little is as yet known about female–female relationships in Guinea baboons. Given the freedom of choice that females in that species have, a similar hominin multilevel system would presumably have made it easy for particular females to stay together or at least in contact. However, a captive study (Maestripieri et al. 2007) suggests complications. A large zoo enclosure contained 11 "harems" with 1–4 females per group. Focal female subjects were randomly selected from five of these unimale groups. These subjects displayed a tendency to initiate contact and grooming *less* often when they had *more* relatives in the group. On the other hand, subjects who had more relatives *outside* their own groups initiated interactions with non-group females more often than those with fewer outside relatives. The researchers noted that the latter pattern resembled behavior in hamadryas baboons reported by Swedell (2002).

Interactions within the groups were both affiliative and agonistic (Maestripieri et al. 2007). Each female received more contact and grooming from other females than from the central male and these interactions were largely symmetrical. On the other hand, group-mates exchanged more threats and avoidance than with non-group females. Overall, group cohesion was maintained by approaches among all members, rather than by male coercion.

9.3.2 Female–Female Relationships in Hamadryas Baboons

Female hamadryas baboons are usually separated from kin through coercive formation and maintenance of one-male units. Nevertheless, females find ways to assert affiliation with some of their relatives. Genetic data suggest that, despite the social system, female hamadryas are more likely to be found in an OMU with maternal relatives than would be expected by chance. At least a third of these maternally related dyads are mother–daughter pairs (Städele et al. 2016). This situation is facilitated by the fact that hamadryas females are generally more philopatric within clans and bands than previously thought, presumably because more takeovers occur within clans and bands than between them (Swedell and Plummer 2019).

It may be that the baboons can recognize maternal kin and can find a way to express choice in at least some takeover situations and/or transfer between specific pairs of units due to a relationship between the two leader males. The situation implies that an ancestral maternal kin bias has been retained in hamadryas society (Städele et al. 2016). This suggests that female relationships like those found in troops could have persisted in a hominin transition to multilevel societies (Fig. 9.2). Observation of a captive colony confirmed the maintenance of kinship bonds by hamadryas females and found that mother–daughter dyads did the most grooming. Among wild hamadryas baboons, females risk punishment from their central males in order to visit females in other OMUs, possibly relatives.

Fig. 9.2 An old female hamadryas baboon after the death of a sister who was her "inseparable companion." She is embraced by an adolescent male. (Photo and information from Steven Gotz. Oakland Zoo)

9.4 Further Evolution of Hominin Female Relationships: Two Scenarios

Baboon analogies indicate that female–female cooperation was limited, if not minimal, in early hominins. It may have been limited to temporary coalitions. However, the underlying affiliations suggest that the roots of human female cooperation may be found in that evolutionary era. A scenario based on hamadryas baboons suggests an intermediate step in the transition to human patterns. This effort can be compared with another recent scenario that ignores baboons entirely.

9.4.1 The Swedell/Plummer Scenario

This scenario for later hominin evolution was based on information described above regarding persistence of female kinship bonds in hamadryas OMUs despite coerced transfers. It envisioned modifications resulting from changes in ecological conditions, subsistence patterns, and costs of reproduction for females (Swedell and Plummer 2012, 2019). The scenario postulates female subgrouping into small social units in which females assisted one another with food procurement as well as care and provisioning of offspring. Kinship bonds would have strengthened such groups, which might have led to multiple generations of related females. These cooperative breeding units may have aggregated around "favored places" as they tracked seasonally changing food sources.

This view of female bonds in relation to subsistence can be connected with a discussion above about co-feeding groups in chacma baboons. King et al. (2011)

found that co-feeding groups were significantly correlated with social bonds (measured by grooming) and not necessarily with kinship. Females joined other females with whom they had bonded, presumably as friends, perhaps because they were age-mates. Similar grouping tendencies in early hominins might have provided the basis for more cooperative food collection and sharing as savanna expansion made good food patches harder to find.

9.4.2 The Newson/Richerson Scenario

Newson and Richerson (2021) presented what they considered to be a new model for the evolution of *Australopithecus*, although there are numerous similarities to the Swedell-Plummer scenario. Both scenarios alluded to the rigors of open country, including widely spaced resources, increased predator danger, and debilitating heat. Both proposed that hominin mothers adapted to these new problems by grouping together to help each other. They imagined the mothers and their young at a "central place" (cf. "favored place") with cover and water that could be defended. Infants would stay there with some mothers while others took turns going out to forage, a concept that some primatologists have labeled the *creche*. Females would have to trust that others would protect and perhaps feed their infants and that food would be shared in order to overcome different levels of success.

Newson and Richerson find this scenario to be "reasonable" because it evolved in sperm whale mothers. There is some irony in using an oceanic analogy to address the problem of how mothers raised their young "in drier habitats." Another anomaly is the use of sperm whales to model a situation in which predator risk has substantially increased. The size of a female sperm whale is far greater than that of any potential predator, with the possible exception of the giant squid. The only other probable dangers are pods of orcas or perhaps multiple sharks. Of course, predators would probably attack infants. But early hominins would have had the same problem in the face of enemies much more like those that confront baboons. Primatologists have made use of cetacean comparisons in the past (as noted in Chap. 2), but these should be juxtaposed with competing (or supplementary) hypotheses derived from primates.

9.4.3 Female Bands with Males

The adaptations of many baboon populations provide analogies for hominins in hot, dry, open habitats. Among these populations there are various analogies that suggest how and why the postulated female foraging bands might have retained one or a few male associates. To begin with, the baboon troop's subtrooping potential models a society in which groups of mothers might have become differentiated while remaining members of the larger community. There are strong affiliative bonds among groups of females within a baboon troop. Many of these relationships are based on kinship, but not all of them.

Hominin females, smaller than males, would have faced a wide array of larger predators. Foraging by a lone individual or a small party of females would probably have been very dangerous, especially in areas where trees were far apart. The baboon analogy suggests that one or a few males in the party would have considerably reduced the risk. Foraging subtroops of chacma and olive baboons often contain several adult females with one adult male. Hamadryas OMUs have one male protector and sometimes an additional male. As sexual dimorphism in hominins declined, the postulated female bands (in either the Newson-Richerson or Swedell-Plummer scenario) might have been more inclined to retain one or a few males to enhance predator defense, perhaps in positions of equality or even subordination.

The Newson-Richerson scenario suggested that the creche pattern created risks of omission: that females at the "home base" might not protect or feed the young of others. Baboon analogy indicates a more dynamic risk: that females at the home base might actually harm the offspring of others, whether through direct attack or "rough handling" (Fig. 9.3). Baboon analogy also suggests a solution: the young might be protected by adult males who were invested in them (Chap. 8).

Newson and Richerson argued that their speculation about a band of mothers is superior to theories of savanna survival based on "pair bonding." Their concept of pair bonding was clearly based on primate species such as gibbons, tamarins, and titi monkeys. None of these or any other pair-bonded primates are savanna dwellers in the same sense as early hominins and baboons. As described in previous sections, baboons provide a different view of long-term relations between males and females

Fig. 9.3 A hamadryas baboon female displays interest in another female's infant. The outcome of such interactions may be affectionate allocare, but it is sometimes "rough handling" (discussed in text). (Photo by Steven Gotz. Oakland Zoo)

in savanna troops, as well as multilevel societies. Special friendships in baboon troops and unimale groups in multilevel baboon societies suggest that close associations of male and female early hominins persisted from woodlands into savannas and from troops into multilevel societies. This view is more in accord with the human pattern that we see in cross-cultural evidence (cf. Chapais 2008).

9.5 Female–Female Competition

Cross-culturally, women behave more competitively than men in same-sex situations. In 91% of societies sampled, female aggression targeted other women (Reynolds 2022). Rucas and Alami (2022) added that rates of aggression between women were only slightly higher for in-laws than for kin. Sisters were the main target in 8% of the societies and mothers/daughters in 10%.

In a stable baboon society, competition is most often manifested in dominance relationships. Overt aggression, the most obvious representation of competition, is in the behavioral repertoire of female as well as male baboons. Physical aggression also occurs among human females, but is often controlled by cultural factors. On the other hand, humans can manifest competitiveness and hostility in verbal aggression.

Women display less physical aggression than men because they risk their children as well as themselves. Evidence from diverse populations shows that the loss of a mother is much more detrimental to the survival of children than the loss of a father (Campbell 2013). Women prefer less dangerous responses to conflict, such as social isolation of targeted individuals (Reynolds 2022). Throughout the world and in all historical periods, physical aggression between women has been less frequent, less lethal, and less injurious than among men.

9.5.1 *Competition for Food*

Affiliation among women often revolves around gathering and processing of food, whether there is a practical need for cooperation or simply an opportunity for socializing. There is also ample opportunity for conflict over food. In a study of the Tsimane of Bolivia, for example, 19% of women's arguments were about food (the rest were about men and social relationships) (Campbell 2013).

There is no cooperative food-getting in baboons and so the relationship among adults is neutral or competitive. Analogies suggest that competitive tendencies in early hominins were affected by food preferences, whether for differences in the quality of foods or for patches with readily available quantities. Among olive baboons at Gombe (Packer 1995), high-ranking females had shorter interbirth intervals, better infant survival, and accelerated maturation of daughters. Researchers attributed these advantages to nutrition. One criterion for dominance ranking was the ability to supplant others from food.

A study of olive baboons on the Laikipia Plateau of Kenya examined the relative importance of food and mates in shaping patterns of aggression between females (Patterson et al. 2021). The most aggressive females were those that were pregnant or lactating, that is, in the stages of reproduction that require the most energy replenishment. There was no evidence that "sexually receptive" females (presumably those in estrus) were more involved in aggression than any others. Neither was there evidence for competition for male caretakers or for reproductive suppression through attacks on other females. The researchers concluded that competition over food played a more important role than competition over mates in this population.

In a stable dominance hierarchy in a baboon troop, grooming may be a tactic to obtain access to resources. Grooming is hygienic, but the fact that it is pleasurable to the recipient makes it socially useful. Females may spend hours of each day in grooming, perhaps because alliances are so important. Subordinates woo dominant females by grooming them. Females without effective allies may transfer from one troop to another, which raises the question of whether alliances are the *cause* of female philopatry rather than an effect (Barton 2000). Baboon evidence suggests that hominin females used such affiliative tactics as well as occasional aggression to pursue dominance and resources.

Barrett et al. (2002) used data from a natural experiment on adult female chacma baboons to test the hypothesis that variation in aggression through time influences patterns of grooming reciprocity within a social group. Due to a change in the competitive regime, they were able to compare data from periods when aggression was high (period 1) and low (period 2). During period 2, the slope of the relationship between aggression and rank was significantly shallower than during period 1 and less aggression was directed at the lowest-ranking females. The researchers attributed this change to reduced effectiveness of dominance as a means of excluding other females from feeding resources.

The reduction in aggression during period 2 was accompanied by an increase in grooming reciprocity between dyads suggesting that high-ranking females no longer attracted grooming by subordinates in exchange for tolerance, and that grooming in period 2 was exchanged for its intrinsic benefits. The loss of rank-related effects on grooming reciprocity in period 2 compared with period 1 further confirmed this. These findings show that female baboons are able to respond flexibly and swiftly to changes in their social circumstances (Barrett et al. 2002). Presumably early hominins were no less adaptable.

9.5.2 Competition for Males

Despite negative findings in some studies (e.g., Patterson et al. 2021, above), there is considerable evidence that female baboons compete for male services and that these behavior patterns parallel findings for human females. In a recent collection of papers, writers examined competitive behavior in human females. The primary article by Reynolds (2022) made several points relevant here, based on broad cross-cultural comparisons and similar findings in industrial societies. She saw "traces" of "ancestral" behavior in modern women. Reynolds placed these patterns in a historical setting: However, several of these female patterns are subject to comparison with other primates. In the discussion below, points made by Reynolds are italicized. Each of them is followed by comments on the paper (which was a "target article") and by relevant statements from other researchers on the subject. Then comparisons are made with baboon behavior. The underlying postulate is, of course, that the similarities might be analogous to behavior patterns in early hominins.

Throughout most of history women had to compete for the "romantic" partners most likely to acquire resources and to make a long-term commitment to provisioning those resources.

A cross-cultural survey found that aggression between women "often" revolved around competition to acquire and retain mates (Campbell 2013). For example, in the Tsimane study cited above, 25% of women's arguments were about men (vs. 19% about food). For girls under the age of 20, men were the leading cause of conflict (early marriage was characteristic of the culture). Women in "every communal condition" compete over material and social resources, including mates (Rucas and Alami 2022).

Male baboons as mates provide services rather than material resources. These are (potentially) protection and parental investment (Chap. 8). Female chacma baboons at Moremi competed for males as mates and as possible protectors of offspring. This competition among chacma females may be especially intense because the threat of infanticide is higher than in other baboon species (Cheney et al. 2012). Within the OMUs of hamadryas baboons, females compete for access to the central male (Kummer 1968).

In large cross-cultural samples, men often prioritize physical attractiveness in romantic partners. Similar studies show that women experienced more distress and jealousy when confronted with attractive rivals. In Western societies, women considered to be attractive were more likely to marry, to marry highly educated and/or higher earning men, and more likely to have children.

Girls who emphasize attractiveness with clothing, makeup, or behavior are likely to be targeted by other girls (Campbell 2013). In baboons, attractiveness is a function of the sexual swelling. Among chacma baboons studied at Tsaobis, "sexually receptive" females received high levels of aggression from others who were in the same condition. Researchers interpreted this to be consistent with competition for mating opportunities (Baniel et al. 2018a). They reported no evidence for food competition.

At Mikumi, Tanzania, female yellow baboons received the highest rates of coalition attacks from other females when in reproductive states most vulnerable to reproductive suppression (Wasser and Starling 1988). Pregnant females were the most frequent attackers in coalitions. Other, less frequent, attackers were cycling females and then lactating females. These attacks were followed by decreased reproductive performance due to longer periods of cycling before conception, possibly due to lingering stress caused by the attacks.

Women are more intolerant of sexual promiscuity than men. One possible explanation is that they perceive a threat to established relationships. Cross-cultural evidence shows that women who are attractive and who engage in casual sex are most successful in poaching mates. Other women are likely to be inclined to punish such women.

Regardless of social class, race, or ethnicity, accusations of promiscuity are a powerful form of verbal attack because such behavior jeopardizes the possibility of having a reliable long-term mate (Campbell 2013).

At Tsaobis, pregnant and lactating chacma baboons were more aggressive to estrus females when (a) the attackers had recently conceived, and (b) when an estrus female was mate-guarded and sexually active with one of their male friends (Baniel et al. 2018b). Reduced probability of conception in targeted females seemed to suggest reproductive suppression against a rival (however, Baniel et al. 2018a stated that there was no evidence for reproductive suppression).

Similar attacks took place among yellow baboons at Amboseli (Samuels et al. 1987). Physical aggression, including biting, resulted in injuries to 14 of 19 females and "some wounds clearly endangered females' lives" (p.788). Aggression included "mob" attacks and 15 of 18 occurred during the "turgescent" phase of the loser (about half of the cycle).

Silk et al. (2004) investigated such female coalitions and found that third-party interventions occurred in no more than 4% of all disputes. Females did not reciprocate agonistic support nor exchange it for grooming. They supported close maternal kin more often than others (Silk et al. 2006b). Sexual competition probably occurred among early hominin females, as it does in humans and baboons. However, the baboon evidence indicates that group action, though occasionally intense, was sporadic and usually included few individuals.

The competitive behavior of women is affected by the operational sex ratio, that is, the preponderance of one sex (men in this case) over the other at a given time. Cross-culturally this is associated with heightened levels of casual sex and mate poaching. In experiments, women who believe they are in such a situation behave more aggressively toward other women who are attractive.

According to Campbell (2013) the local sex ratio "determines" the intensity of mate competition and that ratio has been often badly skewed against girls in urban neighborhoods. Adding to the problem is variance in male resources. Physical aggression is more likely to occur in these settings and mothers sometimes participate in support of daughters. Three reasons for young women to use "physical interference competition" are that mates and resources are in short supply; a husband is shared; and kin are absent (Benenson 2013).

Among chacma baboons at Moremi, rates of aggression increased with greater skew in sex ratio, especially from low-ranking females. The behavior included coalitionary ("mob") attacks by low-ranking females, mostly against estrus females in consort (i.e., attractive females with mates), resulting in some changes in dominance hierarchy. Rates were higher when there was a larger proportion of nulliparous females. These were more likely to challenge older and more dominant females and also attack other cycling females.

Social instability can have a similar effect. During such a period, rates of aggression increased among chacma females. Under these circumstances, the probability of infanticide increases, protection from a single alpha male becomes vital, and opportunities to mate with the alpha male become the focus of competition among females (Baniel et al. 2018a, b; Huchard and Cowlishaw 2011).

When females live in groups with less reproductive skew and little danger of infanticide, as do the olive baboons at Laikipia, access to the alpha male may not provide major fitness benefits. In addition, there may be less competition over access to special friends because females share these friends with relatively few other females. Instead, female fitness may be more strongly affected by access to nutritional resources, as discussed above (Patterson et al. 2021). Differences in the pattern of aggression between these two baboon populations in different habitats seem to suggest that evolution has "finely tuned" female baboons'responses to the challenges that they face in their local environments.

9.6 Summary and Discussion

Female baboons display a variety of relationships with each other that may represent early hominin behavior in ways that suggest parallels with recent humans. These patterns may have provided evolutionary foundations for human female social relationships. First, there are dyadic associations that are often composed of close kin but not limited to this basis. Second, there are groups and networks of females that perform a variety of functions. Whatever their other purposes or functions, these bonds were subject to evolution by natural selection because they increased the fitness of participants through better health, longer life, and more successful reproduction. Stress reduction probably made a major contribution. Research with modern humans shows that all of these positive results of social bonds also apply to women.

Patrilocality has been thought to be a barrier to female affiliation in human evolution. However, this rule of postmarital residence has rarely had a rigid application, especially among the hunter-gatherers who probably represent our Paleolithic ancestors. In any case, women who were compelled to live in their husbands' communities could make friends there and could visit their relatives in other communities.

The ability of females to maintain their bonds has been demonstrated even in the coercive societies of hamadryas baboons. Females find ways to end up in the same OMUs as their female relatives more often than would be expected by chance, despite separations caused by competition among males. If female early hominins lived in a multilevel society like that of Guinea baboons (rather than hamadryas), their freedom would have been greater. If earlier hominins lived in troops like those of the COKY baboons, the foundation for female affiliation would have been initiated in a system of female philopatry (comparable to matrilocality in humans).

Two recently published scenarios have imagined the changing relationships of females in *Australopithecus* and early *Homo*. Both posit increasing cooperation among females, independent of males. The main difference between them is that the Swedell and Plummer scenario, based on hamadryas baboons, was first presented in 2012. The Boyd and Richardson scenario, which ignores baboons, was published in 2021. An important shortcoming of both scenarios, from the viewpoint here, is the assumption that cooperative female defense would have been sufficient in very dangerous circumstances. These bands of mothers would have benefitted from some male presence (this has been called the "hired gun hypothesis"). The male presence need not have been coercive and dominant, as demonstrated by analogies with Kinda baboons and Guinea baboons.

Early hominin females must have competed with one another, on behalf of themselves and their offspring. The baboon perspective indicates that food and males would have been the goals. Food competition was contingent on the abundance of the general food supply and the availability of preferred foods. Males would have been a source of contention everywhere, though also somewhat contingent on the supply. Certain individuals would always have been preferred as mates and (with greater variation) as protectors and caretakers of offspring. Stable dominance hierarchies would have guided competitive regimes in many societies, but occasional physical aggression would have occasionally altered dominance relationships, expressed rivalries, and perhaps resulted in reproductive suppression. In a few instances aggressive females would have joined in small coalitions.

References

Archie EA, Tung J, Clark M, et al. Social affiliation matters: both same-sex and opposite-sex relationships predict survival in wild female baboons. Proc R Soc B. 2014;281:20141271.

Baniel A, Cowlishaw G, Huchard E. Context dependence of female reproductive competition in wild chacma baboons. Anim Behav. 2018a;139:37–49. https://doi.org/10.1017/j.anbehav.2018.03.001.

Baniel A, Cowlishaw G, Huchard E. Jealous females? Female competition and reproductive suppression in a wild promiscuous primate. Proc R Soc B. 2018b;285:20181332. https://doi.org/10.1098/rspb.2018.1332.

Barrett L, Gaynor D, Henzi P. A dynamic interaction between aggression and grooming reciprocity among female chacma baboons. Anim Behav. 2002;73(7):1047–53. https://doi.org/10.1007/anbe.2002.3008.

Barton RA. Socioecology of baboons: the interaction of male and female strategies. In: Kappeler PM, editor. Primate males, causes and consequences of variation in group composition. New York: Cambridge University Press; 2000. p. 97–107.

Benenson JF. The development of human female competition: allies and adversaries. Philos Trans R Soc B. 2013;368:20130079. https://doi.org/10.1098/rstb.2013.0079.

Campbell A. The evolutionary psychology of women's aggression. Philos Trans R Soc B. 2013;368:20130078. https://doi.org/10.1098/rstb.2013.0078.

Chapais B. Primeval kinship, how pair-bonding gave birth to human society. Cambridge MA: Harvard; 2008.

Cheney DL, Seyfarth RM, Fischer J, et al. Factors affecting mortality and reproduction among free-ranging baboons in the Okavango Delta, Botswana. Int J Primatol. 2004;25:401–28.

Cheney DL, Silk JB, Seyfarth RM. Evidence for intra-sexual selection in wild female baboons. Anim Behav. 2012;84:21–27. https://doi.org/10.1016/j.anbehav.2012.03.010.

Cheney DL, Silk JB, Seyfarth RM. Network connections, dyadic bonds and fitness in wild female baboons. R Soc Open Sci. 2016;3:170255. https://doi.org/10.1098/rsos.160255.

Crockford C, Wittig RM, Whitten PL, et al. Social stressors and coping mechanisms in wild female baboons (*Papio hamadryas ursinus*). Horm Behav. 2008;53(1):254–65. https://doi.org/10.1016/j.yhbeh.2007.10.007.

Fox SA, Scelza B, Silk J, Kramer KL. New perspectives on the evolution of women's cooperation. Philos Trans R Soc B. 2022;378:20210424. https://doi.org/10.1098/rstb.2021.0424.

Huchard E, Cowlishaw G. Female-female aggression around mating: an extra cost of sociality in a multimale primate society. Behav Ecol. 2011;22(5) https://doi.org/10.1093/beheco/arr083.

King AJ, Clark FE, Cowlishaw G, et al. The dining etiquette of desert baboons: the roles of social bonds, kinship, and dominance in co-feeding networks. Am J Primatol. 2011;73(8):768–74. https://doi.org/10.1002/ajp.20918.

Kramer KL. Female cooperation: evolutionary, cross-cultural and ethnographic evidence. Philos Trans R Soc B. 2022;378:20210425. https://doi.org/10.1098/rstb.2021.0425.

Kummer H. Social organization of hamadryas baboons, a field study. Chicago: University Chicago Press; 1968.

Maestripieri D, Mayhew JA, Carlson CL, et al. One-male harems and female social dynamics in Guinea baboons. Folia Primatol. 2007;78(1):56–68. https://doi.org/10.1159/000095686.

Newson L, Richerson P. A story of us: a new look at human evolution. New York: Oxford University Press; 2021.

Packer C. Baboon fertility and social status. Nature. 1995;377(6551):689–90. https://doi.org/10.1038/377689a0.

Patterson SK, Strum SC, Silk J. Early life adversity has long-term effects on sociality and interaction style in female baboons. Proc R Soc B Biol Sci. 2021;289(1978):20212244. https://doi.org/10.1098/rspb.2021.2244.

Reynolds TA. Our grandmothers' legacy: challenges faced by female ancestors leave traces in modern women's same-sex relationships. Arch Sex Behav. 2022;51:3225–56. https://doi.org/10.1007/s10508-020-01768-x.

Rucas SL, Alami S. Female–female competition occurs irrespective of patrilocality. Arch Sex Behav. 2022;51:3287–92. https://doi.org/10.1007/s10508-021-02221-3.

Samuels A, Silk J, Altmann J. Continuity and change in dominance relationships among female baboons. Anim Behav. 1987;35:785–92.

Silk JB, Alberts SC, Altmann J. Social bonds of female baboons enhance infant survival. Science. 2003;302:1231–4.

Silk JB, Alberts SC, Altmann J. Patterns of coalition formation by adult female baboons in Amboseli, Kenya. Anim Behav. 2004;77:573–82.

Silk JB, Altmann J, Alberts SC. Social relationships among adult female baboons (*Papio cynocephalus*): I. Variation in the strength of social bonds. Behav Ecol Sociobiol. 2006a;71:183–95.

Silk JB, Alberts SC, Altmann J. Social relationships among adult female baboons (*Papio cynocephalus*): II. Variation in the quality and stability of social bonds. Behav Ecol Sociobiol. 2006b;71:197–204.

Silk JB, Beehner JC, Berman TJ, et al. The benefits of social capital: close social bonds among female baboons enhance offspring survival. Proc Biol Sci. 2009;277:3099–104.

Silk JB, Beehner JC, Berman TJ, et al. Strong and consistent social bonds enhance the longevity of female baboons. Curr Biol. 2010;20:1359–71.

Städele V, Pines M, Swedell L, Vigilant L. The ties that bind: maternal kin bias in a multi-level primate society despite natal dispersal by both sexes. Am J Primatol. 2016;78:731–44. https://doi.org/10.1002/ajp.22537.

Swedell L. Affiliation among females in wild hamadryas baboons (*Papio hamadryas hamadryas*). Int J Primatol. 2002;23(6):1205–26. https://doi.org/10.1023/A:1021170703006.

Swedell L, Plummer T. A papionin multi-level society as a model for hominin social evolution. Int J Primatol. 2012;33:1175–93. https://doi.org/10.1007/s10774-012-9700-9.

Swedell L, Plummer T. Social evolution in Plio-Pleistocene hominins: insights from hamadryas baboons and paleoecology. J Hum Evol. 2019;137:1–15. https://doi.org/10.1017/j.jhevol.2019.102777.

Wasser SK, Starling AK. Proximate and ultimate causes of reproductive suppression among female yellow baboons at Mikumi National Park, Tanzania. Am J Primatol. 1988;16(2):97–121. https://doi.org/10.1002/ajp.1350160202.

Wittig R, Crockford C, Lehmann J, et al. Focused grooming networks and stress alleviation in wild female baboons. Horm Behav. 2008;54:170–7. https://doi.org/10.1017/J.Yhbeh.2008.02.009.

Wrangham R. An ecological model of female-bonded primate groups. Behavior. 1980;75(3/4):262–300. https://doi.org/10.1163/156853980X0044.

Chapter 10
Nonlinguistic Communication

10.1 Introduction

This chapter is concerned with nonlinguistic signals that may have been important to early hominins, as distinctively represented by baboon analogies. The term "distinctively" is important here because of the extent to which baboons are like many other primates and other animals in much of their communication system. The unusual features of baboon communication have special relevance to early hominins.

Molesti et al. (2019) listed 67 "communicative gestures"observed in a captive group of 47 olive baboons, aged 0–25 years. A gesture was defined as "a movement of the body or part of the body, directed to a specific partner or audience." Most of these gestures are represented in Table 10.1. Presumed idiosyncratic behaviors (performed by seven or fewer individuals) have been eliminated in order to simplify this discussion.

Many of the gestures observed by Molesti and colleagues are shared with numerous other primate species (e.g., grooming, embracing, bared-teeth signals) or many other animals (e.g., chase, flee). These provide no more insight into early hominin communication than more broadly comparative studies. Furthermore, some of these widely shared gestures have simple and obvious analogies with humans (e.g., staring as a threat), such that little discussion seems to be needed here. Finally, there are behaviors indicative of anatomical differences between baboons and hominins (e.g., sexual presenting and mounting) that offer only the simplest functional analogies.

Baboon vocalizations seem to offer the most distinctive and significant analogies for early hominins. Barks and grunts in particular are discussed at length in this chapter. A more complex set of communication signals occurs in baboon greetings. Some of these greeting patterns provide rather specific parallels with certain human behaviors, suggesting an evolutionary origin in early hominins. Preliminary work on baboon leave-taking also suggests analogies with humans and early hominins.

G. E. King, *Baboon Perspectives on Early Human Ancestors*,
https://doi.org/10.1007/978-3-031-36769-4_10

Table 10.1 Communicative signals of captive olive baboons

Signal	Partial description	# of Events	# of Subj.	Adult context
Audible				
Ground slapping	Slap ground or other surface	127	17	
Lip smack	Rapid lip movement	753	46	
Object shake	Vigorous shaking of object	108	20	
Teeth grind	Exaggerated chewing motions	33	10	Agonistic
Tactile				
Bite	Literal	19	13	Agonistic
Body contact	Parts of bodies touch	162	37	
Body–body rubbing	Literal	10	10	
Embrace	Wraps arms/legs around other	72	25	
Grab	Establish hold of other	207	32	
Grooming initiation	Literal	482	46	
Hand–body touch	Literal	464	47	
Hand–genitals touch	Touch genitals of other	51	29	
Hand–hand touch	Literal	51	11	
Head push	Briefly pushes other with head	19	12	Affiliative
Head–body rubbing	Gentle rubbing	21	15	Affiliative
Jump on	Bounce on back of other	42	9	
Mating initiation	Male clasps female	51	12	Sexual
Mock bite	Gentle grip with teeth	410	40	
Mount	Mount w/out sexual function	86	28	
Mouth–body touch	Literal	29	16	
Mouth–genitals touch	Literal	66	30	
Mouth–mouth touch	Gentle contact	45	29	
Pull	Grab and pull other's body part	115	23	
Slap	Hit with open hand	44	15	
Visual				
Air bite	Literal	75	24	
Back and forth look	Exaggerated gaze alternation	36	15	
Bared-teeth	Literal	74	25	
Biting threat	Mouth wide, showing teeth	463	41	
Charge	Run at over short distance	153	35	
Chase	High speed pursuit	91	28	
Display	Body shaking, jumping	45	14	
Eyebrow raising	Eyes wide, brow raised	477	42	
Flee	Literal	139	33	
Freeze	Arm on ground, body lowered	51	23	
Give ground	Move away at moderate speed	939	44	
Greeting	Side by side with other	35	16	
Groom present	Shows other part to be groomed	457	44	

(continued)

Table 10.1 (continued)

Signal	Partial description	# of Events	# of Subj.	Adult context
Ground rubbing	Rub ground/other support w palm	60	15	
Head shake	Literal	36	18	
Lift	Rhythmic raise/lower brows	132	36	
Lunge	Sudden intense forward movement	99	33	
Make room	Move part of body away from other	158	40	
Open mouth	Eyes wide, lips over teeth	60	15	
Peer	Intense look into other's eyes	164	37	
Presentation	Hindquarters presented to other	624	42	
Pursed lips	Lips protruded	48	12	
Scalp backward	Scalp/cheek retracted, ears flattened	206	41	
Spread leg	Stretch hind leg back toward other	42	16	
Stare	Literal	102	33	
Stretch arm	Extend arm(s) toward other	222	38	
Tail raising	Tail held straight and vertical	83	27	
Ventral presentation	Stand up, belly/genitals toward other	17	12	Affiliative

Modified from Molesti et al. (2019). Signals recorded from a captive troop of olive baboons, with total number of occurrences and total number of subjects displaying each signal. Behavioral context is noted where a signal appeared *only* in that context. All others were displayed in more than one context. Many adult patterns also were included in the play of immatures

10.2 Vocalizations

Humans have a set of nonlinguistic vocal signals, such as shouts and grunts, that are comparable to the call systems of other primates. A call system is a relatively fixed set of vocal signals used by a species that are conserved in the evolutionary sense, that is, they are little affected by genetic or ecological variation (Fischer 2021; Hammerschmidt and Fischer 2019). The conserved signals in humans that resemble those of other animals are presumably derived from early hominins. Analogies with baboon communication may suggest how these vocalizations functioned in early hominins, and why they were favored by natural selection in the context of hominin evolution.

10.2.1 A Flexible Call System

Vocal communication in all baboons is based on a call system like those of other primates. However, there are distinctive details in baboon systems that have implications for early hominin communication. Hammerschmidt and Fischer (2019)

considered *Papio* a good “model“for assessing links between social system characteristics and vocal communication because of the variation in social systems and social behavior across the genus.

When they analyzed acoustic variation in the call types of several baboon species, they found that the repertoires were composed of the same general types. However, quantitative analysis of acoustic features discerned subtle variations among chacma, olive, and Guinea baboons in two important categories: grunts and loud calls. The *grunts* of baboons are deep rhythmic sounds that are often emitted in rapid sequences. *Loud calls* are just that, calls that can be heard across fairly long distances and that demand attention from others.

Hammerschmidt and Fischer (2019) considered variation in baboon call rates and intensity to be great enough to support significant plasticity in social relationships, mating patterns, and social organization. A broad analogy with early hominins suggests that the hominins could have communicated flexibly with a relatively simple conserved signal system, at least with regard to various social interactions.

Of course, this functional analogy does not necessarily imply that hominins had the same call system as baboons or even a similar one. Nevertheless, there are analogies between some particular kinds of extant human and baboon utterances. As with other kinds of behavior, the hypothesis here is that hominins and baboons separately evolved comparable responses to similar social and environmental problems.

10.3 Loud Calls

Baboon loud calls are mostly barks, somewhat similar to vocalizations of some domestic dogs. Human analogy with baboon barks is loose and may apply only to males. Baboon barks are comparable to human shouts in volume and functions. They demonstrate how vocalizations like shouts might have been important to early hominins.

10.3.1 “Roars” in Humans

Loud, aggressive vocalizations in humans have been called “roars” or “roar-like vocalizations“by Raine et al. (2019). These researchers used playbacks to explore the form and function of such human vocalizations. Listeners accurately judged the upper body strength of vocalizers from aggressive speech and (separately) from roars. They made more accurate judgments based on the roars. These vocalizations could, therefore, be considered “honest signals,” that is, they conveyed the physical reality to listeners independently of the senders’ intent. However, the vocalizations conveyed some exaggeration of the associated feature (in accord with the competitive function of physical strength). This exaggeration effect occurred with male vocalizers and not with females.

Fukumori et al. (2023) researched "angry shouts" along with other loud vocalizations for the practical purpose of identifying threatening situations via public audio surveillance. It is not clear if these vocalizations resemble the roars described by Raine and colleagues. However, the functional connection between such loud vocalizations and social threat does seem to be clear.

Raine et al. (2019) hypothesized a "homology" between humans and other mammals resulting from natural selection for acoustic structure of aggressive vocalizations that communicates strength (i.e., physical fighting ability) to competitors. Such vocalizations are typically low pitched and structurally "noisy." If Raine and colleagues are correct, these aspects of threat vocalization in humans are more than analogies with baboons—they have a common origin in the evolution of mammals. Whether homology or not, an important function of these vocalizations is to settle disputes without dangerous physical combat. Even an animal much stronger than an opponent may suffer a serious injury that can lead to death, for example, a severed artery or an infection.

10.3.2 "Wahoos" in Baboons

Physical fights in a chacma baboon population were observed to cause potentially fatal injuries, but most disputes were resolved by displays in which a distinctive vocalization was prominent (Kitchen et al. 2013). Males produced a loud call that sounds like two syllables. The *wa-* is a kind of bark that is "ingressive" (i.e., emitted with inhalation); the *-hoo* is a kind of grunt (Boë et al. 2018). Male chacma baboons give these calls in response to predators (alarms) and during aggressive displays that sometimes involve chasing other baboons (contest wahoos). Acoustic analysis revealed significant but subtle differences between the alarm and contest calls that can be difficult for humans to distinguish.

For baboons it is adaptive for listeners to discriminate among calls that are given in qualitatively different contexts. This is particularly true for female chacma baboons because of the varied dangers that they face. In playback experiments, females responded for significantly longer durations to alarm than to contest wahoos and only alarm wahoos caused females to flee. Despite the acoustic similarity of the calls, female baboons appeared to associate alarm and contest wahoos with qualitatively different events (Kitchen et al. 2003).

Male perception of the contest vocalization is entwined with social dominance, which is "fiercely contested" (Kitchen et al. 2013). Physical fighting, a costly behavior, is relatively rare because disputes are frequently resolved by displays that include loud, repetitive wahoos. Males of all ranks adjusted their contest behavior based on the relative fighting ability of opponents. Fighting ability is reliably indicated by calling rate, fundamental frequency, and length of the second syllable (*-hoo*) (Kitchen et al. 2003, 2013).

Wahoos occur in baboon species other than chacmas, but they are rare in wild populations of Guinea baboons (Hammerschmidt and Fischer 2019). This is not because they lack the capability; numerous wahoos were recorded in a captive group of the species (Boë et al. 2018). Rather, wahoos are not stimulated in Guinea baboons under natural conditions because of the high level of male–male tolerance and the consequent rarity of aggressive interactions (Chap. 7).

There seems to be an analogy between male "roars" (or "angry shouts") in humans and male wahoo barks in baboons. This applies to function, but may also be relevant to some aspects of vocal quality. A human tendency for a serious argument to devolve into a "shouting match" may be derived from an early hominin pattern of dominance competition.

Guinea baboons share with chacma baboons an adjunct to the wahoo. In chacmas the vocalization is often preceded by "a short series of throaty roar-grunts" (Hammerschmidt and Fischer 2019). Guinea baboons sometimes emit "such roar-grunts" during branch shaking or chasing females. It would be interesting to know if these roar-grunts have any structural similarity to the roars reported for humans. It seems possible that these apparently softer vocalizations are analogous to human attention-getting exclamations such as the American *hey!* and the British *oi!*

10.3.3 Female Barks

The simpler barks of female chacma baboons parallel the predator warning function of wahoos, but not the competitive aspect. Female barks were also associated with separation from other group members (Cheney et al. 1996). This might seem to suggest intent on the part of the caller. However, though the calls did function to maintain contact between dispersed animals, there was apparently no intent to inform others of location. This was demonstrated by the timing of the calls. Females were more likely to give a contact bark in the 5 min after they themselves had called than to utter the vocalization in the 5 min after another female had called. Playback experiments suggested that separated females responded primarily to the contact barks of close relatives rather than other members of the troop.

Fischer et al. (2001a) studied variation in the barks of female chacmas and found a graded continuum from tonal and harmonically rich calls to calls with a "noisier" and harsher structure. Tonal barks were typically given when the signaler was at risk of losing contact with the group or when a mother and infant had become separated (contact barks). The harsher variants were given in response to large predators (alarm barks). Within the alarm bark category, there are significant differences between calls given in response to mammalian carnivores and those in response to crocodiles (Chap. 6), a distinction that in other species has been attributed to referential understanding (but see Fischer 2021).

Fischer et al. (2001b) tested whether wild baboons made the following discriminations among recorded vocalizations of females: (1) clear contact barks versus harsh alarm barks, and (2) clear contact barks versus intermediate alarm barks. Calls were selected according to an analysis of a suite of acoustic parameters. In these experiments, the baboons responded only to the playback of a harsh alarm bark.

Apparently, only this harsh variant was perceived as warranting a response. The researchers hypothesized that baboons' responses were to a large degree influenced by their assessment of context. This distinction seems to parallel the one made by females with regard to male alarm and contest wahoos.

10.4 Grunts

Grunts are deep rhythmic sounds that occur in all baboon species and convey various meanings, mostly positive, ranging from contact maintenance to friendliness to cooperation. Owren et al. (1997) characterized the grunts of chacma baboons as "tonal, harmonically rich vocalizations." Grunts demonstrate the social versatility of a single call type. They occur in both baboons and humans, with similar vocal characteristics and behavioral implications. Because of the similarity in both form and function, grunts are probably the mode of communication most likely to be analogous between humans and baboons.

10.4.1 An Evolutionary Theory of Human Grunts

McCune (2021) presented a theory of hominin language origins based on grunts. Across mammalian species, including humans, grunts are an initial reflex response to autonomic demand (e.g., effort). In some species, including chimpanzees and vervets, grunting was co-opted for communication. McCune noted that chimpanzee and vervet infants are similar to humans in the shift of grunts from effort to communication.

Throughout the first 18 months of life, human infants produce a growing variety of vocalizations. In studies of referential word onset, McCune and colleagues discovered that one prominent laryngeal vocalization, when produced communicatively, was predictive of each child's referential word onset (McCune et al. 2020). They defined this "grunt"in line with nonhuman primate literature as a laryngeal articulation characterized by abrupt glottal onset and short duration. In infants that were followed from 9 to 16 months of age, grunt production occurred in three contexts: physiological effort, focused attention, and communication.

The evolutionary hypothesis is that the connection between a physiological state (physiological challenge or effort) and an accompanying vocalization (the autonomic grunt) has been an engine in the origin of protolanguage(s) in species with sufficient levels of mental representation. The fact that many primate species' call repertoires include a vocal signal related to travel, an effortful activity, is relevant to the grunt/effort hypothesis. This association occurs in gorillas, vervet monkeys, and chimpanzees.

In addition to the primate species cited by McCune, baboons emit grunts before and during group movements (Hammerschmidt and Fischer 2019). In chacma baboons, at least, grunts may be important for initiation of movement across open areas (Owren et al. 1997). Thus, the McCune hypothesis is supported by the primate genus that is probably most closely analogous to early hominins with regard to group movement, especially across open areas.

Beyond that, the baboon analogy demonstrates the co-optation of grunting into other forms of social communications. Owren et al. (1997) reported chacma baboon grunts in two distinct behavioral contexts: initiation of movement and approach to a mother in an attempt to inspect or handle her infant. The grunts appeared to be acoustically distinct and elicited different responses from receivers.

10.4.2 Grunts and Social Interaction

Humans (at least in Western societies) use soft and simple phrases to soothe others, such as *okay* or *there-there*. Whether or not these have any formal relation to ancestral grunts, baboon use of grunts suggests how comparable vocalizations might have evolved in early hominins. All baboons use grunts to facilitate affiliative social interactions (Hammerschmidt and Ficher 2019). They are produced by both sexes, in all age classes, in situations that vary with individual relationships and social organization.

In olive baboons, "decisions" about whether to grunt or remain silent are influenced by the social context, particularly the likely response of a potential partner to the approach (Silk et al. 2016). When a female emits low amplitude grunts after approaching another female, she is less likely to behave aggressively toward the other female, and more likely to be affiliative and to handle the partner's infant. The female baboons are more likely to grunt when they approach lower-ranking females, who may be anxious about the interaction, than in approaching higher-ranking females. They are less likely to grunt after approaching their own mothers and daughters, presumably because behavioral expectations are clear. Similar to olive baboons, quiet, tonal grunts by female chacma baboons mollified lower-ranking females, facilitating friendly social interactions that included inspection and handling of an infant (Cheney et al. 1995; Owren et al. 1997). Taken together, the patterning of grunts in olive and chacma baboon suggests that these vocalizations play an important role in reducing uncertainty in others about an actor's intentions. This facilitates nonaggressive social interactions (Silk et al. 2018).

Males also modify their communication signals in accord with social relationships. In a study of chacma baboons, Palombit et al. (1999) found two patterns of adult male grunts during interactions with females. First, higher-ranking males grunted significantly more often than subordinates when approaching females in most of the females' reproductive states. Second, males grunted more often when approaching females with which social interaction was potentially highly beneficial and/or social interaction was unlikely to occur due to female evasion, that is, estrus females and lactating females. In chacma baboons, male grunts had contrasting effects on the probability of supplanting a female or interacting in an affinitive manner with her. Supplanting of females was just as common when the approaching male grunted as when he did not. Instead, variance in supplanting was better explained by female avoidance of high-ranking and non-friend males than by the male's vocal behavior.

These results suggest that male grunts themselves do not determine the female's response in all situations. Rather, the female's reproductive state and social relationship with the male (i.e., his "friendship" status and/or rank) affect *both* the male's tendency to call to her and the female's tendency to move away from him. In contrast to supplanting, affinitive interaction occurred significantly more often when males grunted than when they silently approached females. Taken together, results suggest that a female chacma baboon's spatial response to a male's approach (stay or leave) depends on her assessment of nonvocal factors, but her "social" response (interact or not) is influenced by the grunts given by the male (Palombit et al. 1999).

In Guinea baboons, as in olive and chacma baboons, the presence of an infant affects social communication. In females, the probability of grunting was higher when the relationship strength was low, but only when an infant was present. Males were also more likely to grunt when an infant was near a female partner (Faraut et al. 2019). In the tolerant society of Guinea baboons (Chap. 7), males often grunted to each other (Maciej et al. 2013a, b).

Faraut et al. (2019) suggested that grunt usage in baboons can be best conceived as a combination of a motivational and a strategic component. The motivational component expresses the increased disposition to interact in an affiliative fashion, while the strategic component refers to the modulation of grunt usage with regard to relationship quality and context. The motivational component appears to be shared between baboon species. The strategic component varies with social organization and places different premiums on the potential benefits of signaling, resulting in variation in grunting patterns between species (Faraut et al. 2019). With the proliferation of hominin species that were contemporary with each other (Chap. 1), and may have had differing forms of social organization, potential analogies like this one increase in significance.

10.4.3 Grunts and Referential Communication

In the study cited above, Owren et al. (1997) described distinctive grunts uttered by chacma baboons in two different contexts: initiation of movement and approach to a mother. The researchers hypothesized this to be a rudimentary capacity for *referential signaling*, that is, communication *about* something. In playback experiments (Rendall et al. 1999), the subjects differentiated between the two kinds of grunts based only on acoustic features and their distinct responses matched the behaviors in naturally occurring situations. However, some responses to playbacks were also affected by context, such as rank differences between callers and the subjects. The researchers concluded that baboon grunts can function in rudimentary referential fashion, but that context and social identity can also affect recipients' responses.

These results suggested that baboons make inferences about the directedness of vocalizations even in the absence of visual cues, and that the nature of prior interactions affects subsequent behavior. When attending to vocal signals, baboons appeared to recognize the signaler's identity and her probable subsequent behavior, and also the target of her attention. The ability to integrate these cues could be interpreted as a first step toward the recognition of other individuals' intentions and motives (i.e., the cognitive capability known as "theory of mind" or "mind-reading").

Fischer (2021) clarified the general issue by applying the distinction between first-order and second-order intentionality. In first-order systems, the intent or motivation is to influence the behavior of the receiver(s). Second-order communication differs in that the signaler intends to convey information. First-order communication is sufficient in the social systems of baboons and other primates because signals that express motivation (e.g., desire to touch an infant) allow the receiver to predict the sender's behavior. This adds to the comparative inference that early hominins could have engaged in complex social behavior with relatively simple vocal systems (cf. Hammerschmidt and Fischer 2019).

10.5 Gestures

There seems to be little description of gestures in wild baboons, especially when compared with the work on chimpanzees. This might be connected with the relative freedom of the arms and hands in chimpanzees compared with the quadrupedal baboons. However, a captive study indicates that baboons have a substantial capacity for gestural communication in a broad sense. Molesti et al. (2019) studied gestural communication in three social groups of captive olive baboons for 1 year. They defined a communicative gesture as any movement of the body or part of the body that is directed to a specific partner or audience. The definition included facial expressions. A gesture could be directed to a partner by gaze, body orientation, or physical contact.

The researchers recorded almost 9000 gestures that they classified into 67 gesture types (Table 10.1). The majority of the types were visual: 39 types, 58% of the repertoire. Of the rest, 24 were tactile and only 4 were audible. Some gesture types were common and others rare. Of the total number of gestures, 74% were performed as an unaccompanied signal while 26% were combined with another gesture. The prominence of visual signals here is consistent with a long-standing hypothesis that associates the modality with increased terrestriality and reduced interference from vegetational barriers. In this context, Molesti and colleagues noted that baboons evolved in environments very much like those of many early hominins.

This baboon evidence supports inference from the *Pan* species that early hominins might have communicated extensively with gestures at close quarters. Longer range communication might have been encouraged by expansion into more open habitats with less vegetational cover. The baboons in the Molesti study inhabited a large open enclosure (with access to shelter in a building).

10.6 Salutations

Salutations are signals that acknowledge an individual's arrival or departure. Greetings are important signals of peaceful intention and trust between males in both baboon and human societies. The baboon evidence suggests that such behavior may have been adaptive in early hominins. Further, some specific behaviors may have evolved in both lineages because of their social effectiveness. Recent work demonstrates the existence of salutations (or, at least, notifications) of departure in baboons. Baboon greetings have been used as a prime example of the concept of *sequence organization* as applied to social interactions of primates.

10.6.1 Greetings

Ritualized greetings are exchanges of nonaggressive signals. They are important in complex societies, such as those of baboons and humans, and are especially elaborate in the multilevel societies of the hamadryas and Guinea baboons. They are common among the males and are thought by some to balance the trade-offs of male co-residence between the risk of aggression and the need for co-existence (Dal Pesco and Fischer 2018). The signals used by baboons include some that specifically resemble gestures used in some human societies, including males touching or grasping one another's genitals to show trust and imply tolerance and willingness to cooperate. Thus, analogy with baboons suggests that the practice of (need for?) greeting originated in early hominins (Wickler 1972).

While ritualized greetings are widespread in the animal kingdom, the behavioral repertoire described in the genus *Papio* is exceptional, as it involves potentially harmful behaviors such as genital fondling. Although greetings are one of the most striking male social interactions in baboons, their function has been subject to dispute. Dal Pesco and Fischer (2018) examined the social behavior of 24 adolescent and adult male Guinea baboons to test whether greetings reflect relationship quality or function to buffer tension. Greetings were ten times more frequent than aggression and twice as frequent as affiliation. Neither dyadic aggression nor tense context predicted greeting rate, discounting the buffering hypothesis. Greetings occurred almost exclusively between males of the same party, even when other parties were around. Within parties, spatially tolerant partners greeted more frequently but those in dyadic relationships did not greet each other because they were usually in proximity.

Although affiliation did not predict overall greeting rate, intense and potentially costly greetings were more likely between males with stronger affiliative relationships. Greetings in Guinea baboons appear to signal commitment among party members, test relationships among spatially tolerant partners, and accentuate relationship strength among highly affiliated males. Looking at the comparative evidence, the researchers concluded that, although ritualized baboon greetings lack the symbolic component of human rituals, they appear to serve similar functions of strengthening in-group affiliation and promoting cooperation (Dal Pesco and Fischer 2018)

The authors also considered comparative evidence from other baboon species (Dal Pesco and Fischer 2020). These ritualized signals differ between species in their occurrence, form, and function. While ritualized greetings are rare in species with the most intense contest competition, the complexity of and risk involved in greeting rituals increase with the degree of male–male tolerance and cooperation. The variety of societies found in this genus, combined with its role as a model for human socioecological evolution, sheds light on the evolution of ritualized behavior in humans.

10.6.2 Sequence Organization

Mondada and Meguerditchian (2022) applied the idea of *sequence organization* to baboon social interactions, with greeting as a prime example. The concept of sequence organization was posited by students of human behavior as a key element in a larger hypothesis of social interaction. This approach considers an interaction to be more than just one action following another. It views each action as making the next one relevant, including alignment or "disalignment" with the prior action. The sequential order displays the way intelligible actions are produced; how they are progressively identified and recognized; and how they are responded to. It establishes and manifests the rights and obligations of the participants. Setting aside cultural meanings, this is a social process that can be observed in baboons.

Mondada and Meguerditchian (2022) performed an analysis that showed how the actions of baboons are finely coordinated and mutually shape each other. In the openings studied, and in particular in the greetings (constituted by the presentation of the hindquarters), baboons manifested a very precise sense of sequence organization. They displayed it by carefully formatting their courses of action, by scrutinizing them, and by expressing whether or not they conform social and normative expectations. This is particularly observable in the orientation to absence of responses in the other individual. While approaching each other, baboons establish and define their mutual positions. This indicates what the encounter is becoming—either an aligned and even affiliative unfolding of actions, or an interaction that is "disaligned" (simply put, trustful or suspicious). The researchers emphasized that this analytical approach from human studies can be readily applied to baboons. They inferred common factors that may derive from parallel evolutionary backgrounds.

10.6.3 Leave-Taking

Behaviors that mark the parting of two individuals are a common and possibly universal feature of human social life. Baehren and Carvalho (2022) studied leave-taking in wild baboons by testing a range of candidate behaviors on video footage. The behaviors that they addressed were (1) self-scratching, (2) eye gaze, and (3) orientation in the direction of parting. Their analysis controlled for interaction duration and individual variation. The results showed that orientation in the direction of parting occurs predominantly before social separation events. It was not associated with solo or proximity separations. This suggested to the researchers that the behavior plays a unique social role that could have evolved for a social function. It may assist in strengthening social bonds or in minimizing aggression, increasing the affiliative nature of interactions or mitigating risky endings.

This work was limited to one species in one locality. The absence of risk from predators or humans in that national park could have been significant. These baboons were more terrestrial than many other populations and this could increase opportunities for such short-term, recurrent social interactions. Such opportunities increase visual proximity, and as such, leave-taking may be a necessary adaptation to social life on the ground.

Terrestrial adaptations, of course, have crucial relevance to early hominins. Differences in aggressive behavior, social cohesion, and female-bondedness (for example) might also affect leave-taking, resulting in variation among species that has not been studied (Baehren and Carvalho 2022). Variation between forest and savanna species or populations might be significant.

10.7 Summary and Discussion

Nonlinguistic communication is any form of signaling between animals that does not involve language in the human sense of the term. It need not be deliberate on the part of the sender. Patterns of baboon communication suggest how early hominins might have communicated with each other and how effective their system might have been. Some features of baboon communication have nothing special to tell us about nonlinguistic signals in early hominins. This is because baboons are like most other primates in these behaviors; for example, mutual grooming and staring threats. However, some baboon patterns provide more specific analogies.

Vocal communication in baboons uses a call system comparable to that of other primates. It contains a limited number of call types that are characteristic of the genus and seem to be strongly constrained by the genes. Nevertheless, it is a flexible and versatile system because each call type has subtle variations and baboons derive information from integrating social context with the signal. Early hominins probably evolved such a system as they adapted to diverse and changing environments and to the advent of multilevel societies.

Two categories of baboon vocalization seem to be especially pertinent to early hominin adaptation: loud calls and grunts. Loud calls are essentially barks that are broadly analogous to human shouts in volume and at least some functions. Female barks vary from tonal calls (in response to separation) to harsher vocalizations (predator alarms). It seems that there is no intent behind the separation calls, so the cognitive basis for some parts of the system is relatively simple. Males have a distinctive two-phase loud call, the "wahoo." Some wahoos are predator alarms while others are emitted during dominance contests and may serve as a substitute for costly aggression. This vocalization is a good example of flexibility in baboon usage of innate signals. The wahoo is prominent among the aggressive chacmas, but rarely occurs in Guinea baboons where there is little dominance competition among males. On the other hand, "roars" seem to be similar among chacmas, Guinea baboons, and humans.

Grunts are acoustically similar between baboons and humans. Baboon grunts are deep rhythmic sounds that are harmonically rich and serve a variety of communicative functions. An evolutionary theory of grunting views it as a response to autonomic demand (e.g., physical effort) that occurs across all mammals, including humans, that has been coopted to be a social signal. In baboons, for example, grunts occur in the effort of beginning travel and have come to communicate in that context and others. The circumstances in which baboons communicate with grunts must have had many parallels in early hominins, given the similar demands of their physical and social environments. For example, a troop of baboons varied their grunts in relation to habitat usage (perhaps reflecting some level of awareness of their surroundings). They used longer grunts in forested areas, probably in response to reduced visibility and sound transmission.

Grunts signal "benign intent" among baboons, facilitating positive social interactions. The variety of these interactions gives rise to hypotheses about the underlying cognitive processes. Baboons seem to integrate the exact form of a grunt with the signaler's identity, its probable subsequent behavior, and the target of the signaler's attention. This may come close to an understanding of the other's mental process (often called "Theory of Mind"). Another study also made inferences about cognitive processes, suggesting that grunts represent the combination of an affiliative motivation with a "strategic" component (modulation with regard to relationship quality and context). Playbacks of grunts were also used to assess the ability of baboons to comprehend the relationships of others. In contrast to chacmas, male Guinea baboons showed little concern about indications that a female was changing partners. Presumably this was because they live in a society where such affiliations are chosen by females.

A captive study indicates that baboons have a substantial capacity for gestural communication in a broad sense that includes facial expressions and any other movement of the body or part of the body directed to a specific partner or audience. This supports inference from the *Pan* species that early hominins might have communicated extensively with gestures at close quarters. Longer range communication might have been encouraged by expansion into more open habitats with less vegetational cover.

Ritualized greetings between males are an important and somewhat specialized form of communication in baboon and human societies. In both taxa they convey trust and cooperativeness, and in both taxa some similar signals are used (the most dramatic being toleration of genital touching). Baboons also seem to recognize the conclusion of interactions. Recent research on one troop reports a consistent signal of leave-taking, but only when there is complete separation from a social situation. Baboon greeting behaviors may be an example of sequence organization, a theory of human interaction that posits a complex relationship between each behavior and the one that follows it.

References

Baehren L, Carvalho S. Yet another non-unique human behaviour: leave taking in wild chacma baboons (*Papio ursinus*). Animals. 2022;12:2577. https://doi.org/10.3390/ani12192577.

Boë L-J, Sawallis TR, Fagot J, Berthommier F, et al. What's up with wahoo? Exploring baboon vocalizations with speech science techniques. In: Boë L-J, Fagot J, Perrier P, Schwartz J-L, editors. Origins of human language: continuities and discontinuities with nonhuman primates. Berlin: Peter Lang; 2018. p. 59–74.

Cheney DL, Seyfarth RM, Silk JB. The role of grunts in reconciling opponents and facilitating interactions among adult female baboons. Anim Behav. 1995;50(1):249–57. https://doi.org/10.1006/anbe.1995.0237.

Cheney DL, Seyfarth RM, Palombit R. The function and mechanisms underlying baboon 'contact' barks. Anim Behav. 1996;52(1):507–18.

Dal Pesco F, Fischer J. Greetings in male Guinea baboons and the function of rituals in complex social groups. J Hum Evol. 2018;125:87–98. https://doi.org/10.1016/j.jhevol.2018.10.007.

Dal Pesco F, Fischer J. On the evolution of baboon greeting rituals. Philos Trans R Soc B. 2020;375:20190420. https://doi.org/10.1098/rstb.2019.0420.

Ey E, Fischer J. Keeping in contact: flexibility in calls of olive baboons. In: Sommer V, Ross C, editors. Primates of Gashaka: socioecology and conservation in Nigeria's biodiversity hotspot. New York: Springer; 2011. p. 361–84.

Ey E, Rahn C, Hammerschmidt K, Fischer J. Wild female olive baboons adapt their grunt vocalizations to environmental conditions. Ethology. 2009;115:493–503. https://doi.org/10.1111/j.1439-0310.2009.01638.x.

Faraut L, Fischer J. How life in a tolerant society affects the attention to social information in baboons. Anim Behav. 2019;152:11–7. https://doi.org/10.1016/j.anbehav.2019.04.004.

Faraut L, Siviter H, Dal Pesco F, Fischer J. How life in a tolerant society affects the usage of grunts: evidence from female and male Guinea baboons. Anim Behav. 2019;153:83–93. https://doi.org/10.1016/j.anbehav.2019.05.003.

Fedurek P, Neumann C, Bouquet Y, et al. Behavioural patterns of vocal greeting production in four primate species. R Soc Open Sci. 2019;6:182181. https://doi.org/10.1098/rsos.182181.

Fischer J. Primate vocal communication and the evolution of speech. Curr Dir Psychol Sci. 2021;30(1):55–60. https://doi.org/10.1177/0963721420979580.

Fischer J, Hammerschmidt K, Cheney D, Seyfarth RM. Acoustic features of female chacma baboon barks. Ethology. 2001a;107:33–54.

Fischer J, Metz M, Cheney D, Seyfarth RM. Baboon responses to graded bark variants. Anim Behav. 2001b;61:925–31.

Fischer J, Higham JP, Alberts S, et al. The natural history of model organisms: insights into the evolution of social systems and species from baboon studies. elife. 2019b;8:e50989. https://doi.org/10.7554/eLife.50989 .

Fischer J, Wegdell F, Trede F, et al. Vocal convergence in a multi-level primate society: insights into the evolution of vocal learning. Proc R Soc B. 2020;287:20202531. https://doi.org/10.1098/rspb.2020.2531.

Fraser O, Plowman AB. Function of notification in *Papio hamadryas*. Int J Primatol. 2007;28:1439–48. https://doi.org/10.1007/s10764-007-9185-x.

Fukumori T, Ishida T, Yamashita Y. RISC: a corpus for shout type classification and shout intensity prediction. arXiv. 2023:2306.04143.

Hammerschmidt K, Fischer J. Baboon vocal repertoires and the evolution of primate vocal diversity. J Hum Evol. 2019;126:1–13. https://doi.org/10.1016/j.jhevol.2018.10.010.

Kemp C, Rey A, Legou T, et al. The vocal repertoire of Guinea baboons (*Papio papio*). In: Boë JL, Fagot J, Perrier P, Schwartz JL, editors. Origins of human language: continuities and splits with nonhuman primates. Oxford: Peter Lang; 2017. p. 15–58.

Kitchen DM, Cheney DL, Seyfarth RM. Female baboons' responses to male loud calls. Ethology. 2003;109:401–12.

Kitchen DM, Cheney DL, Engh AL, et al. Male baboon responses to experimental manipulations of loud "wahoo calls": testing an honest signal of fighting ability. Behav Ecol Sociobiol. 2013;67:1825–35. https://doi.org/10.1007/s00265-013-1592-8.

Maciej P, Ndao I, Patzelt A, et al. Vocal communication in a complex multi-level society: constrained acoustic structure and flexible call usage in Guinea baboons. Front Zool. 2013a;10:58–72. www.frontiersinzoology.com/content/10/1/58

Maciej P, Patzelt A, Ndao I, et al. Social monitoring in a multilevel society: a playback study with male Guinea baboons. Behav Ecol Sociobiol. 2013b;67:61–8.

McCune L. Laryngeal vocalizations: grunts as a gateway to language? Lingua. 2021;264:103161. https://doi.org/10.1016/j.lingua.2021.103161.

McCune L, Lennon EM, Greenwood A. Gestures, grunts, and words: development in a dynamic system. First Lang. 2020;41(3):243–67. https://doi.org/10.1177/0142723720966820.

Molesti S, Meguerditchian A, Bourjade M. Gestural communication in olive baboons (*Papio anubis*): repertoire and intentionality. Anim Cogn. 2019; https://doi.org/10.1007/s10071-019-01312-y.

Mondada L, Meguerditchian A. Sequence organization and embodied mutual orientations: openings of social interactions between baboons. Philos Trans R Soc B: Biol Sci. 2022, 1859;377 https://doi.org/10.1098/rstb.2021.0101.hal-03871646.

Owren MJ, Seyfarth RM, Cheney DL. The acoustic features of vowel-like grunt calls in chacma baboons (*Papio cynocephalus ursinus*): implications for production processes and functions. J Acoust Soc Am. 1997;101:2951–63.

Palombit RA, Cheney DL, Seyfarth RM. Male grunts as mediators of social interaction with females in wild chacma baboons (*Papio cynocephalus ursinus*). Behaviour. 1999;136:221–42.

Raine J, Pisanski K, Bond R, et al. Human roars communicate upper-body strength more effectively than do screams or aggressive and distressed speech. PLoS One. 2019;14(3):e0213034. https://doi.org/10.1371/journal.pone.0213034.

Rendall D, Seyfarth RM, Cheney DL, Owren MJ. The meaning and function of grunt variants in baboons. Anim Behav. 1999;57:583–92.

Seyfarth R, Cheney D. Pragmatic flexibility in primate vocal production. Curr Opin Behav Sci. 2018;21:56–61.

Silk JB, Seyfarth RM, Cheney DL. Strategic use of affiliative vocalizations by wild female baboons. PLoS One. 2016;11(10):e0163978. https://doi.org/10.1371/journal.pone.0163978.

Silk JB, Roberts ER, Stäzdele V, Strum SC, et al. To grunt or not to grunt: factors governing call production in female olive baboons, *Papio anubis*. PLoS One. 2018;13(11):e0204601. https://doi.org/10.1371/journal.pone.0204601.

Wickler W. The sexual code, the social behavior of animals and men. New York: Doubleday; 1972.

Chapter 11
Language Origins

11.1 Introduction

Language in the human sense is a complex system with *phonemes* (bundles of speech sounds) assembled into *morphemes* (meaningful units) and morphemes connected by rules of *syntax*. Baboon analogies suggest that the behavior and cognition of early hominins contained separate elements that eventually came together to form the basis for human language. Some of these elements were favored by natural selection because they supported communication. Others originally had different functions, but were coopted into the communication system.

11.2 Components of Language

Baboons, especially in experimental settings, have provided evidence for numerous hypotheses about the origins of language. A group of scientists based in France has developed the idea that language evolved from a combination of cognitive, neurological, and anatomical components (Fagot et al. 2019). They argue that baboons and humans share a number of the features that are central to language. Some are domain-general cognitive functions. That is, they contribute to a variety of mental processes, some of which can support communicative behavior. Other features are more domain-specific, that is, evolved to function in a communication system. For example, baboons are comparable to humans in their capacities for vocal and gestural production (Chap. 10).

The French group considered baboons an "excellent" model for the study of language evolution because of the "multidimensionality" of knowledge about the genus. They seem to be virtually unanimous in the opinion that the relevant features shared by baboons and humans are homologies that originated in the common

G. E. King, *Baboon Perspectives on Early Human Ancestors*,
https://doi.org/10.1007/978-3-031-36769-4_11

ancestor of apes and monkeys. The main alternative is the view emphasized in this book, that such similarities are more likely to be analogies arising from the common circumstances of baboon and hominin evolution. Of course, the component view of language origins opens the possibility that different components evolved at different times. In any case there is agreement that baboons can be a valuable source for understanding the evolution of language.

A survey by Prieur et al. (2020) demonstrated that many of the prelinguistic components found in baboons also occur in various other primate species. However, the combination of these components in baboons may be unique, at least among monkeys. As with other aspects of behavior discussed earlier, baboons provide the only model in which the patterns are manifested in environments like those that early hominins occupied.

11.3 Language and Concepts

Fitch (2019) made an important argument about the primate origin of language that complements the conventional emphasis on performance. Fitch's premise is that a defining feature of human language is the flexible representation and recombination of *concepts*. From this he infers that precursors for important components of language should be sought in animal cognition rather than animal communication.

Fitch's hypothesis counters a long-standing assumption that limits the comparative approach to the study of language. According to this view, the absence of some feature of human language from animal communication is evidence for an evolutionary discontinuity between the species. Analysis of animal concepts may provide a more complete understanding of an animal's communicative potential and its significance for comparison with human language.

The key point is that an animal communication system typically expresses only a small subset of the concepts that can be mentally represented and manipulated by that species. Thus, if a particular concept is not expressed in a species' communication system, this is not evidence that the species lacks that concept. Exclusive focus on overt signals will lead to underestimation of conceptual abilities and a flawed comparative analysis of language evolution. Therefore, animal cognition provides a crucial (and often neglected) source of evidence regarding the biology and evolution of human language.

Fitch's primate examples are chimpanzees and vervet monkeys (the latter is *Chlorocebus pygerythrus*, previously classified as *Cercopithecus aethiops*). Vervets have received a great deal of attention for a set of referential alarm calls that distinguish leopards, eagles, and snakes (Price et al. 2015). Fitch argues that concepts in vervet cognition go far beyond these calls. Mental representations of vervets include concepts that are undoubtedly shared with many other primates (e.g., *dominant other*) along with concepts that are probably more specific to the genus. For example, vervets have complex spatial representations of their environment and the ability to mentally track the locations of hidden group members. They can socially learn how to access food and rapidly absorb new social preferences about what to eat based on color. None of this cognitive sophistication is in any way detectable in their vocal communication system.

Fitch acknowledges that similar examples could be provided for many other well-studied primates. This could certainly include baboons. For example, vervet alarm calls can be compared to baboon calls that distinguish raptors, mammalian predators, and crocodiles (Chap. 6). Spatial representations of baboons probably go far beyond those of vervets, because baboon troops typically range more widely and encounter a greater variety of habitats. As to concepts not expressed in overt communication, there is evidence from experiments that address baboon cognition.

11.4 Domain-General Functions in Language Origins

According to Fagot et al. (2019) language is a multilevel phenomenon that requires integrative processes. Many of these are domain-general processes that can also fulfill nonlinguistic functions. This point of view seems to articulate well with the argument for concepts, discussed above. Fitch looks for the antecedents of language in adaptive mental processes; Fagot and colleagues seek the roots of language in domain-general mental functions. In both instances, the cognitive phenomena are not necessarily communicative in themselves, much less linguistic. However, they lend themselves to processes of evolutionary integration that produced language. Both approaches look to nonhuman animals for clues to these roots of language. The premise in each case is that early hominins may have had the same capability (perhaps for different reasons) and that the capability could have fed into the early evolution of language.

11.4.1 Memory

Memory is essential for the retention of language basics and for control of specific communications. Long-term memory is crucial for the storage of phonology, semantics, grammatical rules, pragmatics, and many other aspects of language (Fagot et al. 2019). Guinea baboons have displayed "impressive" long-term memory, retaining a large percentage of pictures from samples of thousands for at least a year. These results were comparable to those obtained with a human subject (Fagot and Cook 2006).

Short-term memory is important in language for keeping track of the words in a sentence and for the rapid verbal learning of the words in a lexicon (Fagot et al. 2019). Experiments show that baboons can store a large amount of information in short-term memory and that they can maintain this information in a small temporal range, measurable in seconds (Fagot and de Lillo 2011; Rodriguez et al. 2011). It seems that baboons have weaker working-memory capacities than humans, but that the two taxa have roughly identical long-term memory capacities. Early hominins could probably retain a substantial lexicon of signals, but actual communication exchanges might have been brief.

11.4.2 Categorization

Categorization is an essential process in the acquisition of language (Fagot et al. 2019). At the structural level, syntactic categories (e.g., noun or verb) are fundamental in grammatical structure. At the semantic level, nouns often refer to categories (e.g., *cat*, referring to all cats). Baboons, like many other animals, can form categories according to concrete criteria and some more abstract criteria. Guinea baboons, after training, can assign examples to correct categories; for instance, different alphanumeric characters (Vauclair and Fagot 1996). Olive baboons (*Papio anubis*) classified foods versus nonfoods (Bovet and Vauclair 1998).

Guinea baboons can also form categories based on relational properties (e.g., including openness versus closedness, Barbet and Fagot 2011) and spatial relations (e.g., far/near, Dépy et al. 1998; above/below. Dépy et al. 1999). For early hominins, the ability to make such distinctions might have facilitated communication about the proximity of predators or the height of foods or sleeping branches in trees.

11.4.3 Statistical Regularities

The ability to detect statistical regularities facilitates language acquisition and processing, including categorization (Fagot et al. 2019). Guinea baboons were trained with touch screens to discriminate real English four-letter words from four-letter strings that were not words (Grainger et al. 2012). Further examination of the baboons' strategies with a modeling approach showed that discrimination between the words and nonwords involved learning of particular bigrams or trigrams that were statistically more frequent in the words than in the nonwords (Hannagan et al. 2014). This performance can be accounted for by the baboon's ability to detect the statistical regularities between and among words, and to develop an open-ended representation of the word and nonword categories on that basis (Fagot 2017). In an experiment with spatial cueing, Guinea baboons demonstrated statistical learning mechanisms similar to those of humans (Goujon and Fagot 2013). A similar recognition capability might have been the basis for infusion of symbolic content in the cognitive processes and communication capability of early hominins.

11.4.4 Analogical Reasoning

Analogical reasoning can be considered a form of categorization that is based on abstract relationships (Fagot et al. 2019). Developmental studies in human children have shown a close relationship between analogical reasoning and the acquisition of linguistic labels (Christie and Gentner 2013). Analogical studies in Guinea baboons used a relational matching task based on Fagot and Thompson (2011). First, the subject sees one pair of objects that are either identical or different. Then, two comparison pairs are shown, and the baboon must indicate the stimulus pair with the same relationship to each other as the sample pair. Correct response to the test requires at least some understanding of the relation between relations.

With pairs of shapes as the stimuli, 6 of 29 baboons solved the matching test and 5 of these 6 then transferred this ability to novel sets of shapes. These results suggest that some Guinea baboons have the capacity to categorize stimuli with regard to both concrete and more abstract criteria, an ability that is critical for language (Fagot et al. 2019). In a simpler version of the test, Fagot et al. (2001) showed that two Guinea baboons could solve the problem when represented by same or different icons. Manipulation of such icons suggests the possibility of manipulating words. The limited capability shown here seems to suggest that early hominins might have had the potential for analogical reasoning in a few individuals, but that it was not common enough to be a factor in the communication system.

11.4.5 Pragmatics and Semantics

Campbell and Tyler (2018) asserted that pragmatics and semantics can be considered domain-general features of language because they are involved in many other cognitive functions. In humans, neural regions involved in semantics during language processing overlap with those that underlie object processing and other functions. Cheney and Seyfarth (2014, 2016) described the operation of pragmatics and semantics in the communication of chacma baboons. A baboon, hearing a call from another, builds a mental representation of the call by associating it with the caller's identity (including rank and kinship connections) as well as recent events involving that individual. In this sense, the process is combinatorial and provides a foundation for semantics (Cheney and Seyfarth 2018). The system is adaptive because it contributes to reproductive success in a long-lived species in which individuals depend on strong social bonds with other individuals and on recognition of social relationships between other troop members. This would obviously apply to early hominins.

11.5 Domain-Specific Components: Vocalization and Speech

Baboons and humans share some physical and mental functions that are specific to the evolution of language rather than being domain-general (Fagot et al. 2019). These include vocal capabilities that evolved in connection with nonlinguistic communication (Chap. 10). Given this view of their origins, it seems more appropriate to label them *communication-specific* rather than *language-specific*. There are two major schools of thought as to the primary mover at the beginning of language evolution. One emphasizes vocalizations and the other emphasizes gestures. Baboons provide evidence for both.

11.5.1 Protophones and the Vocalization-First Theory

Protophones are speech-like sounds produced by human infants during roughly the first year of life. These include vowel-like sounds and also syllables and syllable sequences such as *da* and *dada.* Oller et al. (2021) studied such sounds in human infants, as an approach to determining the prime mover in language origins. Protophone production seemed to be endogenous in that (1) the infants produced them at a high rate even when alone and (2) they did not direct the majority of these sounds to a listener when one was present. Additional evidence for an innate basis is that infants born deaf produce protophones at rates comparable to those of hearing babies.

In another study, protophones were at least 35 times more frequent than gestures at the age of 3 months. The ratio declined, but was still greater than 2.5 at the age of 11 months (Burkhardt-Reed et al. 2021). In cases of directed signals (indicated by gaze), protophones were about twice as likely to be directed to a receiver as were gestures (36% vs. 16%).

Oller et al. (2021) took the early prominence of protophones to be evidence for the vocalization-first theory of language origin. They suggested that the adaptive value of these sounds was to project infant wellness to hominin caregivers who were occupied with other activities. One reason for this supposition is that a long period of infant helplessness must have placed more pressure on hominin caregivers. Second, Oller and colleagues assumed that hominins lived in increasingly larger groups than apes and that the alloparenting became a key factor in hominin life. This situation required the infant to broadcast its fitness status to a broader audience, which fits with the *Homo erectus* scenario of Swedell and Plummer (2019), based on hamadryas baboons (Chap. 9).

Other baboons provide an alternative view. Many baboons live in larger groups than most or all apes. In both troops and multilevel societies, the circle of acquaintances may number 200 or more. There is little or no alloparenting in this genus. The key issue, then, is infant helplessness. Two points can be made. First, it is not clear when in hominin evolution this became a crucial issue. Second, insofar as mothers needed help, the fact that male baboons provide a certain amount of care for infants and juveniles suggests that increased male attention to the young might have eased the hominin mother's burden. Protophones might have evolved to signal infant wellness to a concerned male that might otherwise be attending to other business.

11.5.2 Vowel-Like Sounds

Some protophones proved to be the basis for vowels, a vital aspect of language evolution (Boë et al. 2019). Vowels are the core of speech production and are essential to the acoustic value of consonants. This relationship makes possible the formation of morphemes (units of meaning) that can then be arranged into larger utterances. The basic sounds (phonemes) of a particular human language are a distinctive set and are culturally transmitted, but the phonemes in every language are drawn from a universal set of speech sounds that is based on the vocal anatomy and physiology of humans (Boë et al. 2017).

An influential theory (Lieberman et al. 1969) held that early hominins could not have made vowel sounds because the high position of the larynx in the vocal tract limited modification of the vocal tract shape by tongue, lip, or jaw maneuvers. This hypothesis was countered by the demonstration that human babies produce the same range of sounds as adults despite having a high larynx (De Boysson-Bardies et al. 1989). It can now be added that Guinea baboons make vowel-like sounds with a high larynx (Boë et al. 2017).

Boë et al. (2017) recorded spontaneous vocalizations of 15 adult Guinea baboons living in a captive social group. The study focused on five types of vocalizations that contained *formants*, that is, resonance bands that determine the phonetic quality of a vowel. The vocalizations included grunts and wahoos (produced mainly by males), barks and yaks (mainly by females), and copulation calls (only by females). After splitting the wahoos into two syllables, they identified about 1400 "vowel-like segments" (VLSs). A VLS was defined as any continuous part of a vocalization that contained "a consistent and detectable formant structure."

Acoustical analysis of the VLSs revealed at least five distinct classes, distinguished by different tongue positions. Tongue movements were both vertical and horizontal, as is the case in human languages. Two features of the communicative use of these segments also resembled human vowel functions. First, each of two VLSs occurred in two different calls (bark calls and wahoos; male grunts and female copulation calls). Second (in the case of the wahoo) the baboons consistently produced two different VLSs in succession within a single utterance. All of these features together suggest the kind of system that language evolved from, and that the beginning of that evolutionary process might have taken place in early hominins (Boë et al. 2017). The five vowel-like segments covered a large portion of the baboon's vocal space, in a proportion almost equivalent to that found (for instance) in 12-year-old native speakers of American English. Though recorded in captivity, the vocalizations in this study were highly similar to those already described in the wild (e.g., Maciej et al. 2013), which suggests that the analysis is pertinent to the natural behavior of Guinea baboons.

11.5.3 Vowels Versus Consonants

Gannon et al. (2023) compared the social effectiveness of proto-vowels and proto-consonants in an "open plains" environment, using playbacks of orangutan calls that were broadcast on a South African savanna. Their measurements indicated that only the consonant-like calls were effectively perceptible beyond 100 m under these conditions. Given the occupation of such habitats by human ancestors, the researchers inferred that consonants played an early role in the evolution of language.

A study of early hominin auditory capabilities suggests a somewhat different inference. Quam et al. (2015) studied the anatomy of the outer and middle ear in *Australopithecus africanus*. They interpreted the proportions of the bones to indicate "an increased emphasis on short-range vocal communication in open habitats." Vowel-like vocalizations such as baboon grunts may have been of the greatest importance for ongoing communication in these hominins. From this viewpoint, consonants probably became important later in hominin evolution as a basis for more complex communications. Selection for short-range communication might have been related to the formation of more compact groups for predator defense on the savannas.

11.6 The Gesture-first Theory

The gesture-first theory is a hypothesis of language origin that is often placed in opposition to the vocal-first theory. According to this school of thought, gestural communication was the foundation of language and the basic capabilities were later expanded to encompass vocalization. The term "gesture" is sometimes used almost synonymously with communication signals of any kind. However, gesture in other accounts is limited to arm and hand movements.

11.6.1 Human and Primate Evidence

Fay et al. (2022) summarized evidence for the importance of gesture in human life, with the implication that language originated in communication with manual motions. People gesture while speaking in every culture, blind people gesture, and hearing people can (in experiments) communicate successfully with gestures alone. Gestural languages, with the same expressive range as spoken language, emerge rapidly in populations of deaf children and in communities with a high incidence of deafness. Observations like this stimulated the origin of the gesture-first theory in the eighteenth century. A modern revival has been stimulated by scientific research.

Two experiments by Fay et al. (2022) tested the gesture-first theory against the vocal-first theory. In each experiment, one group of human participants recorded gestures or vocalizations that they invented in order to convey meanings specified by the researchers (conventional language was prohibited). A second group viewed the recordings and tried to guess the meanings. The viewers in both experiments were Australian undergraduates. One set of communicators were from the Pacific island of Vanuatu. The second set consisted of ten vision-impaired and ten sighted Australians. Communication success was twice as high for gestured signals than for the vocal signals within cultures, across cultures, and for participants who were severely vision-impaired. The researchers inferred support for the gesture-first theory.

Fay et al. (2022) noted that another reason for revitalization of the gesture-first theory is new evidence from primates. Comparative studies have demonstrated greater flexibility in primate gestures than in vocal signals; experiments have had greater success in teaching primates sign language than vocal language; and similarities have been observed between the naturalistic gestures produced by human children and other primates. Chimpanzees are prominent in all of this work (and famous for it), but baboons have a role to play that is explored in the following sections.

11.6.2 Flexibility and Intent in Gestural Communication

The previous chapter described research by Molesti et al. (2016). They studied spontaneous gestural communication (with a broad definition) in social groups of captive olive baboons for 1 year and recorded almost 9000 gestures that they classified into 67 gesture types. The majority of these types (39) were visual (58% of the repertoire) and only 4 types were audible. The prominence of visual signals here is consistent with the gesture-first theory. The researchers explained this imbalance in terms of baboon evolution in environments like those that were prominent in hominin evolution.

Fay et al. (2022), in arguing for the gesture-first theory, emphasized that primate studies demonstrated greater flexibility in gestures than in vocal signals (see above). In order to explore flexibility in the baboon communication system, Molesti et al. (2016) identified eight contexts in the social life of their subjects: affiliative, playful, agonistic, submissive, sexual, parental, grooming, and “other”. They found that several different gesture types were used in each context. For example, the baboons used about one-third of all types in the sexual and submissive contexts. Flexibility was also represented by the fact that most gesture types appeared in more than one context. On the average a single gesture type occurred in four different contexts.

With regard to another goal of the Molesti study, the baboons fulfilled "the main criteria of intentional communication" in that they performed goal-directed gestures to influence specific target individuals or audiences. This was manifested in four patterns of behavior: orientation toward the recipient, waiting for a response, sensitivity to the recipient's attention, and adjustment of a signal based on the recipient's attention. On average, the baboons performed about 90% of their gestures while looking at recipients; waited for a response in 87% of the interactions; and used 81% of their gestures when the target was paying attention. The baboons adjusted the modality of their gestures in response to lack of attention (for example, shifting from a visual signal to touching).

This systematic study shows at least one baboon species to be consistent with features of gestural communication in the great apes and some other monkeys. Sensitivity to the attention of signal recipients may be the most significant resemblance to findings from experiments with apes, other baboons, and some other monkeys. This growing body of evidence for primate intentionality suggests that intentionality was present in the interactions of early hominins, including but not limited to communicative behavior that may have laid the foundation for language.

11.6.3 Baboons in Gesture Experiments

Molesti et al. (2016) referred to the consistency of their findings with experimental research on baboons and a few other monkeys. This is exemplified by experiments in which olive baboons were taught to point to food rewards (raisins and banana slices), work that produced some conclusions not apparent in the Molesti study. Meunier et al. (2013b) taught the baboons to point to one baited container among others to get a reward from a human. Pointing and gazing varied according to the attentional status of the human (facing toward or away) and her ability to reach the reward. As in the baboons that were spontaneously communicating with each other (Molesti et al. 2016), the baboons in the experiment showed intentionality by their responsiveness to the attentional status of the target individual. The subjects were also able to respond to the added factor of the human's ability to reach the baited container. Since the baboons in this experiment indicated an object to the communication partner, the behavior was interpreted as referential (adding to the discussion in Chap. 10).

Bourjade et al. (2014, 2015, 2019) continued experiments with the olive baboons that had been trained to perform food requesting gestures. In these experiments the human target either faced the baboons or stood in profile to them (rather than with her back toward them). The subjects were (a) tested immediately after training, and (b) tested again 1 year later. Test conditions varied the human cues to attention.

In immediate testing, the profile group baboons gestured toward untrained cues regardless of their relevance for visual communication. They were also less discriminating toward trained versus untrained cues than baboons trained by a human facing them. In delayed testing of the profile group, the number of gestures toward meaningful untrained cues increased. They were able to discriminate the positions of an experimenter's body and adjusted their gestural communication accordingly. The experimenters inferred that intentional gestures tuned to the audience's attention may first develop through associative learning processes. Hard-wired predispositions for recognizing eye gaze as a necessary component of visual attention are apparently not present in olive baboons, at least in the context of interspecific communication.

Bourjade and colleagues suggested that the baboons first learned their gestures as tools in the sense of means to an end, and then turned them into *semiotic* tools (i.e., communicatively meaningful). To express the thesis in more detail, they suggested that the "typical" training or experience equipped the baboons with a set of tools (gestures and coordination with human cues to attention) and conditions (fluency, contingency, congruency of explicit training with implicit learning) that might then scaffold their ability for "understanding," that is, forming and reasoning about expectations and categories.

In sum, we have baboons that were able to learn communication skills that developed into meaningful entities in association with intentionality, audience awareness, joint attention, persistence, and elaboration—ultimately leading to a kind of understanding (Bourjade 2019; Lamaury et al. 2019). This seems to describe the sort of combination of domain-general and domain-specific functions that Fagot et al. (2018) saw as the underpinnings of language. The presence of such potential in baboons suggests that language might have developed very early in hominin evolution and also provides some support for the gesture-first theory.

11.7 Language and Laterality

Laterality of behavior and the brain are characteristic of humans. The great majority of humans have a strong hand preference, which is associated with a difference between the cerebral hemispheres. This includes certain brain structures with a functional connection to language. Baboon evidence suggests that this system might have evolved early in hominin evolution and that it might have been involved in communication from the beginning (Vauclair & Meguerditchian 2018).

11.7.1 Communicative Laterality Versus Manipulative Laterality

Humans are mainly right-handed for many actions, including gestures, and these tendencies are strongly linked to dominance of the left cerebral hemisphere for language functions. In a series of experiments, olive baboons displayed strong tendencies toward laterality in gestural communication. The first experiment reported population-level right-handedness in 60 captive baboons for a species-specific communicative manual gesture called *hand slap*, a threat that consists of quick and repetitive slapping or rubbing of the hand on the substrate. This study added baboons to captive chimpanzees and human children as subjects with a right-hand bias for communicative gesturing (Meguerditchian and Vauclair 2006).

Further research addressed the question of whether or not handedness was a function of the gesture's communicative nature (Meguerditchian and Vauclair 2009). This study focused on two behaviors that had not been previously investigated: a communicative gesture (*food beg*) and a noncommunicative self-touching behavior (*muzzle wipe*) that served as a control. *Food beg* displayed a trend toward right-handedness that significantly correlated with the preferences of the same individuals for *hand slap*. Hand preferences for *muzzle wipe* did not reveal any trend toward bias at the group level or correlation with hand preferences for *food beg* or *hand slap*. These findings were viewed as support for a hypothesized gestural communication system, based on left-hemisphere dominance in the brain, that differs from the system involved in purely motor functions.

Handedness for the *slap* gesture was robust and consistent across time and two study populations (Meguerditchian et al. 2011). Thirty baboons from the earlier experiment were retested for hand preference in the gesture 4 years later, by an observer unaware of the previous data. Twenty-six of them displayed significant continuity in handedness across the time period in question. Replication of the study in 96 novel individuals revealed a degree of population-level right-handedness similar to the one expressed in the first group of 66 subjects.

A closer link to humans was established by comparing baboons with human infants (Meunier et al. 2012). Researchers studied hand preferences for grasping objects or pointing to objects placed at several different spatial positions. In both species, right-hand preference was significantly stronger for the communicative task than for grasping objects. Noting that spatial location could have been a confounding factor in the preceding experiment, Bourjade et al. (2013) compared the consistency of individuals' hand preference with regard to spatial variation of a communicative partner and a food item to grasp. They found more consistent hand preference for communicative gestures than for grasping actions.

Meunier et al. (2013a) reviewed four studies investigating hand preferences for grasping versus pointing to objects at several spatial positions in human infants and three species of primates. There was a strong convergence in the distribution of hand biases for the two kinds of tasks among human infants, baboons, and macaques. Capuchins, a manipulative species of the Americas, diverged. The researchers inferred that left-lateralized language may be derived from a gestural communication system in the common ancestor of macaques, baboons, and humans. However, the close phylogenetic relationship between baboons and macaques weakens this argument. The pattern in question could be an analogy between the hominin and baboon/macaque lineages.

11.7.2 Language and Brain Laterality

The neurological inferences from experiments such as those cited above have been supported by direct investigation of baboon brain structures that correspond to language-related structures in humans (Meguerditchian et al. 2016). Among primates other than the great apes, the baboon is a good model for such research (Fagot et al. 2019) because the baboon brain is on average twice as large as those of other monkeys, including the closely related macaques. It also has greater *gyrification*, that is, the formation of folds in the cerebral cortex. These features are associated with structures that are homologs for those found in humans.

11.7.3 Wernicke's Area

Wernicke's area is a part of the human brain in the temporal lobe that is frequently associated with language. It is part of the auditory association cortex. In the left hemisphere it performs various functions in language processing (Becker and Meguerditchian 2022). A "bank of tissues" called the planum temporale (PT) is the most reliable "landmark" for quantification of this area (Hopkins 2022). Both the surface area of the PT and the volume of the underlying gray matter consistently display significant leftward asymmetry. The asymmetry is present in newborn infants and increases in association with language development.

Marie et al. (2018) used MRI images to quantify the PT area in 96 adult baboons and found population-level leftward asymmetry in size. The same asymmetry occurs in newborn olive baboons and, as in humans, differentiation increases during development (Becker et al. 2020). Baboons also display leftward bias in gray matter volume (Becker et al. 2022b). Such population-level asymmetries have not been found in rhesus macaques, bonnet macaques, or vervets (Hopkins 2022).

Compared to other primates, chimpanzees and (albeit with less evidence) baboons display "the most robust and consistent population-level asymmetry" of leftward bias in the planum temporale (Hopkins 2022). Studies of both species have used multiple research methods and different levels of analysis, applied to both surface area and gray matter volume. These findings suggest that asymmetry of the PT originated in the common ancestor of *Pan* and *Homo*, and was favored by the conditions in which baboons and early hominins evolved.

11.7.4 Other Brain Structures

Broca's area in humans was once considered the center of speech production. It is now known to have extensive connections in the language network of the brain (Becker and Meguerditchian 2022). Broca's area is involved with speech, gesture, syntax, and sign language. Comparison with other species is hampered by the fact that a homolog in other primates is difficult to discern. However, the inferior arcuate sulcus (IAS), which is part of Broca's area in humans, can be located in other primates. The "ventral portion and its depth" delimit the equivalent of the surface of Broca's area in the monkey brain (Meguerditchian et al. 2013).

Becker et al. (2022a) studied 50 olive baboons with in vivo anatomical MRI and found that communicative gesturing is related to the ventral portion of the inferior arcuate sulcus. Both direction and degree of gestural communication's handedness are associated with each other and correlated with contralateral depth asymmetry at this exact position. Baboons that prefer to communicate with their right hand have a deeper left-than-right IAS than those preferring to communicate with their left hand and vice versa. In contrast to handedness for object manipulation, gestural communication's lateralization is not associated with asymmetry in the depth of the central sulcus. This is consistent with previous work that found handedness for manipulative actions to be related to asymmetry in the central sulcus (Margiotoudi et al. 2019).

A variety of other language-related structures have homologs in baboons and perhaps in other primate species, especially chimpanzees. Becker and Meguerditchian (2022) name the planum parietale and the superior temporal sulcus among others. For example, significant depth asymmetry in favor of the right hemisphere was found in a specific portion of the superior temporal sulcus. The same asymmetry in the human brain is considered a landmark of communication and social cognition.

Becker and Meguerditchian (2022) hypothesized that asymmetries for language areas may not have initially evolved for language (cf. Fagot et al. 2019). Rather, each asymmetry could have evolved independently for different cognitive functions, to adapt to "unknown environmental pressures." This could explain the unclear relationship between structural and functional asymmetries related to language areas. As far as "unknown environmental pressures" are concerned, the study of wild baboon ecology may provide clarification in the future.

11.8 Language Learning

Some baboon experiments have been designed to investigate learning capabilities that might have underpinned the beginnings of language. Language combines abstract representations in a process called *compositionality*. This is a mental operation based on implicit recognition that the meaning of an expression is determined by its components and the rules that define their connection. Dautriche et al. (2022) tested Guinea baboons for a sense of compositionality. They chose negation as the key to the work because it is so fundamental to language and because forms of negation had previously been taught to some animals.

In the first experiment, the baboons learned to associate a cue with iconically related referents (e.g., a blue patch referring to all blue objects), and also to the complementary set associated with it (e.g., a blue patch referring to all *non*-blue objects). This was interpreted as showing the ability to comprehend negative compositional representations. In the second experiment the subjects learned to associate complex cues with the complementary object set. A complex cue in the second experiment was composed of the same cue as in the first experiment plus an additional visual element.

In related research, Chemla et al. (2019) administered a pattern extraction task to Guinea baboons. The results showed that the baboons are like humans in having a learning bias that helps them to discover connected patterns more easily than disconnected. For example, an implicit bias of this kind favors learning rules like "contains between 40% and 80% red" over rules like "contains about 30% red" or "100% red." The experimental task was made as similar as possible to a one that had previously been presented to humans, which was argued to reveal a bias responsible for shaping the lexicons of human languages. The baboon experiment involved subjects in a complex computer task that required learning of three rules of connectedness among icons. Of 23 voluntary participants, 9 failed the first condition and 9 eventually learned all three rules. The performance of some baboons in this task suggests that cognitive roots for regularities in the content and logic of human lexicons could have been present among early hominins.

11.9 Summary and Discussion

A theory about the evolutionary origin of language envisions multiple roots consisting of diverse components that ultimately came together to form a unique system of communication in humans. Some of these components of language were domain-general cognitive abilities that performed a variety of functions. These included memory, categorization, analogical reasoning, and detection of statistical regularities. Other functions and capabilities were domain-specific, that is, originally evolved to facilitate communication.

In humans and other primates, vocal and gestural skills are paramount. Primates, including baboons, are considered by many to provide clues to early hominin developments along these lines. Consistent with the domain-general view of language is the idea that animal cognition should be considered because it includes more concepts than animals are able to convey in their communication systems.

Domain-general and domain-specific functions are intertwined in the controversy about the primary platform for language evolution. Some scientists argue that vocalizations were the basis for language while others maintain that gestures provided the foundation. Baboons provide evidence relevant to both sides.

Research on the vocalization side is exemplified by the work on protophones in infants. These speech-like sounds are much more frequent than gestures during the first year of life and are twice as likely to be directed toward a recipient. Among the protophones of infants are vowel-like sounds, antecedents of the vowels that are crucial to the structure of spoken language. Baboons can pronounce vowel-like sounds despite throat anatomy that supposedly prevented it. Thus, sound articulation necessary for speech could have existed in very early hominins. This is a case where baboons seem to be more relevant than chimpanzees. Despite a low larynx like humans, chimpanzees are not known to produce vowel-like sounds. Baboons have a high larynx and yet have been shown to produce a variety of vowel-like sounds in their natural communication.

With respect to the adaptive significance of protophones, it was hypothesized that they signaled infant well-being (or lack of it) to caretakers among early hominins who were busy with other tasks. It was further speculated that protophones became increasingly important as group size increased because of the involvement of multiple caretakers in a cooperative system. Baboons provide valuable models for the postulated social situation in that troops tend to be large, and multilevel societies of even greater size and complexity occur in two species. Baboon societies illustrate various possible distractions from childcare that would have made vocal signals of well-being adaptive for caretakers and their young. The idea that larger hominin societies entailed "cooperative breeding" can be linked to the evolutionary scenario of Swedell and Plummer (2019), based on hamadryas baboons (Chap. 9).

Gesture-first theorists point to the ubiquity of gestures in human communication and the fact that gestures can convey meaning effectively without speech. In experiments that pitted gestures against vocalizations, communication success was twice as high for gestures between people from different cultures, and for vision-impaired recipients.

Baboons display a substantial repertoire of spontaneous communicative gestures that are flexible and variable. In both spontaneous behavior and experimental situations, baboon gestural communication displays the kind of domain-general functions that are vital to language: intentionality, persistence, and elaboration. They also manifest capabilities more specific to communication, such as awareness of a partner's attentional status. Experimenters postulated a sequence in which the baboons learned gestures as simple "tools" and transformed them into semiotic tools as a basis for reasoning. This could be analogous to the development of language in hominins, and the existence of such capabilities in baboons suggests that language, expressed through gestures, might have emerged at a very early stage in hominin evolution.

Some proponents of the gesture-first theory have done extensive work on handedness in gesturing and associated laterality of brain structures. They have found that baboons display a hand preference in gestural communication that does not appear in the manipulations of objects. The distinction also occurs in human infants with regard to pointing and grasping. In a comparative study, baboons and closely related macaques converged with human infants while capuchin monkeys diverged.

There is some evidence that connects baboon handedness with parts of the brain that are homologs with brain structures that belong to the language network in humans. This is particularly true of the planum temporale, which borders Wernicke's area. Chimpanzees and baboons are alone (so far) in displaying a robust leftward bias in the planum temporale, according to diverse analyses of surface area and volume.

Experiments have probed further into baboon cognitive capabilities that can be related to language. One such area is compositionality, the ability to combine abstract representations (e.g., not + blue). In two complex experiments, baboons demonstrated this ability with regard to the abstract concept of negation. Another study indicated that the baboons had a learning bias that facilitated the discovery of connected patterns, favoring cognitive rules that link two variables (e.g., *between x and y* rather than *approximately x* or *all x*). Taken together, this research suggests capabilities for abstraction in early hominins that would have provided the basis for syntax and lexical generation.

In some experiments, large numbers of baboons performed well. This suggests that the language components under consideration might have been common in early hominins. In other experiments, only a few baboons responded correctly to tests. This seems to suggest that the language components in question might have existed in a few early hominin individuals. These components would not have contributed to the communication system at that time, although they might have spread later and contributed to the evolution of language.

References

Barbet I, Fagot J. Processing of contour closure by baboons (Papio papio). J Exper Psychol Anim Behav Proc. 2011;37(4):407–419. https://doi.org/10.1037/a0025365.

Becker Y, Meguerditchian A. Structural brain asymmetries for language: A comparative approach across primates. Symmetry. 2022;14(5):876. https://doi.org/10.3390/sym14050876.

Becker Y, Sein J, Velly L, et al. Early left-planum temporale asymmetry in newborn monkeys (*Papio anubis*): a longitudinal structural MRI study at two stages of development. NeuroImage. 2020;227:117575. https://doi.org/10.1016/j.neuroimage.2020.117575.

Becker Y, Claidière N, Margiotoudi K, et al. Broca's cerebral asymmetry reflects gestural communication's lateralization in monkeys (*Papio anubis*). eLife Sciences. 2022a;11. https://doi.org/10.7554/eLife.70521.

Becker Y, Loh KK, Coulon O, Meguerditchian A. The Arcuate Fasciculus and language origins: disentangling existing conceptions that influence evolutionary accounts. Neurosci Biobehav Rev. 2022b;134:104490. https://doi.org/10.1016/j.neubiorev.2021.12.013.

Boë LJ, Berthommier F, Legou T, et al. Evidence of a vocalic proto-system in the baboon (*Papio papio*) suggests pre-hominin speech precursors. PLoS One. 2017;12:e0169321. https://doi.org/10.1371/journalpone.0169321.

Boë LJ, Sawallis TR, Fagot J, et al. Which way to the dawn of speech? Reanalyzing half a century of debates and data in light of speech science. Sci Adv. 2019;5:eaaw3916.

Bourjade M, Meunier H, Blois-Heulin C, Vauclair J. Baboons' hand preference resists to spatial factors for a communicative gesture but not for a simple manipulative action. Dev Psychobiol. 2013;55(6):651–61. https://doi.org/10.1002/dev.21128.

Bourjade M, Meguerditchian A, Maille A, et al. Olive baboons (*Papio anubis*) adjust their visual and auditory intentional gestures to the visual attention of others. Anim Behav. 2014;87:121–8.

Bourjade M, Canteloup C, Meguerditchian A, et al. Training experience in gestures affects the display of social gaze in baboons' communication with a human. Anim Cogn. 2015;18:239–50. https://doi.org/10.1007/s10071-014-0793-5.

Bourjade M, Gaunet F, Maugard A, Meguerditchian A. Manipulating social cues in baboon gesture learning: what does it tell us about the evolution of communication? Anim Cogn. 2019;22:113–25. https://doi.org/10.1007/s10071-018-1227-6.

Bovet D, Vauclair J. Functional categorization of objects and of their pictures in baboons (Papio anubis). Learning and Motivation. 1998;29(3):309–322. https://doi.org/10.1006/lmot.1998.1009.

Burkhardt-Reed MM, Long HL, Bowman DD, et al. The origin of language and relative roles of voice and gesture in early communication development. Inf Behav Dev. 2021;65:101648. https://doi.org/10.1016/j.infbeh.2021.101648.

Campbell KL, Tyler LK. Language-related domain-specific and domain-general systems in the human brain. Curr Opin Behav Sci. 2018;21:132–7. https://doi.org/10.1016/j.cobeha.2018.04.008.

Chemla E, Dautriche I, Buccola B, Fagot J. Constraints on the lexicons of human languages have cognitive roots present in baboons (*Papio papio*). Proc Natl Acad Sci USA. 2019;116:14926–30. https://doi.org/10.1073/pnas.1907023116.

Cheney DL, Seyfarth RM. The evolution of language from social cognition. Curr Opin Neurobiol. 2014;28:5–9.

Cheney DL, Seyfarth RM. The origin of meaning in animal signals. Anim Behav. 2016;124:1–8.

Cheney DL, Seyfarth RM. Pragmatic flexibility in primate vocal production. Curr Opin Behav Sci. 2018;21:56–61.

Christie S, Gentner D. Language helps children succeed on a classic analogy task. Cogn Sci. 2013;38:383. https://doi.org/10.1111/cogs.12099.

Dautriche I, Buccola B, Berthet M, et al. Evidence for compositionality in baboons *(Papio papio)* through the test case of negation. Sci Rep. 2022;12:19181. https://doi.org/10.1038/s41598-022-21143-1.

De Boysson-Bardies B, Halle P, Sagart L, Durand C. A cross-linguistic investigation of vowel formants in babbling. J Child Lang. 1989;16:1–17.

Dépy D, Fagot J, Vauclair J. Comparative assessment of distance processing and hemispheric specialization in humans (Homo sapiens) and baboons (Papio papio). Brain Cogn.1998;38:165e182. https://doi.org/10.1006/brcg.1998.1027.

Dépy D, Fagot J, Vauclair J. Processing of above/below categorical spatial relations by baboons (Papio papio). Behav Proc. 1999;48: 1–9. https://doi.org/10.1016/S0376-6357(99)00055-8.

Fagot J. Orthographic processing in animals: implications for comparative psychologists. Learn Behav. 2017;45:209. https://doi.org/10.3758/s13420-017-0267-0.

Fagot J, Cook R. Evidence for large long-term memory capacities in baboons and pigeons and its implications for learning and the evolution of cognition. Proc Nat Acad Sci. 2006;103:17564–7.

Fagot J, de Lillo C. A comparative study of working memory: immediate serial spatial recall in baboons (*Papio papio*) and humans. Neuropsychologia. 2011;49:3870–80.

Fagot J, Thompson RKR. Generalized relational matching by Guinea baboons (*Papio papio*) in two by two-item analogy problems. Psychol Sci. 2011;22:1304–9.

Fagot J, Wasserman E, Young M. Discriminating the relation between relations: the role of entropy in abstract conceptualization by baboons and humans. J Exper Psychol Animal Behav Proc. 2001;27:316–28.

Fagot J, Malassis R, Medam T, Montant M. Comparing human and nonhuman animal performance on domain-general functions: towards a multiple bottleneck scenario of language evolution. In: Boë JL, Fagot J, Perrier P, Schwartz JL, editors. Origins of human language: continuities and splits with nonhuman primates. Oxford: Peter Lang; 2018. p. 329–65.

Fagot J, Boë L-J, Berthomier F, et al. The baboon: a model for the study of language evolution. J Hum Evol. 2019;126:39–50. https://doi.org/10.1016/j.jhevol.2018.10.006.

Fay N, Walker B, Ellison TM, et al. Gesture is the primary modality for language creation. Proc R Soc B. 2022;289:20220066. https://doi.org/10.1098/rspb.2022.0066.

Fitch TW. Animal cognition and the evolution of human language: why we cannot focus solely on communication. Philos Trans R Soc B. 2019;375:20190046. https://doi.org/10.1098/rstb.2019.0046.

Gannon C, Hill RA, Lameira AR. Open plains are not a level playing field for hominid consonant-like versus vowel-like calls. Sci Rep. 2023;13:21138. https://doi.org/10.1038/s41598-023-48165-7.

Goujon A, Fagot J. Learning of spatial statistics in nonhuman primates: contextual cueing in baboons (*Papio papio*). Behav Brain Res. 2013;247:101–9.

Grainger J, Dufau S, Montant M, et al. Orthographic processing in baboons (*Papio papio*). Science. 2012;336:245–8.

Hannagan T, Ziegler JC, Dufau S, et al. Deep learning of orthographic representations in baboons. PLoS One. 2014;9:e84843. https://doi.org/10.1371/journalpone.0084843.

Hopkins WD. Neuroanatomical asymmetries in nonhuman primates in the homologs to Broca's and Wernicke's areas: a mini-review. Emerg Top Life Sci. 2022;6(3):271. https://doi.org/10.1042/ETLS20210279.

Lamaury A, Cochet H, Bourjade M. Acquisition of joint attention by olive baboons gesturing toward humans. Anim Cogn. 2019;22:567–75. https://doi.org/10.1007/s10071-017-1111-9.

Lieberman PH, Klatt DH, Wilson WH. Vocal tract limitations on the vowel repertoires of rhesus monkey and other nonhuman primates. Science. 1969;164:1185–7.

Maciej P, Ndao I, Hammerschmidt K, Fischer J. Vocal communication in a complex multi-level society: constrained acoustic structure and flexible call usage in Guinea baboons. Front Zool. 2013;10:58. https://doi.org/10.1186/1742-9994-10-58.

Margiotoudi K, Marie D, Claidière N, et al. Handedness in monkeys reflects hemispheric specialization within the central sulcus. An in vivo MRI study in right- and left-handed olive baboons. Cortex. 2019;118(5):203. https://doi.org/10.1016/j.cortex.2019.01.001.

Marie D, Roth M, Lacoste R, et al. Left brain asymmetry of the planum temporale in a non hominid primate: redefining the origin of brain specialization for language. Cereb Cortex. 2018;28:1808–15.

Meguerditchian A, Vauclair J. Baboons communicate with their right hand. Behav Brain Res. 2006;171:170–4.

Meguerditchian A, Vauclair J. Contrast of hand preferences between communicative gestures and noncommunicative actions in baboons: implications for the origins of hemispheric specialization for language. Brain Lang. 2009;108:167–74.

Meguerditchian A, Molesti S, Vauclair J. Right-handedness predominance in 162 baboons for gestural communication: consistency across time and groups. Behav Neurosci. 2011;125:653–60.

Meguerditchian A, Vauclair J, Hopkins WD. On the origins of human handedness and language: a comparative review of hand preferences for bimanual coordinated actions and gestural communication in nonhuman primates. Dev Psychobiol. 2013;55:637–50.

Meguerditchian A, Marie D, Love SA et al (2016) Human-like brain specialization in baboons: an in vivo anatomical MRI study of language areas homologs in 96 subjects. In: Roberts SG et al (eds) The evolution of language: proceedings of the 11th international conference (EVOLANG11). https://doi.org/10.17617/2.2248195.

Meunier H, Vauclair J, Fagard J. Human infants and baboons show the same pattern of handedness for a communicative gesture. PLoS One. 2012;7:e33959. https://doi.org/10.1371/journalpone.0033959.

Meunier H, Fagard J, Maugard A, et al. Patterns of hemispheric specialization for a communicative gesture in different primate species. Dev Psychobiol. 2013a;55:662–71. https://doi.org/10.1002/dev.21151.

Meunier H, Prieur J, Vauclair J. Olive baboons communicate intentionally by pointing. Anim Cogn. 2013b;16:155–63. https://doi.org/10.1007/s10071-012-0558-y.

Molesti S, Vauclair J, Meguerditchian A. Hand preferences for unimanual and bimanual coordinated actions in olive baboons (*Papio anubis*): consistency over time and across populations. J Comp Psychol. 2016;130:341–50.

Oller DK, Ramsay G, Bene E, et al. Protophones, the precursors to speech, dominate the human infant vocal landscape. Philos Trans R Soc B. 2021;376:20200255. https://doi.org/10.1098/rstb.2020.0255.

Price T, Wadewitz P, Cheney D et al. Vervets revisited: A quantitative analysis of alarm call structure and context specificity. Sci Rep. 2015;5:13220. https://doi.org/10.1038/srep13220

Prieur J, Barbu S, Blois-Heulin C, Lemasson A. The origins of gestures and language: history, current advances and proposed theories. Biol Rev. 2020;95:531–54. https://doi.org/10.1111/brv.12576.

Quam R, Martinez I, Rosa M, et al. Early hominin auditory capacities. Sci Adv. 2015;1:e1500355. https://doi.org/10.1126/sciadv.1500355.

Rodriguez JS, Zvrcher NR, Barlett TQ, et al. CANTAB delayed matching to sample task performance in juvenile baboons. J Neurosci Methods. 2011;196:258–63.

Swedell L, Plummer T. Social evolution in Plio-Pleistocene hominins: Insights from hamadryas baboons and paleoecology. J Hum Evol. 2019;137:1–15. https://doi.org/10.1016/j.jhevol.2019.102667.

Vauclair J, Fagot J. Categorization of alphanumeric characters by Guinea baboons: Within- and between-class stimulus discrimination. Cahiers de Psychologie Cognitive. 1996; 15(5):449–462.

Vauclair J, Meguerditchian A. The gestural origin of language and its lateralization: theory and data from studies of nonhuman primates. In: Kern S, Gayraud F, Marsico E, editors. Emergence of linguistic abilities: from gestures to grammar. Newcastle upon Tyne: Cambridge Scholars Publishing Ltd; 2018. p. 43–59.

Chapter 12
Cognition

12.1 Introduction

Baboons may offer some potential insights into the mental capabilities, and even some specific mental processes, of early hominins. Behaviors of wild and captive baboons suggest relevant hypotheses, and recent experiments have explicitly addressed baboon–hominin comparisons. Baboons are significant in this regard for at least two reasons. First, they provide a baseline or minimum case for a number of hominin cognitive functions. If baboons can do something, it may be that even the earliest hominins had similar capability. Second, baboons have adapted their behavior in ways that are strongly parallel to hominins under the same conditions: diverse and increasingly dry habitats; and variation from troops to complex multilevel societies. These parallel contexts may have selected for similar cognitive mechanisms. Various aspects of cognition played important roles in the previous two chapters on nonlinguistic communication and language. This chapter elaborates on some of them and adds others.

12.2 Quantification

Quantification takes various forms in animals. It may be straightforward counting or it may be more abstract accretion of information, as in probability and statistics. Quantification facilitates responses to a variety of environmental and social situations. Research with baboons was the first to suggest that humans and other animals naturally experience similar environmental pressures for representing small numerical values. This is important because environmental pressure to represent small numerosities is a possible causal factor in the evolution of cognitive systems for numerical representation (Piantadosi and Cantlon 2017).

G. E. King, *Baboon Perspectives on Early Human Ancestors*,
https://doi.org/10.1007/978-3-031-36769-4_12

Cantlon et al. (2015) reported that two captive olive baboons exhibited cognitive ability that was algorithmically and logically similar to human counting. The subjects were given the task of choosing between two food caches. First, they saw one cache baited with a number of food items, one item at a time. Then the process was repeated with a second cache. At the point when the second set was approximately equal to the first set, the baboons spontaneously moved to choose the second set even before that cache was completely baited. The researchers presented an analysis with the conclusion that the baboons compared quantities by using an approximate counting algorithm that was incremental, iterative, and condition-controlled. This "proto-counting" algorithm was argued to be structurally similar to formal counting in humans and thus representative of an evolutionary precursor to human counting.

A different experiment with caches tested one-to-one correspondence in captive olive baboons (Koopman et al. 2019). This is a logical rule that is important in counting and representing exact number. It is the understanding that two sets are equal if each item in one set corresponds to exactly one item in the second set. The baboons were given a quantity discrimination task where two caches (unlike the previous experiment) were baited with *different* quantities of food. When the quantities were baited in a manner that highlighted the one-to-one relation between them, the baboons performed significantly better than when one-to-one correspondence cues were not provided. Early hominins may have had such intuitions about numerical equality. This mental ability could have facilitated social negotiations about matters such as dividing a number of small prey animals that were perceived as additive units.

Piantadosi and Cantlon (2017) used data from the troop movements of olive baboons in a study cited below (Strandburg-Peshkin et al. 2015, in "Initiation and Direction of Movement") to compare various models of quantitative decision-making. Their analysis led to the conclusion that numerical representation is the mechanism by which baboons cognitively track and tally votes during social decision-making. The decisions made by these naturally behaving animals relied specifically on numerical representations that may have key homologies (or analogies) with human number representations.

These findings suggest the types of problems that shaped the evolution of hominin numerical cognition, in this case to monitor social behavior during collective movements. Democratic decision-making is thus one utility of numerical representation in the primate lineage and may have been of special importance to early hominins as they began to traverse open country in large troops. All of this research points to sensitivity to numerical value in baboons that is comparable to that of humans. By at least one measure the natural numerical sensitivity of wild baboons is comparable to that of 3-year-old human children (Piantadosi and Cantlon 2017).

Bryer et al. (2021) included two baboon species in a survey of quantitative sensitivity in 49 studies that covered 33 bird and mammal species. They used a "novel Bayesian model that combined phylogenetic regression with a model of number psychophysics and random effect components" to calculate the Weber fraction for each species. The Weber fraction is a measure of a subject's precision in cognitive quantity representation. Lower fractions indicate better performance in discriminating quantities. For example, 0.2 means that an animal can reliably discriminate 10 versus 12 items. Olive baboons scored 0.97 and hamadryas baboons scored 0.63. The hamadryas score was better than that of gorillas (0.76) and two macaque species and matched the performance of capuchin monkeys.

These results suggest a significant degree of quantity representation in baboons. They also seem to suggest a difference between olive baboons (0.97) and hamadryas baboons (0.63), and therefore (possibly) another case of two alternative *Papio* analogies for early hominins. Bryer and colleagues cited a study of olive baboons as an example of the abilities they were trying to measure (Strandburg-Peshkin et al. 2015).

One more conclusion of this research was that quantitative sensitivity in primates was positively correlated with domain-general cognition scores. Bryer and colleagues suggested that quantitative precision might involve both generalized and specialized functions. Whatever the particular neural pathways, early hominins probably counted the numbers in various aggregates, perhaps a quantity of small animals killed or potential enemies in a neighboring troop.

12.3 Reasoning by Exclusion

Baboon capability for reasoning by analogy was discussed in the previous chapter in connection with its role in language. There is also evidence that baboons are capable of inferential reasoning by exclusion. Schmitt and Fischer (2009) tested olive baboons in object choice experiments with varied information given to the subjects about food located in 1 of 2 cups. All seven subjects located the reward when they had previously seen it with both cups open. With partial information (only one cup was opened), four baboons apparently inferred the location of the reward by reasoning. The other three may have adopted the alternative strategy of avoiding the empty cup.

Marsh et al. (2015) extended knowledge of this capability by testing a hamadryas baboon along with a member of the closely related macaque genus and two New World species. They used the traditional two-way object choice task and a more complex three-way object choice task. Both Old World species were able to infer by exclusion in the three-object task while the New World species did poorly. A specific application of this information to the behavior of early hominins is not apparent. However, it is further evidence for diverse cognitive abilities in early hominins, perhaps in relation to domain-general functions.

12.4 Cognitive Flexibility

Flexibility and innovation were crucial features of hominin evolution. The significance of baboons in this regard is suggested by an experiment that compared hamadryas baboons with rhesus macaques (*Macaca mulatta*). The subjects were naive immatures of both sexes from both species that had been raised in group cages (Anikaev et al. 2020). They were given a test battery aimed at evaluating general cognitive ability. Specifically, they were given tests for exploratory activity and learning ability. More baboons than rhesus were active in performing the tasks and they were less likely to decrease activity as the tasks became more complex. The higher level of activity was strongly correlated with test success. The researchers noted that these results were consistent with earlier findings on manipulative ability and attributed the differences between the species to behavioral plasticity and adaptability. They postulated that these qualities contributed to the evolutionary success of baboons. Similar qualities could have meant evolutionary success for early hominins as they evolved under conditions much like those facing baboons.

Cognitive flexibility in nonhuman primates is traditionally measured in the laboratory with the conceptual set shifting task (CSST). Gullstrand et al. (2022) used this task to continuously test 24 Guinea baboons over a period of about 10 years. The task involved the presentation of three stimuli on a touch screen, all made from three possible colors and three shapes. The subjects had to touch the stimulus containing the stimulus dimension (e.g., green) that was constantly rewarded until the stimulus dimension changed. Analysis of perseveration responses, scores, and response times collected during the last 2 years of testing (approximately 1.6 million trials) indicated that the baboons had developed an "expert" form of cognitive flexibility.

Another important finding was that the performance was age-dependent: developing in juveniles, optimal in adults, declining in middle age, and strongly impaired in the oldest group. A direct comparison with data collected for some of the same baboons performing the same task revealed that the performance of all age groups had improved after 10 years of training, including those now old. Gullstrand and colleagues inferred that cognitive flexibility in humans has a long evolutionary history.

A *cognitive set* can be either helpful or harmful in problem solving. A large set of similar problems may be solved mechanically by applying a single-solution method. However, efficiency might be sacrificed if a better solution exists and is overlooked. These alternatives were the subject of a nonverbal computer task to compare 104 humans and 15 Guinea baboons (Pope et al. 2015). A substantial difference was found in the ability to break cognitive set. The majority of humans were highly impaired by set while the baboons were almost completely unaffected. Analysis of the human data revealed that children (aged 7–10) were significantly better able to break set than adolescents and adults. Flexibility like that of baboons might have been adaptive for early hominins. The deficit in the human response might come from an evolutionary decline due to changing circumstances, or from ontogenetic decline due to lack of use in those changed circumstances.

12.5 Exploratory Tendencies

The previous section alluded to testing of immature hamadryas baboons for exploratory tendencies (Anikaev et al. 2020). Exploratory tendencies were also tested in adult male hamadryas baboons, as well as adult males of three other monkey species: green monkeys (*Cercopithecus aethiops*) and two macaque species (Anikaev et al. 2021; 2023a, b). Hamadryas baboons and long-tailed macaques displayed the highest levels of exploratory activity with regard to tactile, oral, visual, and olfactory contacts with the target object (a multicolored plastic cube). Such exploratory behavior among early hominins could have led to tool using and tool-making innovations (long-tailed macaques are unusual among monkeys in the extent of tool use, such as cracking open shellfish for food). It should be noted that the species in the tests varied widely in the number of participants: hamadryas 57, long-tailed macaques 14, rhesus macaques 23, and green monkeys 11.

In the experiment just cited, hamadryas baboons and long-tailed macaques displayed the greatest amount of variation among individuals. In a more limited experiment with skill learning by mature males, hamadryas baboons demonstrated higher learning ability, training level, and repeatability than rhesus macaques. However, the species were similar in having a relatively low level of intraindividual variation in learning ability. The combined results of these experiments seem to suggest a highly speculative scenario in which innovation was led by relatively few individuals but readily learned by many others.

Some additional dimensions of exploratory tendencies, in a different *Papio* species under more natural conditions, emerged from a field experiment with chacma baboons by Martina et al. (2021). They quantified individual differences under varied conditions as to the probability of exploring two types of stimuli for two cognitive tests: a set of colored paper bags (in an associative learning test) and a blue cardboard square (in a second-order conditioning test). Results for both stimuli were similar in two respects. First, juveniles were more likely than others to contact the stimuli and they explored them for longer periods of time. Second, individuals involved in another activity at the time of testing were less likely to contact the stimuli and had lower exploration times than others.

Juveniles and females with a high propensity to use social information (i.e., scrounging food from others) were more likely to contact the paper bags and had shorter latencies than adult males. These results suggest that any tests of exploratory behavior (and inferences to early hominins) should include all age-sex categories, time available from other activities, and types of stimuli.

12.6 Two More Experiments

Two other important aspects of general cognition in baboons are represented by one experiment each. *Metacognition* is the ability of an organism to evaluate its own states of knowledge and engage in appropriate information-seeking when lack of knowledge is detected. This capability would probably have been highly adaptive for early hominins and baboons experiencing diverse and changing environmental conditions. Malassis et al. (2015) assessed metacognition in three Guinea baboons by requiring them to use a touchscreen to report on the location of two target stimuli that had previously appeared briefly on a grid. They could either respond directly or use a "Repeat" key to review the target stimuli. In Experiment 1, the baboons used the Repeat key more frequently in difficult trials and transferred this use of the key to novel conditions. Two baboons showed higher accuracy when they declined using the key compared to baseline trials in which the key was not available, suggesting accurate metacognitive monitoring judgments. The same two baboons were consistently faster at reporting the targets' locations after a repetition of the stimulus. Allowing for individual variation, results support the hypothesis of metacognition in some baboons.

The *affect-as-information hypothesis* predicts that the positive or negative valence of mood differentially affects processing of environmental details. It was originally formulated in relation to human learning. Marzouki et al. (2014) tested Guinea baboons for this factor in learning. Six Guinea baboons, living in a social group, had free access to a computerized visual search task. Trials that immediately followed spontaneously expressed emotional behavior were analyzed. Response times following negatively valenced behaviors were slower than those following neutral and positively valenced behaviors, respectively. Thus, moods affected baboon performance in highly automated tasks in the same way as in humans during tasks with much higher cognitive demands. The researchers postulated a common adaptive mechanism by which moods influence performance in various ecological contexts. The baboon analogy suggests that this human mental bias has roots in early hominin behavior.

12.7 Ecological Factors in Cognition

A comparative study by DeCasien et al. (2017) covered a large sample of primates (more than 140 species). They arrived at two general conclusions that have special relevance to baboons as models for early hominins. One has to do with diet. There is a key difference between strepsirrhines ("lower" primates) and haplorrhines ("higher" primates). The latter group includes hominins and baboons. Among haplorrhines, omnivorous species tend to have larger relative brain size than do folivores. The researchers attributed this to the fact that omnivorous haplorrhines often hunt vertebrates as well as collecting invertebrates. Their example was the capuchins of the Americas; however, baboons (along with chimpanzees) hunt larger and sometimes more elusive prey than any other primates (Chap. 5).

DeCasien and colleagues also postulated an important role for environmental variation. They cited studies indicating a positive correlation between relative brain size and survival in mammal species that were introduced to new environments and inferred that "long-term environmental variability could select for behavioral versatility." Baboons and early hominins in Africa experienced geographic and temporal variability across a similarly wide range of environments (Chap. 2) and both taxa are noteworthy for behavioral flexibility.

12.7.1 Foraging and Home Range

DeCasien and colleagues suggested that foraging for fruit affects brain size, since frugivores tend to have larger brains than folivores across all primates. The feeding ecology of forest baboons suggests that frugivory might have had some impact on the earliest hominins (Chap. 4). However, frugivory was unlikely to be a major factor for hominins or baboons as they expanded into increasingly open and dry habitats where they would have encountered less fruit. Cognitive abilities were more likely to be selected by the need to exploit plant foods that were more difficult to obtain and process, such as underground storage organs (Chap. 4).

A broad survey of primates by Powell et al. (2017) supported DeCasien and colleagues in finding stronger and more robust associations with ecological factors related to foraging than with social patterns. However, inclusion of additional variables and datasets led to the conclusion that home range size rather than diet was the most consistent correlate of brain size. This tentative conclusion was complicated by variation between datasets, suggesting that the effects of diet and home range are hard to separate. The researchers postulated that diet and ranging might form an adaptive "syndrome" because more frugivorous diets are strongly associated with more patchily distributed resources and larger home ranges. Here again, forest baboons fit the pattern, but the expansion of baboons and hominins brought them into contact with different pressures. Analogy with savanna baboons indicates a continuation of large home ranges in early hominins but in relation to the distribution of different foods (Chap. 4).

Vidal-Cordascoa et al. (2020) argued that home range area does not reflect time spent in moving and, therefore, leaves this question open: Could the effort involved in movement have affected brain size evolution in primates. They applied a large comparative dataset of extant primate species and phylogenetic comparative methods to the problem. The specific aim was to test the influence of daily movement on primate brain sizes while controlling for other behavioral and ecological factors. The study produced a significant correlation between daily movement and brain mass that was not explained by home range, diet, social group size, or body mass. The researchers concluded that, viewed on an evolutionary timescale, longer daily movement distance is not a constraining factor for the energetic investment in a larger brain. Instead, increased mobility could have contributed to the evolution of increased brain mass. This conclusion seems particularly relevant to the evolution of baboons and hominins in open habitats.

12.8 Manipulation of Stones

In relation to the origins of hominin tool use, chimpanzees and a few monkey species have (rightfully) received a great deal of attention. However, baboons have been reported to manipulate stones in relevant patterns that differ from those of the other tool-using primates. These baboon behaviors are all the more interesting in light of recent suggestions that the earliest use and manufacture of stone tools by hominins (in the Oldowan cultures) may have been less complex than is usually assumed.

12.8.1 Stone Tools in the Early Oldowan

Shea (2017) argued that lithic evidence dating before 1.7 mya reflects only occasional stone tool use, "much like that practiced by nonhuman primates." He noted the production of flake tools as a distinction between hominin and primate stone manipulation; however, this gap has narrowed with the report of flakes like those in the Oldowan produced by the stone using activities of some monkeys (Proffitt et al. 2023). Shea placed the onset of habitual stone tool use at about 1.7 mya, perhaps correlated with increased hominin "logistical mobility" (possibly carrying things such as raw materials or carcasses).

Plummer and Finestone (2018) examined remains from archeological sites dating between 2.6 and 2 mya and reached conclusions consistent with Shea's argument. They perceived a shift from low-density artifact scatters in a narrow set of depositional contexts to denser concentrations that included abundant fossils. The earlier sites seem to represent briefer periods of occupation and perhaps more variation in competence. Raw materials display less selectivity and were transported over shorter distances (usually a few meters and never more than 100 m). The capabilities of hominins before this time period were presumably more limited. Baboons are among the primates that show us what might have been happening in that earlier period.

12.8.2 Stone Manipulation by Baboons in Feeding

Wild baboons display sporadic object manipulations, some not seen in other primates, that suggest precursors to tool use in early hominins. Some chacmas selectively overturn rocks to feed on underlying invertebrates, moving certain rocks and ignoring others. Postulating a cost–benefit trade-off, Mare et al. (2021) investigated the choices made by these "clever baboons." Using optimal foraging theory, they predicted that the baboons would prefer to move medium-sized rocks, a trade-off between the effort needed for larger rocks and insufficient prey under smaller rocks. Recording and analysis of 666 rocks moved by baboons and 619 unmoved rocks verified the hypothesis: the baboons preferred medium-sized rocks. However, the researchers were surprised to find that the baboons did not prefer rounded rocks that could be rolled in order to use less energy than needed to flip flat rocks. Instead, the baboons lessened the needed effort by flipping flatter rocks along the shortest axis. As an additional aid, the baboons tended to move rocks downslope (Mare et al. 2019).

Goodall et al. (1973) observed one instance in which an olive baboon picked up a "fairly large stone" and used it to "forcefully rub" her muzzle to rid it of sticky juice from seeds that she had eaten. This was a modified version of a common behavior in the troop: the baboons often rubbed their mouths against large rocks or tree trunks. Similarly, one baboon used "a piece of maize kernel" to wipe blood from his mouth. This paper also mentioned earlier reports of baboon tool use for which no details had been given. Baboons in one wild troop used rocks to break open hard-shelled fruits and one baboon crushed a scorpion with a stone and then ate it.

12.8.3 Use of Stones As Weapons by Baboons

One form of baboon tool use provides a good example for the idea that "natural history" in the form of anecdotes should not be ignored (Strum 2019). After years of sporadic accounts from diverse sources, primatologists reported that three troops in southern Africa dropped and threw stones at them from the tops of cliffs on numerous occasions after moving to positions in relation to the observers that indicated intent (Hamilton et al. 1975). Individual baboons threw multiple stones and some worked vigorously to free rocky material from the canyon walls after loose stones had been exhausted. The size of the stones suggests selection for the purpose. Mean size for a measured sample was about 6.5 by 4 inches (~16.5 cm × 10 cm) and mean weight was about 3 ounces (~90 g). The report from southern Africa elicited accounts from other scientists about similar incidents in Kenya (Pickford 1975) and Sudan (Pettet 1975).

Critics of the original report argued that the humans were too far away to be struck by the stones, and they pointed out that some baboons throw sticks during displays without understanding their potential as weapons (Cheney and Seyfarth

2007). However, Pettet (1975) stated that his baboons displayed good aim much of the time. Agitation of the baboons at the presence of perceived enemies might account for some of the errant throws. Some of the underlying mechanisms for this moving and rolling of rocks might be the same as, or similar to, the movement of rocks for subsistence purposes, including the downslope orientation (see above). Some further support for aimed throwing by baboons comes from 12 reports of such behavior in zoos (Kortlandt and Kooij 1963, cited by Goodall et al. 1973).

12.8.4 Possible Construction of Terrestrial Shelters by Early Hominins

Isbell et al. (2018) raised the question of how hominins protected themselves at night when they eventually had to sleep on the ground, and suggested that they surrounded themselves with *bomas* constructed from thorny branches. Although chimpanzee nesting techniques seem highly relevant to such a speculation, baboon use of stones suggests a complementary factor. As noted in the previous section, baboons sometimes use stones in what may be intentional defense against presumptive predators (humans). If early hominins chose hills or other high points for ground sleeping, they might have deterred predators with rocks thrown from behind their shelters. In yet another case of complementary evidence, chimpanzees have been shown to use hills as observation points for watching neighboring communities. Early hominins might have taken the next step by constructing nighttime shelters on the hills to give them the advantage over predators. To make one more speculative leap: accumulated stones not needed as ammunition might have been used to augment or anchor the branches of the bomas.

12.9 Group Travel Decisions

To survive in complex and changing environments, early hominins must have had an effective grasp of space and time. An issue that seems like almost equal parts ecological and social cognition is the matter of travel decisions for the group. Whether living in troops or multilevel societies or in both successively, early hominins must have traveled in groups.

Conflicts of interest about where to go and what to do were primary challenges of group living. Did early hominins need "strong leaders" or were their decisions "democratic" in some sense? How were decisions transmitted throughout the group? What are the differences, if any, between initiation of travel and group organization during travel? Were the cognitive mechanisms sophisticated or did simple ones suffice?

Questions like these bear on the possible evolution of hominin patterns that may have important implications for modern humans. Relevant spatial comprehension includes a general grasp of surroundings and awareness of particular goals, such as concentrations of food. Retention of such knowledge requires effective memory. These are cognitive issues, whether travel direction is a group phenomenon or determined by one or a few individuals.

12.9.1 Cognitive Maps

A *cognitive map* is a mental representation of the environment. Summarizing a great deal of earlier work by many scientists, de Raad and Hill (2019) described two kinds of mental map found in animals and applied these concepts to the chacma baboons that they studied. The *topological map* is a route-based view of the environment that links various features (e.g., food trees) by learned travel routes. The *Euclidean map* is a quantitative representation that contains more detailed information about the spatial relations among landmarks. It allows calculation of direction and distance to places that are not in view and facilitates the use of direct routes, as well as the ability to take novel routes and shortcuts.

Chacma baboons in the study by de Raad and Hill (2019) were like other primates in that their foraging movements were goal-directed and guided by a cognitive map. Their behavior clearly indicated topological spatial awareness in at least two respects (de Raad and Hill 2019). First, they traveled through a dense network of repeated routes, approaching their goals from a small number of the same directions. Second, when leaving one site the initial travel direction was significantly different from the direction to the next travel goal. Other data hinted at Euclidean spatial awareness but were inconclusive.

An important point is that the functional difference between the two types of cognitive map is minimized when a topological map has a large number of landmarks. As noted in the previous chapter, experiments with Guinea baboons revealed long-term memory capacity on a par with that of humans. Other experiments showed that baboons can assimilate directional categories, such as up/down and far/near. Early hominins probably could have adapted to most habitats using topological maps of their surroundings.

12.9.2 Initiation and Direction of Movement

In the past, it was often assumed that baboon travel direction was determined by a particular individual (usually thought to be the alpha male) or a category of individuals (the dominant males). Rowell (1966) raised early doubts about these assumptions, describing olive baboons in which females were influential. Accumulating research demonstrates greater variation and subtlety. One of the recent developments is that much of group travel behavior involves relatively low levels of cognition.

Systematic observation of a chacma baboon troop in South Africa found that about 75% of the adults made start attempts and about a third of these were by females (Sané and Zinner 2008). There was no general sex difference in the probability of success, although the alpha male played a slightly more prominent role. The researchers concluded that timing of departure and initial direction resulted from a "partially shared consensus" or "distributed leadership." Female participation might have been limited by concern for their young: those with dependent offspring were "particularly rare" among the initiators. However, the absence of baboon predators in the area casts some doubt on this interpretation.

Comparable research on a chacma baboon troop in the Namibian desert found both similarities and differences (King et al. 2011). It differed from the South African research in finding no sex difference in total start attempts, but was similar in finding no effect of sex on successful initiations. Here again, the alpha male played a somewhat greater role than others, making more successful initiations than expected by chance. Given the levels of dominance and aggression in chacma baboons (Chap. 7), even the slight prominence of the alpha male might not be characteristic of other baboon species.

In mobile social groups, influences driving group movement can vary between democratic and despotic. The arrival at any single pattern of influence is thought to be underpinned by both environmental factors and group composition. Beginning with those theoretical premises, Bonnell et al. (2017) sought to identify the specific patterns of influence driving travel decision-making in a chacma baboon troop. They used spatial data to extract individual patterns of movement and scaled these estimates of individual-level bias to the level of the group. This provided the basis for constructing an influence network and assessing its emergent structural properties. The results indicated that there is heterogeneity in movement bias: individual animals respond consistently to particular group members, and higher-ranking animals are more likely to influence the movement of others. This seems like partial confirmation of the old idea that dominant individuals are leaders. However, this is just one factor in a complex phenomenon.

The heterogeneity in movement bias resulted in a group-level network structure that consisted of a single core and two outer shells. The presence of a core suggests that a set of highly interdependent individuals drove routine group movements. In other words, heterogeneity at the individual level seems to produce group-level influence structures. The researchers concluded that study of movement patterns in mobile social groups can add to the exploration of both how these structures develop (i.e., mechanistic aspects) and what consequences they have for outcomes at the individual and group levels (i.e., functional aspects).

In a study of olive baboons in Kenya, using GPS collars on members of one troop, Strandburg-Peshkin et al. (2015) found no differences in sex or dominance with regard to either attempted or successful initiations. Their subjects made shared decisions based on numbers, being more likely to follow a larger group of initiators. On the other hand, the baboons were less likely to follow any of the initiators if there were a large number of them indicating many different directions. When conflicts arose over the direction of movement, baboons chose one direction over the other when the angle between them was large, but they compromised if the angle was small. According to the researchers, these results are consistent with models of collective motion, suggesting that democratic collective action emerging from simple rules is widespread, even in complex, socially stratified societies. Bonnell et al. (2017) suggested that their results, described above, modified the picture presented by Strandburg-Peshkin and colleagues. However, they studied a different species and, as noted above, the higher levels of dominance and aggression in male chacma baboons might translate into greater influence on troop movement.

Guinea baboons, despite their multilevel social structure, resembled other *Papio* species in departure behavior (Montanari et al. 2021). Roughly two-thirds of attempts were led by males and one-third by females. The females were almost as successful as the males (81–87%). Social status seemed to be immaterial in that the success of bachelor males slightly exceeded that of central males (92–84%).

In hamadryas baboons, the other multilevel species, departures are generally controlled by the leaders of one-male units. However, hamadryas baboons may not be entirely different from other *Papio* species: anecdotal evidence from some studies indicates that some attempted departures were thwarted by the failure of females to follow (Montanari et al. 2021). Among the males, only about 60% of attempts were successful. The "amoeba-like" process suggests the kind of choices made by olive baboons. The available evidence for yellow baboons suggests that they display the same general trend toward "partially-shared consensus" that characterizes other *Papio* species (Montanari et al. 2021).

12.9.3 Group Organization During Travel

The second travel issue for early hominins and baboons is group organization while moving. This entails group cohesion and the spatial distribution of group members. Early reports described a consistent order of progression (e.g., DeVore and Washburn 1963) that included fewer dominant males in front, followed by pregnant and estrus females, with dominant males in the center along with young juveniles and females with infants. Less organized groups were quickly reported (e.g., Rowell 1966). Recent research, some using new technology, has discerned more complicated patterns.

Bonnell et al. (2017), in the work on chacma baboons described above, depicted a system in which group movement was organized from individual biases. Individual relations were also important in a study of olive baboons wearing GPS collars. Individual movements were best predicted by the actions of 4–6 "neighbors" (Farine et al. 2016). The "local rule" was to follow neighbors, with some bias toward particular individuals termed "affiliates." The most predictive affiliates were those that were generally in spatial proximity. The result was troop cohesion throughout the day. Emergence of group movement from individual movements may at least partly explain the fact that other studies found no difference in movement dynamics between small and large baboon troops (Farine et al. 2016).

An additional factor in the decisions of individuals is locomotor compromises with others, which were discerned in a study of olive baboons (Harel et al. 2021). Size variation is related to varied locomotor capabilities. Many troop members adjusted their stride frequencies in relation to the size of their nearest neighbors, which contributed to troop cohesion. However, individuals at the front did not make such adjustments. This would tend to maintain large adult or subadult males at the front, as depicted in some of the early reports.

Allowing for the distinctive social organization of Guinea baboons, their movement dynamics are consistent with those of the troop species (Montanari et al. 2021). Members of the same primary unit travel together, which might result from the same local rule that characterizes the other species. The most consistent feature of moving groups was the presence of bachelor males at the front. This might result from the kind of compromise factors described for olive baboons (Harel et al. 2021, above). Bachelor males would be larger than females and not concerned with adjusting to affiliated females or young.

12.10 Summary and Discussion

This chapter added to the discussion of domain-general capabilities of baboons in the previous chapter. It then described domain-specific capabilities related to behavioral ecology in the range of environments occupied by both extant baboons and early hominins.

Baboons are like many other animals in having an elementary ability to count, but may go beyond most other animals in some respects. Two different experiments with food caches tested baboon numerosity. In one of these the baboons conformed to the logical rule of one-to-one correspondence, which is the understanding that two sets are equal if each item in one set corresponds to exactly one item in the second set. In the other experiment, the baboons chose the second cache as soon as it was filled to the same level as the first. Analysis concluded that the baboons used an algorithm that is structurally similar to the one that humans use in formal counting. A very different study examined the travel behavior of baboons and concluded that the decision-making involved was based on numerical reasoning like that used by humans.

Quantification capabilities of baboons also include the extraction of statistical regularities from the environment, which provides important information. Experiment indicates statistical learning mechanisms that are similar to those of humans. Both may have originated in similar complex and changing environments.

The previous chapter discussed the possibility of analogous reasoning in baboons. Experiments also suggest that baboons can use inferential reasoning by exclusion. Olive baboons solved the two-cup problem in a standard test. Hamadryas baboon solved a three-cup problem (as did a macaque) while several New World monkeys did poorly with it.

Hamadryas baboons did better than macaques in a battery of tests aimed at general cognitive ability, although the difference rested more on behavioral dispositions than on reasoning ability. More of the hamadryas subjects engaged with the tasks and they were less likely to decrease participation as the tasks became more difficult. This result points to the role of emotions in cognition. Emotions also affected task performance by Guinea baboons: spontaneous activity in a computerized visual search task varied with moods, in parallel with human responses on other tasks.

Cognitive flexibility was suggested by an experiment that compared young hamadryas baboons and rhesus macaques in the performance of tasks related to exploratory activity and learning ability. More baboons than rhesus were active in performing the tasks and they were less likely to decrease activity as the tasks became more complex. The higher level of activity was strongly correlated with test success. The researchers noted that these results were consistent with earlier findings on manipulative ability and attributed the difference to behavioral plasticity and adaptability. They postulated that these qualities contributed to the evolutionary success of baboons. Similar qualities could have meant evolutionary success for early hominins as they evolved under conditions much like those facing baboons.

Metacognition would probably have been highly adaptive for early hominins and baboons experiencing diverse and changing environmental conditions. In an experiment with three Guinea baboons, the subjects used a touchscreen to report on the location of two target stimuli that had previously appeared briefly on a grid. The baboons used a "Repeat" key more often in difficult trials and transferred use of the key to novel conditions. Two baboons showed higher accuracy when they declined by using the key, compared to baseline trials in which the key was not available. The researchers inferred that this was due to accurate metacognitive judgments. The same two baboons were consistently faster at reporting the targets' locations after a repetition of the stimulus.

There is evidence that primate brain size is correlated with ecological factors that were especially important in baboon and early hominin evolution: decline in frugivory, larger home ranges, and/or more travel time. Routine movement of baboons through their home ranges implicates cognitive maps. These are not as sophisticated as Euclidean maps, but memory for a large number of landmarks considerably narrows the difference between the two. Experiments with Guinea baboons have revealed long-term memory capacity on a par with that of humans (though short-term memory does not display comparable effectiveness). Other experiments show that baboons can assimilate what they see into useful categories in ordinary life, such as food/nonfood, up/down, far/near. Ability to categorize pictures and alphanumeric characters may represent a capacity for more abstract cognition.

Initiation and direction of group movement are largely a matter of the quasi-democratic process of distributed leadership (partially shared consensus), although socially dominant individuals sometimes play a larger role than others. Group organization during movement may arise from relatively simple cognitive processes. Some research shows that baboons follow nearest neighbors and/or affiliates, and cohesion may result from some individuals adjusting their stride to match others. Travel decisions encompass both the ecological and the social sides of baboon and hominin life.

Baboons display manipulations of stone that have not been reported for other monkeys or chimpanzees. Baboons move rocks while foraging in ways that suggest some cognitive underpinnings for tool use. They select medium-sized rocks that are likely to cover edible invertebrates but are not too difficult to move; they prefer flat rocks that they can flip along the shortest axis; and they flip the rocks downhill. Baboons also drop and throw rocks as weapons against presumed predators (humans), apparently selecting sizes likely to be damaging. One observer credited them with good aim. Aimed throwing is also reported for zoo baboons. Assuming that chimpanzees have not always been the tool-users that they are today, the baboon observations suggest what the earliest foundation for hominin tool use might have been like.

References

Anikaev AE, Chalyan VG, Meishvili NV. Parameters of experimental activity in the study of the cognitive abilities of rhesus monkeys (*Macaca mulatta*) and baboons (*Papio hamadryas*). Bull Exper Biol Med. 2020;168:793–6. https://doi.org/10.1007/s10517-020-04804-x.

Anikaev AE, Meishvili NV, Chalyan VG, Anikaeva EN. Study the differences between the parameters of learning and exploratory activities in adult male rhesus monkeys (*Macaca mulatta*) and baboon hamadryas (*Papio hamadryas*). Bull Exper Biol Med. 2021;172(9):380–4. https://doi.org/10.1007/s10517-022-05397-3.

Anikaev AE, Chalyan VG, Meishvili NV, Anikaeva EN. Studies of the ability to carry out exploratory activity in lower monkeys. Neurosci Behav Physiol. 2023a;53(7):1229–41. https://doi.org/10.1007/s11055-023-01519-0. (translated)

Anikaev AE, Meishvili NV, Chalyan VG, Anikaeva EN. Analysis of the ability to skill formation [sic] and stability of the formed skill in mature male rhesus monkeys (*Macaca mulatta*) and hamadryas baboons (*Papio hamadryas*). Bull Exper Biol Med. 2023b;175(2):244–7. https://doi.org/10.1007/s10517-023-05849-4.

Bonnell TR, Clarke PM, Henzi SP, Barrett L. Individual-level movement bias leads to the formation of higher-order social structure in a mobile group of baboons. R Soc Open Sci. 2017;4:170148. https://doi.org/10.1098/rsos.170148.

Bryer MAH, Koopman SE, Cantlon JE, et al. The evolution of quantitative sensitivity. Philos Trans R Soc B. 2021;377:20200529. https://doi.org/10.1098/rstb.2020.0529.

Cantlon JF, Piantadosi ST, Ferrigno S, et al. The origins of counting algorithms. Psychol Sci. 2015;26(6):853–65. https://doi.org/10.1177/0956797615572907.

Cheney DL, Seyfarth RM. Baboon metaphysics: the evolution of a social mind. Chicago: University of Chicago Press; 2007.

de Raad AL, Hill RA. Topological spatial representation in wild chacma baboons (*Papio ursinus*). Anim Cogn. 2019;22:397–412. https://doi.org/10.1007/s10071-019-01253-6.

DeCasien AR, Williams SA, Higham JP. Primate brain size is predicted by diet but not sociality. Nat Ecol Evol. 2017;1(5):112. https://doi.org/10.1038/s41559-017-0112.

DeVore I, Washburn SL. Baboon ecology and human evolution. In: Howell FC, Bourliere F, editors. African ecology and human evolution. New York: Aldine; 1963. p. 335–67.

Farine DR, Strandburg-Peshkin A, Berger-Wolf T, et al. Both nearest neighbors and long-term affiliates predict individual locations during collective movement in wild baboons. Sci Rep. 2016; 6:27704. https://doi.org/10.1038/srep27704.

Goodall (van Lawick) J, van Lawick H, Packer C. Tool-use in free-living baboons in Gombe National Park, Tanzania. Nature. 1973;241:212–3.

Gullstrand J, Claidière N, Fagot J. Age effect in expert cognitive flexibility in Guinea baboons (*Papio papio*). Behav Brain Res. 2022;434:114043. https://doi.org/10.1016/j.bbr.2022.114043.

Hamilton WJ III, Buskirk RE, Buskirk WH. Defensive stoning by baboons. Nature. 1975;256:488–9. https://doi.org/10.1038/256488a0.

Harel R, Loftus JC, Crofoot MC. Locomotor compromises maintain group cohesion in baboon troops on the move. Proc R Soc B. 2021;288:20210839. https://doi.org/10.1098/rspb.2021.0839.

Isbell LA, Bidner LR, Van Cleave EK, Matsumoto-Oda A. GPS-identified vulnerabilities of savannah-woodland primates to leopard predation and their implications for early hominins. J Hum Evol. 2018;118:1–13. https://doi.org/10.1016/j.jhevol.2018.02.003.

King A, Sueur C, Huchard E, Cowlishaw G. A rule-of-thumb based on social affiliation explains collective movements in desert baboons. Anim Behav. 2011;82(6):1332–45. https://doi.org/10.1016/j.anbehav.2011.09.017.

Koopman SE, Arre AM, Piantadosi ST, Cantlon JF. One-to-one correspondence without language. R Soc Open Sci. 2019;6:190495. https://doi.org/10.1098/rsos.190495.

Kortlandt A, Kooij M. Protohominid behavior in Primates. Symp Zool S Lond. 1963;10:61–88.

Malassis R, Gheusi G, Fagot J. Assessment of metacognitive monitoring and control in baboons (*Papio papio*). Anim Cogn. 2015;18:1347–62. https://doi.org/10.1007/s10071-015-0907-8.

Mare C, Landman M, Kerley GIH. Rocking the landscape: Chacma baboons (*Papio ursinus*) as zoogeomorphic agents. Geomorphology. 2019;327:504–10. https://doi.org/10.1016/j.geomorph.2018.11.028.

Mare C, Landman M, Kerley GIH. How should a clever baboon choose and move rocks? Biotropica. 2021;53:162–9. https://doi.org/10.1111/btp.12857.

Marsh H, Vining AQ, Levandoski EK, Judge PG. Inference by exclusion in lion-tailed macaques (*Macaca silenus*), a hamadryas baboon (*Papio hamadryas*), capuchins (*Sapajus apella*), and squirrel monkeys (*Saimiri sciureus*). J Comp Psychol. 2015;129(3):256–67. https://doi.org/10.1037/a0039316.

Martina C, Cowlishaw G, Carter AJ. Individual differences in task participation in wild chacma baboons. Anim Behav. 2021;172:73–91. https://doi.org/10.1016/j.anbehav.2020.11.020.

Marzouki Y, Gullstrand J, Goujon A, Fagot J. Baboons' response speed is biased by their moods. PLoS One. 2014;9(7):e102562. https://doi.org/10.1371/journal.pone.0102562.

Montanari D, O'Hearn W, Hambuckers J. Coordination during group departures and progressions in the tolerant multi-level society of wild Guinea baboons (*Papio papio*). Sci Rep. 2021;11(1):21938. https://doi.org/10.1038/s41598-021-01356-6.

Pettet A. Defensive stoning by baboons (reply). Nature. 1975;256:549. https://doi.org/10.1038/258549a0.

Piantadosi ST, Cantlon JF. True numerical cognition in the wild. Psychol Sci. 2017;28(4):462–9. https://doi.org/10.1177/0956797616686862.

Pickford M. Defensive stoning by baboons (reply). Nature. 1975;256:549–50. https://doi.org/10.1038/258549c0.

Plummer TW, Finestone EM. Archeological Sites from 2.6–2.0 Ma: Toward a deeper understanding of the Early Oldowan. In Schwartz JH (ed) Rethinking human evolution. MIT Press; 2018. https://doi.org/10.7551/mitpress/11032.001.0001.

Pope S, Meguerditchian A, Hopkins WD, Fagot J. Baboons (*Papio papio*), but not humans, break cognitive set in a visuomotor task. Anim Cogn. 2015;18(6):1339–46. https://doi.org/10.1007/s10071-015-0904-y.

Powell LE, Isler K, Barton RA. Re-evaluating the link between brain size and behavioural ecology in primates. Proc R Soc B. 2017;284:20171765. https://doi.org/10.1098/rspb.2017.1765.

Proffitt T, Reeves JS, Braun DR et al. Wild macaques challenge the origin of intentional tool production. Sci Adv. 2023;9(10): eade8159. https://doi.org/10.1126/sciadv.ade8159.

Rowell TE. Forest living baboons in Uganda. J Zool Lond. 1966;149:344–64.

Sané S, Zinner D. To follow or not to follow: Decision making and leadership during the morning departure in chacma baboons. Anim Behav. 2008;75(6):1995–2004. https://doi.org/10.1016/j.anbehav.2007.12.012.

Schmitt V, Fischer J. Inferential reasoning and modality dependent discrimination learning in olive baboons (*Papio hamadryas anubis*). J Comp Psychol. 2009;123(3):316–25. https://doi.org/10.1037/a0016218.

Shea J. Occasional, obligatory, and habitual stone tool use in hominin evolution. Evol Anthropol. 2017;26(5):200–17. https://doi.org/10.1002/evan.21547.

Strandburg-Peshkin A, Farine DR, Couzin ID, Crofoot MC. Shared decision-making drives collective movement in wild baboons. Science. 2015;348:1358–61. https://doi.org/10.1126/science.aaa5099. PMID: 26089514

Strum SC. Why natural history is important to (primate) science: a baboon case study. Int J Primatol. 2019;40(4):596–612. https://doi.org/10.1007/s10764-019-00117-7.

Vidal-Cordascoa M, Rodríguez-González L, Prado-Nóvoa O, et al. Daily distance traveled is associated with greater brain size in primates. Folia Primatol. 2020;91:654–68. https://doi.org/10.1159/000510782.

Chapter 13
Social Cognition

13.1 Introduction

The concept of social cognition encompasses a broad range of processes underlying the ability to identify, mentally represent, and respond to other individuals and groups, their behaviors, intentions, and relations (Hopper et al. 2018). It covers many aspects of human cognition, most of which potentially have roots in early hominin evolution. These include generalized capabilities such as perception, memory, and attention, as well as more specific phenomena such as social categorization, imitation, and Theory of Mind. In a broad sense, social cognition also encompasses emotions. Despite a traditional division between cognitive and affective processes, they now can be seen as a set of closely connected and interdependent processes (Shkurko 2020).

Baboons have played an important role in the development of social cognition studies because the complexity of their societies was quickly apparent, even from limited field studies of savanna baboons in the early days of research on the genus. The multilevel societies of hamadryas baboons quickly followed in the literature. More recently, greater diversity in baboon societies has become known through studies of Guinea baboons and Kinda baboons. In the growing knowledge of baboon social variation, there are illustrations of basic concepts of social cognition as well as analogies for the evolution of particular features of social cognition in early hominins. Some aspects of social cognition in baboons were introduced in the preceding three chapters. Others are discussed below.

G. E. King, *Baboon Perspectives on Early Human Ancestors*,
https://doi.org/10.1007/978-3-031-36769-4_13

13.2 Theories of Social Cognition

Social cognition has been conceptualized in several different ways that overlap with one another. With regard to comparison of baboons and early hominins, three hypotheses are especially pertinent: Machiavellian Intelligence (MIH), Social Brain (SBH), and Embodied Social Brain (ESBH). The first two share the basic premise that the evolution of cognition in hominins and other primates (as well as many other mammals) has been driven primarily or entirely by social complexity. The ESBH introduces a critical perspective that emphasizes brain–body interaction and is more open to ecological as well as social factors in cognition.

The main point to be made in the following discussion, aligned with the purpose of this book, is to demonstrate the importance of baboons in theorizing about social cognition. As usual here, the capabilities of baboons are taken to represent behavior that was possible for early hominins. Some of the examples alluded to in this section are treated in more detail later in the chapter.

13.2.1 The Machiavellian Intelligence Hypothesis

A crucial development in the study of social cognition was the publication of the book *Machiavellian Intelligence* (Byrne and Whiten 1988a, b). It has been called a "keystone" for the field (Hopper et al. 2018). The basic concept is that primate cognitive abilities have been shaped by complex social environments rather than technical or ecological problems. The term "Machiavellian" might seem to limit the hypothesis to competitive or agonistic interactions, but the intent was and is much broader.

The Machiavellian Intelligence Hypothesis (MIH) does encompass competition, and early research focused on *tactical deception* among baboons (Whiten and Byrne 1988). This is the ability to mislead others for personal gain by occasional false use of a normally honest behavior. It was considered "a particularly sensitive yardstick for the depth of Machiavellian intelligence a species can display." Field study of chacma baboons, for example, found several types of tactical deception that used devices such as screaming as if being attacked and exaggerated staring as if seeing a predator. These deceptive behaviors distracted others from food or diverted them from attacks.

Though the initially obvious cases of tactical deception involved agonistic and competitive behavior, the MIH is also concerned with cognition that underlies coordination and cooperation. A recent example (Hopper et al. 2018) is a study of baboon decisions about group movements that was described in the previous chapter. Investigation of the cognitive basis for coordination led to the conclusion that "democratic" collective action could be based on relatively simple behavioral rules (Strandburg-Peshkin et al. 2015). In this case, close examination of the phenomenon suggested a simpler cognitive foundation than others had envisioned.

On the other hand, the capacity for more complicated forms of cooperation seems to have been demonstrated by experiments with Guinea baboons. Pairs of the baboons were tested on side-by-side computers that yielded food rewards under varying conditions. Confronted with a partner who failed to cooperate to obtain a reward for both participants, the excluded individual tried to manipulate the other or went looking for another partner.

13.2.2 The Social Brain Hypothesis

The Social Brain Hypothesis (SBH) argues that the need to live in large social groups selected for increased brain size and, by extension, the cognitive capacities needed to ensure that these groups remain functional and cohesive (Dunbar and Shultz 2007; Shultz and Dunbar 2010). The SBH posits that primates solved the ecological problem of predation risk through the evolution of group-living, and then solved the problem of competition among individuals in social groups by evolving large brains and complex cognitive capacities. Evidence of the hypothesis is not necessarily tied directly to brains. Instead, the objective is to establish the existence of cognitive capacities that only a large brain can support (e.g., understanding of third-party relations). This is one example of extensive overlap in the topics of concern for the SBH and the MIH. In fact, the two hypotheses have been equated (Dunbar 1993, 2003). The difference is that the SBH emphasizes the relationship between social behavior and a more detailed view of the physical brain.

Broadly speaking, the thesis is that the cognitive demands of social relationships within a large group select for a larger brain. More specifically, selection is for the executive functions centered in the neocortex. The relationship between neocortex volume and group size is particularly strong in primates. However, this is just a first approximation. It represents a deeper relationship with behavioral indices of social complexity such as coalition formation, tactical deception, and social play (Dunbar and Shultz 2007).

The group-size proxy can be used for inferences about early hominins. Aiello and Dunbar (1993, 2003) used established scaling relationships to calculate neocortex volume from the cranial capacity of fossil skulls. The group sizes derived from the neocortical estimates were "bracketed" by observed group sizes of extant chimpanzees and humans, which means roughly 10–200 individuals. This is virtually the same range of variation as among the troops of COKY baboons and bands of hamadryas baboons.

Baboons have smaller brains than chimpanzees or early hominins, but they have larger brains than most other monkeys. They have used these brains to survive in the same range of habitats as those occupied by early hominins and not matched by chimpanzees. The apparent similarity in group sizes suggests that baboons encountered social challenges similar to those of early hominins while they engaged in that expansion and perhaps solved the problems in similar ways.

13.2.3 The Embodied Social Brain

Barrett et al. (2021) have advocated an alternative interpretation of the social brain concept that is less concerned with the size of the brain and more focused on the question of whether regions of the primate brain were specialized for sensing and responding to particular kinds of bodily social stimuli—facial expression, eyes gaze, head and body orientation, and biological motion. This view of the social brain, introduced by researchers such as Lesley Brothers and David Perrett in the years around 1980, aimed to establish what particular circuits of the primate brain were doing, and whether these were dedicated to a specific category of objects—other animate beings—as distinct from the broader category of physical objects.

Barrett et al. (2021) gave three interrelated reasons for the revival of this approach:

1. *Group size and brain structures:* Recent comparative work has questioned the link between group size and neocortex size and has demonstrated the importance of non-cortical areas, particularly the cerebellum, in primate brain evolution.
2. *Brain–body relationship:* Recent theory holds that brains evolved first and foremost to control bodies. From this it is inferred that cognition is better conceived of as a set of processes that mediate the adaptive control of bodies in dynamic, unpredictable environments. This contrasts with the traditional "disembodied" view of cognition as a purely brain-based process involving mental representations of the outside world.
3. *Neural reuse:* This concept suggests that much local neural structure is evolutionarily (and developmentally) conserved, but combined and recombined in different ways to perform diverse functions.

Rather than looking for human-like cognitive representations in the neocortex of primates, ESB advocates a research program that would attempt to understand how both human and nonhuman cognition emerge from the reuse of systems that have evolved for embodied sensory-motor control.

Here again, baboons provided the illustrative example though this one is hypothetical. At the beginning of the day some baboons are on the sleeping cliffs grooming, others are already foraging, and a few are beginning the daily journey. A baboon must make decisions concerning hand and foot placement as she descends from the cliff. There are also decisions about which other baboons she can safely approach along the route. On the ground, there are decisions about where to forage and this means monitoring location and activities of nearby baboons. While foraging, the baboon must decide where to move as others approach or move further away. Simultaneously, she is coordinating her hand movements as she picks and processes food items. Like every other baboon, the exemplar is usually in action and responding in real time to a flow of social and environmental stimuli. Options like these are strongly influenced by biomechanical constraints and environmental factors. Therefore, they cannot be considered clean-cut abstractions that exist in an independent mental realm, as is supposedly implied by the SBH.

Embodied decisions respond to constant change in available actions and relevant variables (e.g., success probability, action cost). The baboon (or early hominin) continuously acquires sensory information about relevant affordances as part of ongoing activity, with no temporal distinction between choice and implementation. Such decisions do not require a central executive of any kind. They depend on which of the reciprocally connected sensorimotor networks are the first to "commit" to a given action strongly enough to pull in the rest. With regard to brain evolution, the implication is that more attention should be paid to parietal regions associated with the size and structure of social groups as well as demands of the foraging environment and the degree of terrestriality (Barrett et al. 2021). Neurological research on primates supports the view that various brain structures are involved in social behavior (Platt et al. 2016).

13.2.4 Proximate Factors in Brain Size

The theories cited above all assume that brain size (whether in whole or in part) evolved by natural section. Some baboon evidence suggests the involvement of proximal factors as well as direct selection. A study of captive olive baboons in groups ranging in size from 2 to 63 found that average brain volume was proportional to group size (Meguerditchian et al. 2021). Variation in the size of the enclosures had no such effect, removing one important confounding effect. Increased brain size was largely a function of white matter (although gray matter showed some effect). White matter plays an important role in connecting parts of the brain that are basic to social cognition. The researchers inferred that the connection between group size and brain size supports the evolutionary hypothesis of the social brain. They also noted that the context of their experiment requires explanation in terms of plasticity.

The implication for early hominins is that natural selection might have had a dual effect, favoring intrinsic brain size and also the capacity to increase brain size in response to group size. Consideration of the alternative (or complementary) baboon models (Chap. 7) suggests the following possibilities: (1) troop size might have increased with expansion into more open country in relation to predator pressure and/or other factors (such as the beginnings of cooperative hunting); (2) total community size might have increased with the development of increasingly complex multilevel societies; or (3) both.

13.2.5 Cultural Intelligence

Proponents of the Cultural Intelligence Hypothesis argued that there are two major problems with the Social Brain Hypothesis (van Shaik et al. 2012). The first problem attributed to the SBH is that it cannot account for grade shifts, where species or other taxa have significantly different brain sizes than predicted by social organization. For example, relatively small-brained spotted hyenas display cognitive abilities and social organization similar to that of baboons. Second, the SBH cannot account for the fact that species with high socio-cognitive abilities also excel in general cognition. For birds and mammals, van Shaik et al. (2012) proposed to integrate the social brain hypothesis into a broader framework that they called cultural intelligence. This hypothesis stresses the high costs of brain tissue, general behavioral flexibility, and the role of social learning in acquiring cognitive skills.

The phrase "cultural intelligence" implies a level of social learning beyond that of baboons and early hominins, probably at the level of *Homo erectus*. However, the behavioral flexibility and learning capabilities of baboons suggest that early hominins provided a fertile field for the evolution/development of cultural intelligence. Baboons display high levels of behavioral flexibility and learning capability, even when compared to closely related species such as rhesus macaques.

Anikayev et al. (2022) tested adult male hamadryas baboons and rhesus macaques for learning ability and exploratory behavior. One task required an individual with freedom of choice to learn that food was always in one of two consistently placed containers. The exploration experiment presented the subject with the novel stimulus of a multicolored plastic cube. The baboons learned the location of the food significantly more quickly than did the macaques. They exceeded the macaques in exploration of the novel object in terms of duration of contact and the diversity of investigative and manipulative behaviors.

These results agreed with a series of prior experiments that culminated with Anikayev et al. (2021). In that paper they explicitly compared the baboons to early hominins. They concluded that ecology was the main drive of cognitive adaptation in open country because of needs such as memory for the location of resources (see Chap. 12 here). However, they also acknowledged the significance of social complexity and accepted the proposition that ecology and social organization interacted with each other.

13.3 Cognition in Social Interactions

Students of the evolution of primate and hominin social cognition have examined a variety of specific behavioral patterns. They do not necessarily place them exclusively in any of the general theories described in the preceding section.

13.3.1 Tactical Deception

Tactical deception occurs when an individual is able to use an "honest" act from its normal repertoire in a different context to mislead familiar individuals. Most primate groups are so intimate that any deception is likely to be subtle and infrequent.

Whiten and Byrne (1988) gathered accounts of deceptive behavior in various primates and classified them in terms of the function they perform. For each class, they sketched the features of another individual's state of mind that an individual acting with deceptive intent must be able to represent, thus acting as a "natural psychologist." In an 18-month field study of chacma baboons, the authors and P. Henzi observed eight instances of apparent tactical deception, which they classified into four types.

Type 1. A juvenile screams, falsely representing an attack, which results in the diversion of others; this makes available to the juvenile the USOs that the others have been digging up.

Type 2. A juvenile engages in exaggerated "looking," usually indicative of a predator or another baboon troop, with the result that an adult male is distracted from an attack.

Type 3. An adult female uses unneeded aid solicitation gestures to manipulate an adult male. In one instance the male was distracted from attacking the female. In another, the male left a food patch that he had appropriated from the female.

Type 4. A single instance in a transitional situation where male A had become dominant with respect to access to females, but male B continued to lead the troop. Male B led others, including A, away from a food patch and then circled around to occupy the patch by himself.

Hiding is another form of deceptive behavior. Whiten and Byrne (1988) cited the description by H. Kummer of an adult female hamadryas who gradually shifted her seated position over a distance of 2 m until she was hidden from the leader male by a large rock. There, she groomed a subadult male—behavior that the leader would not tolerate. Whiten and Byrne (1988) regarded such behavior as "sufficiently fine tuned" to require crediting the agent with mental representation of the target's attentional state.

Although baboon deceptive behavior played an important role in the development of social cognition studies, it now seems that they are matched by other monkeys and over-matched by great apes. Regarding this particular topic, baboon analogy demonstrates the lowest common denominator that might have existed in the earliest hominins living in troops (this assumes that the last common ancestor did not have the full range of cognitive capabilities found in extant chimpanzees).

13.3.2 Cooperation

Humans are "strategic cooperators" in the sense that they make decisions on the basis of costs and benefits in order to maintain high levels of cooperation. This capability may have played a key role in human evolution (Formaux et al. 2023). Wild baboons seem to cooperate in some ways, as when males respond as a group to predators or occasionally engage in simultaneous hunts. However, such behaviors can also be explained as independent but parallel responses of individuals. Guinea baboons seem to display the capability for more complex forms of cooperation, at least in experiments.

Formaux et al. (2023) tested Guinea baboons for their cognitive ability to engage in cooperative behavior under seminatural conditions. The subjects were a small troop with freedom to move around an enclosure. They were presented with a test apparatus that they could use voluntarily to gain small food rewards. Two adjacent computer screens allowed participants to obtain rewards for themselves and partners. Each individual could make choices based on the behavior of partners.

During experiment 1, eight individuals reached predefined criteria of at least 80% prosocial choice in one block of 50 trials when a partner was present. They displayed flexibility when the contingencies of the task were reversed. Unchanged response to the reversal during a ghost control phase confirmed that the presence of a partner was essential to their behavior. In the second experiment, the reward structure was changed so that an individual could no longer receive a direct reward. In this circumstance, individuals made the prosocial choice if the partner had previously made a prosocial choice.

During the first, less demanding experiment, the baboons only used partner choice: changing partners more frequently when the partner did not make the prosocial choice. In the more demanding second experiment, prosocial baboons developed two more strategies when paired with a previously non-prosocial partner: They more frequently chose the selfish stimulus, and they were more likely to not respond at all, interrupting the trial and leading to a partner change. In brief, they used direct reciprocity and partner choice to develop and maintain high levels of cooperation.

The researchers inferred that their subjects had the cognitive capacities to adjust their level of cooperation strategically, using a combination of partner choice and partner control strategies. They concluded that such capacities were probably present in our ancestors and would have provided the foundations for the evolution of typically human forms of cooperation (Formaux et al. 2023).

13.3.3 Pointing

A pointing gesture creates a referential triangle that incorporates a distant object into the relationship between the signaler and the gesture's recipient. Pointing, long assumed to be specific to the human species, emerges spontaneously in captive chimpanzees and can be learned by monkeys. Meunier et al. (2013) tested olive baboons for understanding and use of learned pointing behavior. Specifically, they asked whether the behavior was conditioned and dependent on reinforcement or whether the baboons understood it as a mechanism for manipulating the attention of a partner.

Nine subjects had been trained with operant conditioning to exhibit pointing. The experiment tested their ability to communicate intentionally about the location of an unreachable food reward in three different contexts, varying with regard to a human partner's attentional state. In each context, the experimenters quantified the frequency of communication signals, including gestures and gaze alternations between the distal food and the human partner. They found that the baboons were able to modulate their manual and visual communicative signals as a function of the experimenter's attentional state. They concluded that baboons can intentionally produce pointing gestures and understand that a human recipient must be looking at the pointing gesture for them to perform their attention-directing actions. Even at a rudimentary level, such capability would have enhanced early hominin cooperative behavior in important areas such as predator defense and hunting.

13.3.4 Social Facilitation

Social facilitation, which occurs in a wide variety of animal species, is a phenomenon in which one or more individuals manifest certain behaviors in the presence of conspecifics regardless of relevance to any other circumstances. Huguet et al. (2014) tested baboons for the cognitive basis of such social facilitation. The subjects freely engaged in computer-based conflict response tasks that required cognitive control for successful performance. The results indicated that the social presence not only enhanced dominant responses, but also depleted resources for cognitive control. As a result, the baboons experienced greater cognitive conflicts, were less able to inhibit an older learned action in favor of a new one, and were also less able to take advantage of previous experience.

According to the researchers, these findings explain why inappropriate behaviors are not easily suppressed by primates when acting in social contexts. If these "inappropriate" behaviors are maladaptive, natural selection may favor greater cognitive control to overcome the facilitated responses. The researchers hypothesized that such a demand for greater cognitive control in social groups might have been a factor in the evolution of human intelligence (Huguet et al. 2014). A speculation: an evolved tendency to resist group influences on behavior might be one factor in cognitive dissonance in extant humans.

13.4 Self in Society

Baboon analogies suggest how early hominin individuals might have perceived themselves in relation to other individuals in society, both conceptually and emotionally. These analogies are reinforced by comparison with contemporary humans.

13.4.1 Possession and Ownership

Nancekivell et al. (2019) theorized that human understanding of ownership depends on a naïve concept of ownership that emerges early in development. They drew on research from multiple disciplines to suggest that the phenomenon emerges in childhood, develops across the lifespan, and may be universal despite variation across cultures and history. These researchers alluded to an experiment in which hamadryas baboons seemed to respect the rights of a possessor even if lower ranking than a potential poacher.

Sigg and Falett (1985) investigated the usefulness of concepts of possession and property in analyzing the relationship between hamadryas baboons and objects. Their subjects were placed in five experimental situations involving possession of fruits or a food container. Results with both desirable objects indicated that dominant males were controlled by an "inhibition" against taking food away from lower-ranking companions. Also, given the choice, males preferred neutral food cans over cans previously used by the partner. The significance of possession in these experiments was highlighted by the results of a different test, when food pieces were thrown between two partners. In these tests, the dominant never allowed the subordinate to take a piece.

Nancekivell et al. (2019) expressed reservations about the significance of this case and others: "… further work is needed to test whether any of these examples reflect possession of a naïve theory of ownership and to test alternative accounts. For example, many of these findings could be explained by animals showing respect for temporary physical possession of objects, and cost–benefit analyses of whether attempts to take others' possessions are likely to be worthwhile."

Advancing knowledge of baboons suggests another qualification, based on additional results reported by Sigg and Falett (1985). The inhibition they described appeared only in male–male dyads. In male–female and female–female dyads, the extent of "respect" was contingent on rank difference and the type of food. This can be compared to the "respect" that wild male hamadryas baboons display toward one another regarding their respective female associates (Chap. 8). Recognition of possession may be an adaptation to the hamadryas system of one-male units that are nested within larger social groupings, and may be an analogy for social evolution in hominins (Evans et al. 2022). The study by Sigg and Falett hints at the possibility that respect for possession of objects might have been generalized from respect for possession of females, at least in males. Recognition of possession among females could have a completely different origin and set of cognitive correlates.

If the hamadryas analogy is correct, it suggests that the concept of ownership (or right of possession) evolved in male early hominins to minimize conflicts over females. However, we now know that Guinea baboons live in a similar multilevel system based on unimale groups in which males are tolerant of one another and females choose their male associates (Chaps. 7 and 8). Early hominins in such a system would not have had the same pressure to evolve a system of possession rights with regard to females and would have taken a different (not yet explicated) path to concepts of possession and ownership.

13.4.2 Personality

Personalities, variable cognitive/behavioral tendencies across individuals, have been demonstrated in a wide range of primates and other animals. Baboons are no exception. The significance of personality in baboons suggests that this was a factor in the social behavior of early hominins. In the previous chapter, it was suggested that behaviors such as innovation and exploration might have been affected by variable tendencies toward leadership and/or readiness to learn from others.

A chacma baboon project was devoted to personality. In a 7-year study of 45 females, Seyfarth et al. (2012) identified "three relatively stable personality dimensions, each characterized by a distinct suite of behaviors." These were labeled Nice, Aloof, and Loner. Nice females were friendly to all other females, often grunted to subordinates, and had strong social bonds and stable preferences for top partners. Aloof females were more aggressive and grunted mainly to superiors. Loner females were often alone and relatively unfriendly. The baboons seemed to have some recognition of these variations. They approached Nice females at high rates and approached the others at much lower rates. The different personality types varied somewhat in their responses to social challenges: male immigration and the danger of infanticide; and the death of a close relative.

A different project explored the following hypotheses with negative results: (1) that human observers become a "neutral" stimulus and (2) that this habituation process is "equal" across group members (Allan et al. 2020). Based on flight initiation distance, the baboons viewed the observers as a high-ranking social threat rather than a neutral stimulus. Habituation was not equal across group members. There were repeated individual differences that were more important than contextual factors (such as habitat) in determining the distance at which baboons reacted to the observers by visually orienting and/or moving away. A strong correlation between visual and displacement tolerance indicated that this was a personality trait.

Bracken et al. (2022) used high resolution GPS data to investigate personality and plasticity in the movement of chacma baboons across natural and urban environments in a South African city. With regard to personality, the baboons displayed individual differences in movement metrics. Individuals that traveled straighter paths on average, traveled even straighter paths in urban space. Those that increased their step length and decreased their residency times the most in urban space were high-ranking individuals.

A study of olive baboons explored individual differences in coping style and stress reactivity. As in many other primatological studies, these researchers used a "personality-like framework" derived from the human personality literature (Pritchard and Palombit 2022). Coping style and stress reactivity were quantified using observer ratings in individually targeted field experiments. Three personality trait factors emerged: Neuroticism, Assertiveness, and Friendliness. Personality trait differences showed little association with coping style, but Neuroticism was predicted by stress reactivity.

Taken together, the baboon personality research suggests that any of the larger early hominin groups contained a variety of personalities with differing responses to social situations. Interactions of these individuals may have affected the adaptive behavior of all concerned. The extent to which these personalities are heritable is not clear. In one study of baboons, close female kin did not have personalities that were more alike than those of nonrelatives (Platt et al. 2016).

13.4.3 Emotions

The research on personality in baboons seems to imply that emotions are important in shaping social behavior, for example, friendliness, aggressiveness, anxiety, and fear. Delgado et al. (2023) noted that studies showed stress-relieving affiliative behaviors among wild female baboons are linked to decreased cortisol responses and relatively long life. They took this to be an evolutionary perspective on how human individuals form and maintain strong social networks, which has become a significant public health priority. Delgado and colleagues reviewed psychological and neural mechanisms that enable people to connect with each other to alleviate the consequences of stress and isolation. Central to this process is the experience of rewards derived from positive social interactions, which encourage the sharing of perspectives and affective states.

One of the neurological mechanisms underlying emotion in humans is hemispheric specialization. This is manifested in asymmetries of facial expressions that mainly indicate right hemisphere dominance. Wallez and Vauclair (2011) extended this research to olive baboons with recordings of two affiliative behaviors (lipsmack and copulation call) and two agonistic ones (screeching and eyebrow raising). This study provided evidence for right hemisphere specialization in the production of some baboon vocal and facial expressions of emotion. There was no indication that dominance status or sex had any influence on the results. The researchers considered the results as indicative of "neurophysiological and neuroanatomical homologies between baboons and humans in the cortical control of emotional vocal and facial expressions." Of course, analogy is the alternative possibility.

13.5 Social Information

Social information includes information about others and information from others. The topic of grunts (Chap. 10) was one of many topics that led to some discussion of social information. This section provides further comparison of social information in baboons and humans, with implications for early hominins.

13.5.1 Social Comparison

Comparison of self to others is an important characteristic of human social life and may have been a component of social organization in early hominins. Dumas et al. (2017) used a computerized task presented in a social context to explore the psychological mechanisms of social comparison in humans and baboons. They found that the effects of social comparison on a subject's performance were guided both by similarity (same versus different sex) and by task complexity. Comparing oneself with a better-off other (upward comparison) increased performance when the other was similar rather than dissimilar, and a reverse effect was obtained when the self was better (downward comparison). Furthermore, when the other was similar, upward comparison led to a better performance than downward comparison. The beneficial effect of upward comparison on baboons' performance was only observed during the simple task. Humans and baboons responded in comparable ways, depending on whether the other in the experiment was similar or dissimilar to the subject and whether the other was better or worse off. The researchers inferred that humans and baboons shared cognitive mechanisms for social comparison.

13.5.2 Information About Others

Chacma baboons display sophisticated knowledge about relationships in their troops (Cheney and Seyfarth 2007). Various studies show that they behave in accord with relationships in the social hierarchy; track the consortship status of mating pairs; and respond to conflicts by selectively aiding unrelated individuals who have been grooming partners (summarized by Fischer et al. 2019). Playback experiments with wild chacma troops have demonstrated how such social traits influence the attention structure of individuals. They respond strongly to vocalizations of apparent intruders represented by the playbacks. In relation to other troop members, they respond strongly to rank reversal consortship break-ups that are simulated by playbacks.

Guinea baboons differ, apparently on the basis of greater gregariousness and spatial tolerance with little or no concern for dominance. They show more interest in the vocalizations of other group members than those of outsiders, treating them as sources of information about current associations. Researchers inferred from such variations that the "value" of types of social information may differ from one species to another (Faraut and Fischer 2019; Fischer et al. 2019). Thus, any choice of analogies for hominin relationships must follow from the choice of the likely social structure. These results have potential implications for the evolution of social cognition in hominins, since hominins probably lived in troops and in multilevel societies at various stages.

13.5.3 Information from Others

Informativeness—defined as reduction of uncertainty—is central to human communication (Reboul et al. 2022). It allows the rapid dissemination of novel information among individuals (Carter et al. 2016). Reboul et al. (2022) investigated the sensitivity of baboons to informativeness in a series of experiments. On a computer screen, they manipulated the informativity of a cue relative to a response display. The baboons were allowed to anticipate answers or wait varying amounts of time for a revealed answer. Anticipations increased with informativity, while response times to revealed trials decreased. Further experiments reduced rewards for anticipation responses (to 70%) with the result that the link between anticipations and informativeness disappeared. However, the link between informativeness and decreased response times for revealed trials persisted. Additionally, in all experiments, the number of correct answers in revealed trials with fast reaction times increased with informativeness. The researchers concluded that baboons are sensitive to informativeness as an ecologically sound means to tracking reward.

Carter et al. (2016) considered the limitations on informativeness in baboons in a social setting. An individual's ability to use information is likely to be dependent on phenotypic constraints operating at three successive steps: acquisition, application, and exploitation. They identified phenotypic constraints at each step: peripheral individuals in the proximity network were less likely to acquire and apply social information, while subordinate females were less likely to exploit it successfully. Social bonds and personality also played a limiting role along the sequence. As a result of these constraints, the average individual acquired social information on less than 25% of occasions and exploited it on less than 5 percent of occasions. This study highlighted the sequential nature of information use and the fundamental importance of phenotypic constraints on this sequence. Early hominins may have gained some benefit from sensitivity to informativeness, but were probably subject to limitations like those of baboons.

13.5.4 Culture?

One of the most important aspects of information transfer in humans is cultural traditions, transmission of behavior patterns from one generation to the next. The manufacture of stone tools has often been taken as evidence of culture in early hominins. However, there is now some doubt that the earliest efforts represent culture. Snyder et al. (2022) performed an experiment with 25 humans who were naïve with regard to stone knapping techniques. All of them learned the techniques individually, producing and using core and flake tools. If the earliest stone tools do not represent culture in hominins, this leaves an open question as to what the earliest forms of culture might have been. Beyond material culture, there is the question of what social patterns were likely to become cultural in early hominins. Chimpanzees provide an abundance of evidence, but baboons display at least one instance that differs from those of the apes.

Perhaps the best documented case of baboon culture is that of the peaceful chacmas, because the evidence for transmission is clear. Sapolsky and Share (2004) reported a case in which half of the males in a troop succumbed to tuberculosis under circumstances which dictated that the more aggressive males died and atypically unaggressive individuals survived. A decade later, males in that troop were still unaggressive. Due to dispersal, none of the males who survived the epidemic remained in the troop; therefore, new males joining the troop must have adopted the unique social pattern. Features of this male culture included high rates of grooming and affiliation with females, a relaxed dominance hierarchy, and physiological measures suggesting less stress among low-ranking males. All of this occurred among members of the species reputed to be the most aggressive baboons. Models that explained transmission of this cultural pattern centered on the treatment of incoming males by resident females.

Olive baboons at Kekopey in Kenya, over the course of two studies, displayed the development of hunting traditions (Strum 1975). Hunting males spent more time away from the troop, traveled farther (up to 3.2 km from the troop periphery), and spent up to 2 h hunting whether successful or not. Relay chases by the hunters began as accidents but were quickly adopted as strategy, resulting in greater success.

Other cases are only candidates for culture because they are unusual patterns shared within a group or population and absent from the rest of the species. Two of these have been described earlier in different contexts (Chaps. 4 and 5). The most distinctive is the fishing behavior of chacma baboons in a desert canyon (Hamilton and Tilson 1985). They obtained fish from drying pools by various means that included wading into the water to grope for live fish under boulders, and slapping the water at the edges of pools to stun nearby fish. Consumption of particular plant foods, including certain toxic plants, might be local cultural traditions; however, intergenerational transmission has not been verified and availability has not been eliminated as the determining factor.

13.6 Summary and Discussion

In a broad sense, social cognition encompasses all mental processes involved in an individual's reaction to other members of its social group and interaction with them. Cognitive processes may simply be inferred from behavior, but many attempts have been made to relate these processes to the structure and function of the brain. Social cognition in humans and primates has been the subject of several theories with many overlaps and a few crucial differences.

The Machiavellian Intelligence Hypothesis (MIH) and the Social Brain Hypothesis (SIH) have a great deal in common and are sometimes equated with one another. Both postulate that primate cognitive evolution has been driven mostly or entirely by the requirements of living in complex societies, rather than ecological pressures. Both connect primate cognitive evolution with relatively large brains and especially with the size of the neocortex. The Embodied Social Brain Hypothesis (ESBH) criticizes the MIH and SIH for separating mental processes from the actions of the body: mental processes are constantly adjusting to movement of the body in relation to varied substrates and to changes in the immediate physical and social environments. The ESBH is more open to ecological effects on cognition and argues that brain structures other than the neocortex, for example, the cerebellum, should receive more attention. All three of these hypotheses have used baboon examples to illustrate key points, which suggests that the roots of much human social cognition go back to the evolution of early hominins.

The Cultural Intelligence Hypothesis argues that the other theories of cognitive evolution fall short in failing to account for two facts. First, species with very different brain sizes can have very similar social organization (e.g., spotted hyenas and baboons). Second, species with high levels of social cognition also have high levels of general cognition. The CIH proposes a broader perspective on cognitive evolution that emphasizes the costs of brain tissue, behavioral flexibility, and the role of learning in the acquisition of cognitive skills. The degree of behavioral flexibility and social learning in baboons suggests that this view of cognitive evolution can also be rooted in early hominins.

There has been no attempt here to resolve differences among these hypotheses. In accord with the theme of this book, the point was to demonstrate the significance of baboons in the development of the theories. This, of course, suggests that the theories apply to early hominins as well as to extant humans.

This view is supported by a variety of specific hominin–baboon analogies. Some have been presented in the preceding chapters because the level of communication in baboon societies entails social cognition. Some were noted as examples of the precepts of the social cognition theories. Some were treated in more detail in the latter parts of this chapter: tactical deception, pointing, cooperation, social facilitation, social comparison, possession and ownership, personality, and emotions.

Social information is important to baboons, as it is to humans. Examples vary somewhat with species. Chacmas show that they are aware of the dominance hierarchy, mating pairs, and conflicts. They respond strongly to intruders simulated by vocal playbacks. Guinea baboons have little or no concern for signs of dominance relationships. They are more interested in vocalizations from within their own group rather than from outsiders, treating the former as knowledge about affiliations. Whether the emphasis is on dominance or affiliation, all baboons attend to information about their own position in the group. In experiments, baboons react much like humans to comparisons with others who are similar or dissimilar and better-off or worse-off. Researchers inferred similar cognitive mechanisms for social comparison.

Informativeness, defined as reduction of uncertainty in communication, allows rapid dissemination of novel information. This is a key factor in social and behavioral flexibility. In experiments, baboons displayed sensitivity to informativeness, but within social limits. Sensitivity to informativeness would have facilitated hominin adaptation to changing environmental and social conditions. Baboon studies also suggest the limits on such sensitivity in early hominins, perhaps prior to the major behavioral changes after 2 mya.

Culture entails the acceptance of novel information and its transmission from one generation to the next. The evidence for culture in baboons is small compared to what has been reported for some other monkeys and for apes. Nevertheless, there is enough to show that culture is compatible with the baboon way of life and that culture could have arisen among early hominins with many similarities to baboons. Outstanding examples are the unique peace culture in a troop of chacma baboons and the fluctuating predatory patterns of a troop of olive baboons. Consumption of unusual plant products by certain troops or populations might also be examples.

References

Allan ATL, Bailey AL, Hill RA. Habituation is not neutral or equal: Individual differences in tolerance suggest an overlooked personality trait. Sci Adv. 2020;6:eaaz0870. https://doi.org/10.1126/sciadv.aaz0870.

Anikayev AE, Chalyan VG, Meishvili NV, Anikaeva EN. Comparison of learning ability in rhesus macaques (*Macaca mulatta*) and hamadryas baboons (*Papio hamadryas*) in the light of creating a model of primate cognitive evolution. Neurosci Behav Physiol. 2021;51(9):1304–11. https://doi.org/10.1007/s11055-021-01194-z.

Anikayev AE, Meishvili NV, Chalyan VG, Anikaeva EN. Study the differences between the parameters of learning and exploratory activities in adult male rhesus monkeys (Macaca mulatta) and baboon hamadryas (*Papio hamadryas*). Bull Exper Biol Med. 2022;172(3):381–4. https://doi.org/10.1007/s10517-022-05397-3.

Barrett L, Henzi SP, Barton RA. Experts in action: why we need an embodied social brain hypothesis. Philos Trans R Soc B. 2021;377:20200533. https://doi.org/10.1098/rstb.2020.0533.

Bracken AM, Christensen C, O'Riain MJ, et al. Personality and plasticity in Cape chacma baboons' movement across natural and urban environments. SSRN. 2022; https://doi.org/10.2139/ssrn.4137696. (preprint)

Byrne R, Whiten A. Machiavellian intelligence: social expertise and the evolution of intellect in monkeys, apes, and humans. Oxford University Press; 1988a.

Byrne R, Whiten A. Tactical deception of familiar individuals in baboons. In: Byrne R, Whiten A, editors. Machiavellian intelligence: social expertise and the evolution of intellect in monkeys, apes, and humans. Oxford University Press; 1988b. p. 205–10.

Carter AJ, Tico MT, Cowlishaw G. Sequential phenotypic constraints on social information use in wild baboons. eLife Sci. 2016;5:e13125. https://doi.org/10.7554/eLife.13125.

Cheney DL, Seyfarth RM. Baboon metaphysics: the evolution of a social mind. Chicago: University of Chicago Press; 2007.

Delgado MR, Fareri DS, Chang LJ. Characterizing the mechanisms of social connection. Neuron. 2023;111:3911. https://doi.org/10.1016/jneuron.2023.09.012.

Dumas F, Fagot J, Davranche K, Claidière N. Other better versus self better in baboons: an evolutionary approach of social comparison. Proc R Soc B. 2017;284:20170248. https://doi.org/10.1098/rspb.2017.0248.

Dunbar RIM. Dunbar RIM. 1993. Coevolution of neocortex size, group size and language in humans. Behav Brain Sci. 1993;16:681–735.

Dunbar RIM. The social brain: Mind, language, and society in evolutionary perspective. Annu Rev Anthropol. 2003;32:163–81. https://doi.org/10.1146/annurev.anthro.32.061002.093158.

Dunbar RIM, Shultz S. Understanding primate brain evolution. Philos Trans R Soc B. 2007;362:649–58. https://doi.org/10.1098/rstb.2006.2001.

Evans KD, Swedell L, Chowdhury S. Respect for males amid suppression of females: selective use of aggression and fitness correlates in the male-dominated society of hamadryas baboons. J Hum Evol. 2022;165:103151. https://doi.org/10.1016/j.jhevol.2022.103151.

Faraut L, Fischer J. How life in a tolerant society affects the attention to social information in baboons. Anim Behav. 2019;152:11–7. https://doi.org/10.1016/j.anbehav.2019.04.004.

Fischer J, Higham JP, Alberts S, et al. The natural history of model organisms: Insights into the evolution of social systems and species from baboon studies. eLife Sci. 2019;8:e50989. https://doi.org/10.7554/eLife.50989.

Formaux A, Sperber D, Fagot J, Claidière N. Guinea baboons are strategic cooperators. Sci Adv. 2023;9(43):eadi5282. https://doi.org/10.1126/sciadv.adi5282.

Hamilton WJ III, Tilson RL. Fishing baboons at desert waterholes. Am J Primatol. 1985;8:255–7. https://doi.org/10.1002/ajp.1350080308.

Hopper LM, van de Waal E, Caldwell C. Celebrating the continued importance of "Machiavellian Intelligence" 30 years on. J Comp Psychol. 2018;132(4):427–33. https://doi.org/10.1037/com0000157.

Huguet P, Belletier C, Barbet I, et al. Cognitive control under social influence in baboons. J Exp Psychol. 2014;143(6):2067–73. https://doi.org/10.1037/xge0000026.

Meguerditchian A, Marie D, Margiotoudi K, et al. Baboons (*Papio anubis*) living in larger social groups have bigger brains. Evol Hum Behav. 2021;42:30–4. https://doi.org/10.1016/j.evolhumbehav.2020.06.010.

Meunier H. The pertinence of studying neuroethology in nonhuman primates for human behavior in groups and organizations. Org Res Meth. 2019;22(1):250–74. https://doi.org/10.1177/1094428118756741.

Meunier H, Prieur J, Vauclair J. Olive baboons communicate intentionally by pointing. Anim Cogn. 2013;16:155–63. https://doi.org/10.1007/s10071-012-0558-y.

Nancekivell SE, Friedman O, Gelman SA. Ownership matters: people possess a naïve theory of ownership. Trends Co Sci. 2019;23(2):102–13. https://doi.org/10.1016/j.tics.2018.11.008.

Platt ML, Seyfarth RM, Cheney DL. Adaptations for social cognition in the primate brain. Philos Trans R Soc B. 2016;371:20150096. https://doi.org/10.1098/rstb.2015.0096.

Pritchard AJ, Palombit RA. Survey-rated personality traits and experimentally measured coping style and stress reactivity, in wild baboons. Am J Primatol. 2022:e23429. https://doi.org/10.1002/ajp.23429.

Reboul A, Mascaro O, Claidière N, Fagot J. Are monkeys sensitive to informativeness: An experimental study with baboons (*Papio papio*). PLoS One. 2022;17(7):e0270502. https://doi.org/10.1371/journal.pone.0270502.

Sapolsky RM, Share LJ. A pacific culture among wild baboons: its emergence and transmission. PLoS Biol. 2004;2(4):0534–41. http://biology.plosjournals.org

Seyfarth RM, Silk JB, Cheney DL. Variation in personality and fitness in wild female baboons. Proc Natl Acad Sci. 2012;109:16980–5.

Shkurko A. Social cognition. In: Shackelford TK, Weekes-Shackelford VA, editors. Encyclopedia of evolutionary psychological science; 2020. https://doi.org/10.1007/978-3-319-16999-6_3146-1.

Shultz S, Dunbar R. Encephalization is not a universal macroevolutionary phenomenon in mammals but is associated with sociality. Proc Natl Acad Sci USA. 2010;107(50):21582–6. https://doi.org/10.1073/pnas.1005246107.

Sigg H, Falett J. Experiments on respect of possession and property in hamadryas baboons (*Papio hamadryas*). Anim Behav. 1985;13(3):978–84.

Snyder WD, Reeves JS, Tennie C. Early knapping techniques do not necessitate cultural transmission. Sci Adv. 2022;8(27):eabo2894. https://doi.org/10.1126/sciadv.abo2894.

Strandburg-Peshkin A, Farine DR, Couzin ID, Crofoot MC. Shared decision-making drives collective movement in wild baboons. Science. 2015;348:1358–61. https://doi.org/10.1126/science.aaa5099. PMID: 26089514

Strum SC. Primate predation: Interim report on the development of a tradition in a troop of olive baboons. Science. 1975;187:755–7.

van Shaik CP, Isler K, Burkart JM. Explaining brain size variation: from social to cultural brain. Trends Cogni Sci. 2012;16(5):277–84.

Wallez C, Vauclair J. Right hemisphere dominance for emotion processing in baboons. Brain Cogn. 2011;75(2):164–9. https://doi.org/10.1016/j.bandc.2010.11.004.

Whiten A, Byrne RW. The manipulation of attention in primate tactical deception. In: Byrne R, Whiten A, editors. Machiavellian intelligence: social expertise and the evolution of intellect in monkeys, apes, and humans. Oxford University Press; 1988. p. 211–23.

Epilogue: The Once and Future Baboon

To borrow a phrase, the baboon model for early hominin behavior has had a long strange journey. It has been lauded, reviled, and revived. It has been buffeted by paleoanthropological and political currents. It has gone from relatively straightforward field observations to increasingly sophisticated theoretical concepts and technological methods.

Viewed as a competitor with other species for the purpose of comparison with early hominins, baboons have proven to be more relevant than most. Of particular importance, as discussed at various points throughout this book, baboon–hominin comparisons are readily coordinated with inferences from the other leading source of such information—*Pan*. Mutual support between the approaches is exemplified by the fact that both genera face predation by leopards and can respond aggressively. In some respects, baboonanalogies suggest modifications of chimpanzee inferences. Stone tools, for instance, may have had a very early origin as weapons against predators, dropped or rolled from heights as in several baboon populations. Finally, baboons are probably more relevant to early hominins than *Pan* in some important respects, such as adaptations to life on open savannas and in multilevel societies.

Baboon behavioral research seemed to start well with fieldwork by Eugene Marais (1969), who observed free-living chacma baboons "shortly after" the First World War. Marais's interpretations and theories (especially with regard to the "human psyche") are questionable, but he made a sincere attempt to accurately record the behavior of wild baboons. To begin with, he understood and made clear that his main study troop had probably been influenced by human presence before he began observations. He inferred this from briefer observations of more remote troops. Marais' report anticipated discoveries made (or remade) decades later. One such discovery was that the baboons obtained water in a dry area by digging up a species of plant with a water-storing USO. Another was the use of stone and anvil to break open baobab fruits. Not anticipated was the digging of grooves in soil to divert and cool drinking water from a thermal spring. In this case Marais noted his

G. E. King, *Baboon Perspectives on Early Human Ancestors*, https://doi.org/10.1007/978-3-031-36769-4

uncertainty as to the cognitive implications for the baboons, because many of the grooves were dug in the wrong direction. Finally, he seems to have performed one or more field experiments regarding learning capabilities: "… if an infant arboreal baboon is given to a mountain troop … it grows up with the complete knowledge necessary for it to exist in its new environment" (1969: 68).

Rather than continue and improve on this beginning, baboon research foundered in faulty work with captive animals. For example, a female olive baboon who showed an aptitude for tool use was abused: "(I) had to treat her roughly for which reason she bit me well and truly. During the following days she bit me again as well as others, for which reason I gave her a good beating" (Bolwig 1961: 147). This cruelty, however, pales next to the disaster in the London Zoo that resulted from ignorance of natural behavior. Monkey Hill was supposed to be an outdoor improvement on the cages that had housed primates at the Zoo. In May 1925, 97 Hamadryas baboons arrived from Africa and were placed on the Hill. They were supposed to be all males, but 6 (of 97) were females (Burt 2006). Fighting for the females began immediately. Thirty females were added in, along with five immature males. Half of the new females were dead within 8 weeks, caught between desperately fighting males. Of fifteen infants born on the Hill, only one survived. Some were determined to have died of strangulation due to being held too tightly during fights. Possibly there were cases of deliberate infanticide.

These results could have been predicted with knowledge of the natural behavior and demography of hamadryas baboons. However, the anatomist who was in charge of the colony considered the behavior of the captives to be natural. He held fieldwork in contempt as unscientific. This is not surprising in view of his own experience of fieldwork, observing chacma baboons in South Africa in 1930. His short observation periods were often made in the context of baboon hunts and were curtailed by attempts to shoot females for "anatomical purposes" (Burt 2006).

Though other wild primates (e.g., howler monkeys) were studied scientifically in the following decades, baboons did not receive such attention until the 1950s. Beginning in that period, anthropologists Sherwood Washburn and Irven DeVore approached baboons in Kenya with an explicit interest in the reconstruction of human evolution (Washburn & DeVore 1961). K.R.L. Hall, a psychologist, started his work on baboons with social vigilance in chacmas, and then expanded to other aspects of social behavior, including ecological influences. He carried this knowledge with him when he turned to laboratory research to test hypotheses about the evolution of primate society. G. Stolz and L.P. Saayman also studied troops of chacma baboons in South Africa, considering demography and ecology as well as social behavior (Stolz and Saayman 1970).

Then the 1960s-flavored backlash began (see Chap. 3). Baboons were too dimorphic to compare with hominins and to do so was sexist. Baboons were really forest animals and could not be compared with savanna hominins. Then it was found that the earliest hominins lived in forests or woodlands, so they should not be compared with savanna baboons. In any event, baboons were vicious and should not be compared to intrinsically peace-loving humans.

Meanwhile, the paleoanthropological rollercoaster produced a great deal of evidence to show that early hominins were significantly more dimorphic than modern or recent humans. Some might have been as dimorphic as baboons. Furthermore, though hominins had indeed originated in forests or woodlands, they had expanded to occupy a wide range of habitats, including dry and open savannas, throughout southern, eastern, and northern Africa.

Continuing research on wild baboons showed that they were not particularly vicious compared to most other primates and that the various species of the genus varied significantly in aggressiveness and the degree of sexual dimorphism. The recently studied Kinda baboons seem to differ from all other baboon species in that males do not compete for dominance and access to females; instead, they queue up to move into such positions as their troop tenure goes on. Such discoveries generate alternative hypotheses as to the behavior of early hominins. Finally, though baboons were originally thought of as associated with savannas, primatologists showed that they occupy every kind of habitat in which hominins existed, from forest edges to open savannas.

Probably the most important and valid criticism of early fieldwork on baboons was that male behavior was overemphasized and females neglected. This was remedied by a new generation of baboon researchers (e.g., Rowell 1966; Seyfarth 1976; Strum 1987), many of them women. It became clear that females were the heart of the troop, with a social structure of their own. Though males could dominate females, females had ways of coping with male aggressiveness. One of the most relevant to hominin evolution was the discovery of Special Friendships (Ransom 1981; Smuts 1985). This was scientific progress, not a political or philosophical revolution. The significance of males was recognized, but females were now given increasing attention.

Another major development was the recognition of multilevel societies in hamadryas baboons (Kummer 1968), a phenomenon comparable to the typical societies of humans. At the base of hamadryas societies are unimale groups, which males form by obtaining and retaining mates through coercion of the females and sometimes by violence against other males. Here again, males received a great deal of (perhaps disproportionate) attention, and again, later researchers added information about the often subtler behaviors and relations of females (e.g., Swedell 2006). The hamadryas baboon system has been applied to detailed reconstruction of early hominin society (Swedell and Plummer 2012, 2019).

For some critics, these findings might revive concerns about male dominance and aggression. However, despite occasional violent competition between male hamadryas baboons, continuing research has demonstrated considerable tolerance and cooperation among them (Evens et al. 2022). Furthermore, Guinea baboons provide an alternative model for early hominins. They are organized into multilevel societies that are structurally comparable to those of hamadryas; however, Guinea baboons differ sharply in the higher degree of male–male tolerance and affiliation. In addition, unimale groups of Guinea baboons are formed through female choice of mates and the females are free to move from one mate to another (Petersdorf et al. 2019). This is one of the most dramatic examples of variation among baboon species, presenting alternative models or analogies for reconstruction of early hominin behavior.

What is the future of baboons and baboon research? While many primates are threatened or endangered, baboons seem to be doing relatively well. To paraphrase something a colleague said many decades ago, when he returned from his research in the field—if global disaster occurs, two mammals will survive: leopards and baboons.

According to the New England Primate Conservancy (https://neprimateconservancy.org/) Guinea baboon conservation status is "Near Threatened." They have been seriously affected by habitat loss, hunting, and persecution as pests. Another possible factor affecting Guinea baboons is that they are more bound to permanent water sources than the other species: "A rule of thumb for this species is they are always found near a permanent water source, from freshwater springs and marshes to rock pools and mountain rivers." The other baboon species are rated "Least Concern." All but the Kinda baboons are much more numerous than Guinea baboons.

Nevertheless, all baboons face similar dangers (African Wildlife Foundation, www.awf.org/wildlife-conservation/baboon) and populations within species may be at risk currently or in the near future, even if whole species are not. Given the variation known to exist in baboons, the loss of any population is a potential scientific disaster. The main threat to baboons is habitat loss due to overgrazing, agricultural expansion, settlement expansion, and irrigation projects. In addition, baboons are often intentionally poisoned because they are considered pests. Some are hunted for their skins, especially hamadryas. Use of baboons in laboratories and medical research has also increased.

For practical as well as humanitarian reasons, baboon projects (like other primate research projects) have expanded outreach to local human populations. For a long time, local people have provided invaluable services as research assistants, guards, and rangers. More recently, there has been increasing involvement of local people in conservation efforts, including mutual exploration of compromises between the needs of baboons and neighboring humans. Research projects have been joined by scientists who are citizens of the host countries.

Barring disaster, the future of baboon research seems relatively bright. There are ongoing long-term projects devoted to the ecology and behavior of all currently recognized baboon species. These projects are briefly surveyed in the Appendix.

References

Bolwig N. An intelligent tool-using baboon. S Afr J Sci. 1961;57:147–52.

Burt J. Solly Zuckerman: the making of a primatological career in Britain, 1925–1945. Stud Hist Phi Biol Biomed Sci. 2006;37:295–310. www.elsevier.com/locate/shpsc

Evens KD, Swedell L, Chowdhury S. Respect for males amid suppression of females: selective use of aggression and fitness correlates in the male-dominated society of hamadryas baboons. J Hum Evol. 2022;165:103151. https://doi.org/10.1016/j.jhevol.2022.103151.

Kummer H. Social organization of hamadryas baboons, a field study. Chicago: University of Chicago Press; 1968.

Marais E. The soul of the ape. Atheneum; 1969.

Petersdorf M, Weyher A, Kamilar JM, et al. Sexual selection in the Kinda baboon. J Hum Evol. 2019;135(S37) https://doi.org/10.1017/j.jhevol.2019.07.007.

Ransom TW. Beach troop of the Gombe. Lewisburg PA: Bucknell University Press; 1981.

Rowell T. Forest living baboons in Uganda. J Zool Lond. 1966;149:344–64. https://doi.org/10.1016/j.jhevol.2019.06.006.

Seyfarth RM. Social relationships among adult female baboons. Anim Behav. 1976;24:917–38.

Smuts BB. Sex and friendship in baboons. New York: Aldine; 1985.

Stolz G, Saayman LP. Ecology and behaviour of baboons in the Northern Transvaal. Ann Transvaal Museum. 1970;26:99–14.

Strum SC. Almost human: a journey into the world of baboons. New York: Random House; 1987.

Swedell L. Strategies of sex and survival in female hamadryas baboons: through a female lens. Upper Saddle River NJ: Pearson/Prentice Hall; 2006.

Swedell L, Plummer T. A papionin multilevel society as a model for hominin social evolution. Int J Primatol. 2012;33(5):1165–93. https://doi.org/10.1007/s10764-012-9600-9.

Swedell L, Plummer T. Social evolution in Plio-Pleistocene hominins: Insights from hamadryas baboons and paleoecology. J Hum Evol. 2019;137 https://doi.org/10.1016/j.jhevol.2019.102667.

Washburn SL, DeVore I. The social life of baboons. Sci Amer. 1961;204(6):62–71.

Appendix: Ongoing Long-Term Baboon Research Projects

Introduction

Long-term studies of animals reveal behavioral patterns that are not apparent in shorter studies. They also lend statistical validity to patterns that have provided limited data in shorter studies. At least one ongoing long-term field project is currently devoted to each of the six commonly accepted species of *Papio*. Several more such studies have been mounted in the past and are now concluded. In some of these cases the data are still being processed.

Yellow Baboons

Amboseli Baboon Research Project (https://amboselibaboons.nd.edu/)

In 1963, Jeanne and Stuart Altmann searched for a baboon study site and settled on the Maasai-Amboseli Game Reserve (later Amboseli National Park) in Kenya. In 1971, they established the research project. Susan Alberts joined the field team and eventually became a director. Diverse research topics include feeding ecology, nutrition, behavioral endocrinology, kin recognition, sexual selection, aging research, functional genetics, hybridization, parasitology, and relations with other species. The project has tracked hundreds of known individuals in several social groups over the course of their entire lives, accumulating life history data for more than 1500 baboons. Among discoveries from these data: (1) The presence or absence of fathers influences the maturation rates of their offspring: young baboons who grow up without a father reach adulthood more slowly. (2) Maternal dominance rank has pervasive effects on the physiology of sons: low ranking mothers have sons

G. E. King, *Baboon Perspectives on Early Human Ancestors*,
https://doi.org/10.1007/978-3-031-36769-4

with higher baseline stress levels. (3) Close social relationships among females improve the survival probability of their offspring. The last few decades have seen remarkable changes to the Amboseli ecosystem and a number of interesting responses to these changes in the baboons' ecology and behavior.

Olive Baboons

Uaso Ngiro Baboon Project (http://www.baboonsrus.com/)

This was originally the Gilgil Baboon Project, named for a village near the private ranch where the research was first done. During this period, major discoveries were made concerning the predatory activities of the baboons (Chap. 5 in this book). Voluminous data confirmed and elaborated on earlier work documenting the central place of females in troop organization and the formation of male–female friendships (Chaps. 6 and 7 here). From its inception, the director of the project has been Dr. Shirley Strum, now Professor Emerita of Anthropology at the University of California, Berkeley.

When research at Gilgil became untenable, Director Strum oversaw translocation of three troops to the Laikipia Plateau in 1984. Research continued in this high-altitude savanna/bushland about 100 km north of Mt. Kenya. Drought is common and presents a challenge for the baboons. The effects of severe drought can be compared between these olive baboons and the yellow baboons of Amboseli.

Gashaka Primate Project (https://www.ucl.ac.uk/gashaka/)

Gashaka-Gumti is Nigeria's largest National Park. The southern sector (Gashaka) is a mosaic of woodland, lowland and gallery forest, montane forest, and grassland. The founder and Director of the Gashaka Primate Project is Volker Sommer, who is Professor of Evolutionary Anthropology at University College London. The Director of Baboon Research is Caroline Ross, Reader in Evolutionary Anthropology at Roehampton University, London. Her research interests include primate socioecology, life-history evolution, and human–wildlife conflict.

Baboon studies at Gashaka include auditory communication signals. The Gashaka baboons are unusual in that they thrive in both open woodland savanna and "rather thick" forest. This natural experiment is interesting because similar flexibility may have been responsible for the evolutionary success of early hominins. For example, depending on which setting they are in, the baboons vary in the use of their basic communication signals.

Gombe Stream Research Center (https://janegoodall.ca/what-we-do/africa-programs/gombe-stream-research-centre/)

Gombe Stream National Park in Tanzania, where Dr. Jane Goodall first studied wild chimpanzees, is a biodiverse and vibrant ecosystem. Since 1972, the Research Center has conducted baboon field studies led by Dr. Tony Collins. The project has amassed more than 40 years of data, collected mainly by local Tanzanian field assistants and supervising scientists.

The primary focus has been on the ecology and life histories of the baboons, but other topics include disease transmission, conservation, and long-term ecological changes. Among the important discoveries is that female baboons engage in aggressive competition that is linked to reproductive constraints. Population studies contribute to overall understanding of ecosystem health, especially because the baboons are an important prey species for chimpanzees. Data analysis is carried out at the JGI Center for Primate Studies on the campus of the University of Minnesota, directed by Dr. Anne Pusey.

Chacma Baboons

Tsaobis Baboon Project

The Tsaobis Baboon Project is situated on the edge of the Namib Desert in central Namibia. The project is hosted by Tsaobis Nature Park and affiliated with the Gobabeb Namib Research Institute. It was established in 2000 by the Institute of Zoology, the research arm of the Zoological Society of London. The three project directors represent a collaboration of three international research institutions: Guy Cowlishaw at the Institute of Zoology, ZSL; Alecia Carter at the Department of Anthropology, University College London; and Elise Huchard at the Institute for Evolutionary Sciences, a mixed research unit from the University of Montpellier and French National Centre for Scientific Research (CNRS). The Project runs an internship program to provide training to the next generation of Namibian ecologists and conservation scientists. Research on the Tsaobis baboons began in 1990, and has been ongoing on an annual basis since 2000.

The aim of the Tsaobis Project is to carry out fundamental research in behavioral and population ecology, using desert baboons as a model. The work also has a conservation theme in its attention to the effects of complex social structure on population dynamics and extinction. The main subjects are three troops of chacma baboons comprising a total of more than 160 animals. Each baboon is individually recognized, of known age, and of known relatedness to other individuals in the population.

Okavango (https://www.sas.upenn.edu/~seyfarth/Baboon%20research/)

Chacma baboons were studied in the Okavango Delta of Botswana from 1992 to 2008, ending with forced termination. The project is ongoing in the sense that analysis of the findings continued after fieldwork was terminated. The principal investigators were D.L. Cheney and R.M. Seyfarth, but many others were engaged in important research at this site. The emphasis was on behavior, communication, and cognition of animals living under natural conditions.

Important findings included long-term causes of mortality: the majority of deaths among adult females and juveniles were due to predation, while infants were more likely to die of infanticide. There were strong seasonal effects on birth and mortality, with the majority of conceptions occurring during the period of highest rainfall. Mortality due to predation and infanticide was highest during the 3-month period when peak flooding caused the group to be scattered and constrained to move along predictable routes. Predation and infanticide counterbalanced the slight reproductive advantage of dominant females that resulted from shorter interbirth intervals and more rapid infant growth rates. Infanticide affected high-ranking and low-ranking females more than middle-ranking females, while predation affected females of all ranks relatively equally. As a result, there were few rank-related differences in estimated female lifetime reproductive success.

Kinda Baboons: Kasanka Baboon Project (http://kasankababoonproject.com/)

The Kasanka Baboon Project, founded by Dr. Anna Weyher in 2010, studies the social behavior and ecology of Kinda baboons in Kasanka National Park of Zambia. It was the first project to investigate this species in the wild. These baboons are smaller and more slender than other species in the genus and they display distinctive features of social behavior. The KBP has grown into a comprehensive organization that conducts innovative biological research, reduces poaching, educates and employs the local community, and helps local girls go to school.

Kasanka researchers suggest that Kinda baboons may provide new insights into the selective environments that affected early hominins. The species displays distinctive patterns of relationship among males and between males and females. Kinda males exhibit a queuing pattern of dominance acquisition rather than the contest-based pattern found in other baboon species. Alpha male tenures are relatively long. Grooming interactions are commonly initiated by males and terminated by females. Males groom females more often and longer than the reverse, regardless of the female's reproductive state or the presence of an infant (factors that affect male–female relationships in other baboon species).

Hamadryas Baboons: Filoha Hamadryas Project (https://qcpages.qc.cuny.edu/ANTHRO/Web_Pages/swedell/filoha.html)

The FHP is co-directed by Dr. Larissa Swedell and Dr. Shahrina Chowdhury. It studies the behavioral biology of hamadryas baboons in Ethiopia. The Filoha site is located in Awash National Park at the Filoha hot springs (the name derives from the Amharic terms for hot water). The primary focus of research over the years has been one band numbering about 200 individuals. Several other bands have been studied intermittently. Each band uses multiple cliffs as sleeping sites, one of which is the cliff located at Filoha.

The Filoha Hamadryas Project began with Larissa Swedell's doctoral research in 1996–1998, which focused on behavioral ecology and the reproductive and social strategies of females. Further work has elaborated on elements of the male-dominated multilevel social system, including mechanisms of female acquisition and loss, the relationship between ecology and social grouping patterns, modes of dispersal, and patterns of kinship within and among social units.

Most recently, research has focused on (1) sexual conflict, in particular the coercive behavior of hamadryas males, its costs for females, and ways in which females might mitigate these costs; and (2) the adaptive value of social relationships in hamadryas society, including the ways in which leader males may benefit from the presence of follower males in their units. FHP is collaborating with other research teams on projects related to parasite ecology, functional morphology, and the gut microbiome. Active data collection at Filoha has been temporarily suspended due to the COVID-19 pandemic and the Ethiopian Civil War.

Guinea Baboons: Simenti Project (https://www.dpz.eu/en/unit/cognitive-ethology/research/guinea-baboon-research.html)

In 2007, under the auspices of the Cognitive Ecology Laboratory of the German Primate Center, a long-term study of wild Guinea baboons began. The study community ranges near the field station Centre de Recherche de Primatology (CRP) Simenti, which is located in the Niokolo Koba National Park in Senegal. The head of the Project is Prof. Dr. Julia Fischer and the Senior Scientist is Dr. Dietmar Zinner. Several study groups are part of a community of more than 400 individuals with a home range of almost 25 km^2 around Simenti. Several individuals are fitted with radio or GPS collars. More than 500 identified baboons are included in the long-term database started in 2010.

The project began with the expectation that comparative analyses would provide insights into the selective forces that shape primate social evolution. Topics of study have included social and communicative behavior, ecology, and population genetic structure. Guinea baboons live in a nested multilevel system with female-biased dispersal where males form "strong bonds" that can last for several years. Average relatedness was significantly higher between strongly bonded males, suggesting that kin biases contribute to the social preferences of males. Researchers think that this system presents "intriguing parallels" with the organization hypothesized for early human societies.

Long-term study provided an important test of the evolutionary significance of male relationships. Although strongly bonded males supported each other more frequently during conflicts, this did not promote reproductive success as in other baboons. Males who spent less time socializing with other males were associated with a higher number of females and sired more offspring.

In contrast to hamadryas baboons, females choose their mates and are free to move from one to another. The prominent role of females was also apparent in the pattern of group movement. Although adult males made two-thirds of the attempts at initiating departure, the rate of success was the same for both sexes.

With regard to ecology, most Guinea baboon populations, including the one at Simenti, live in more productive habitats than hamadryas baboons. Since both live in multilevel societies, this difference contradicts simplistic ideas about the evolutionary relation between ecology and social system.

Index

A

Active defense, 54, 110–112, 114, 115, 117, 119
Alternative hypotheses, 113, 263
Analogies, 23, 35, 47, 48, 50–55, 57–60, 63–66, 71, 77–79, 85, 86, 88, 91–93, 95, 101, 102, 107, 110, 113–115, 118, 119, 134, 138, 141–143, 147–150, 155, 157, 158, 160, 163, 169, 170, 173, 174, 176–179, 184, 187, 189–192, 194, 195, 197, 199, 203, 204, 214, 224, 225, 228, 229, 241, 247, 249, 250, 252, 253, 256, 261, 264
Aquatic animals, 95
Arboreality, 7, 37, 38
Ardipithecus, 2, 7, 21, 22, 30, 31, 35, 37–39, 41, 64, 65
Australopithecus, 2–4, 7–12, 21, 22, 32, 37–39, 41, 64, 65, 70, 73, 79, 87, 110, 120, 162, 177, 210
Avoidance, 101, 106, 108–110, 119, 120, 127, 128, 142, 175, 195

B

Baboons, 1, 2, 8, 12–23, 29–42, 47, 49–60, 63–80, 85–97, 101, 102, 105–120, 125–144, 147–165, 169–184, 187–201, 203–219, 223–238, 241–257, 261–264, 267–272
Barks, 68, 73, 74, 80, 108, 115, 187, 190–193, 200, 209
Body mass, 3, 4, 15–16, 19, 22, 29, 38–42, 54, 57, 60, 103, 107–108, 119, 137, 144, 157, 229
Brain, 4–6, 12, 22, 34, 55, 57, 70, 77, 79, 80, 113, 126, 213–216, 218, 228–230, 237, 242–246, 255, 256

C

Call systems, 189–190, 199
Carnivorans, 101, 102, 104–105, 107–114, 119, 120
Conceptual frameworks, 47–60
Cooperation, 96, 126, 134, 141–144, 163, 170, 179, 184, 193, 198, 242, 243, 247–248, 256, 263

D

Deception, 242, 243, 246–247, 256
Domain-general, 203, 205–208, 213, 217, 218, 225, 236

E

Early *Homo*, 8, 11–12, 30, 37, 65, 70–72, 78, 88, 91, 103, 110, 115, 162
Eclectic feeding, 65
Ecological cognition, 228–230, 256
Environmental stresses, 56

G. E. King, *Baboon Perspectives on Early Human Ancestors*,
https://doi.org/10.1007/978-3-031-36769-4

F

Families, 58, 60, 102, 126, 141, 147, 149–151, 154, 155, 158, 163–165, 170
Faunivory, 85, 89, 90, 95, 96
Female affiliation, 169–174, 183
Female aggression, 179
Female competition, 179–183
Female cooperation, 176
Female dyadic bonds, 173
Female networks, 172–174, 183
Fishing, 88, 255
Fruits, 6, 13, 56, 57, 65, 67–71, 73, 75, 76, 79, 87, 129, 229, 231, 250, 261

G

Gestures, 187, 196–197, 200, 208, 210–215, 217, 218, 247–249
Grasses, 31, 33, 58, 68–73, 76, 79, 91
Greetings, 139, 140, 158, 187, 188, 197–198, 201
Group travels, 159, 232–236
Grunts, 187, 189–191, 193–196, 200, 209, 210, 252

H

Hominins, 1–13, 15, 19–23, 29–42, 47–55, 57–60, 63–80, 85–97, 101–120, 125–138, 141–144, 147, 148, 150, 151, 154, 155, 157, 158, 160, 162–164, 169–171, 173–184, 187, 189, 190, 192–197, 199, 200, 203–215, 217–219, 223–233, 235–238, 241–243, 245–247, 249–254, 256, 257, 261–264, 268, 270
Hunting, 53, 78, 91–94, 96, 97, 102, 104, 106, 108, 126, 142, 144, 245, 249, 255, 264

I

Infanticide, 148, 153, 154, 159, 161, 162, 164, 181, 183, 251, 262, 270
Insects, 71, 85–87, 90, 95

L

Languages, 114, 193, 199, 203–219, 223, 225
Last common ancestor (LCA), 1, 5, 6, 21, 55, 60, 95, 113, 134, 155, 247
Laterality, 213–216, 218
Leave-taking, 187, 199, 201

M

Machiavellian intelligence, 242–243, 255
Male–female bonds, 134, 150, 153
Mating effort, 136, 151, 154, 155
Meat, 85–97, 106, 107, 136
Men in groups, 125, 141–142, 144
Models, 1, 5, 19, 20, 22, 34, 38, 42, 47–55, 57–59, 63, 87, 95, 112, 113, 115, 125–131, 133–135, 137, 142–144, 147, 148, 154, 156, 157, 159, 160, 163, 177, 190, 198, 203, 204, 215, 218, 224, 228, 235, 245, 255, 261, 263, 264, 269
Multilevel society, 129–134, 143, 158–160, 163, 183

P

Paleoenvironments, 8, 34, 37, 66, 78
Papio, 12, 13, 15–22, 50, 68, 78, 86, 89, 133, 134, 142, 143, 153, 154, 156, 164, 174, 190, 197, 206, 225, 227, 235, 267
Papiophobia, 52–58
Parental effort, 155
Personalities, 251–252, 254, 256
Philopatry, 133–134, 143, 144, 170, 174, 180, 183
Privacy, 160, 163, 165
Protein prioritization, 69

R

Raptors, 51, 114–115, 120, 205

S

Scavenging, 90, 92–94, 96, 102, 104, 105, 107
Scenarios, 47–49, 51–52, 59, 65, 67, 69, 87, 114, 133, 134, 143, 157, 176–179, 184, 208, 218, 227
Selective feeding, 66, 79
Sexual coercion, 161, 162, 164
Sexual dimorphism, 3–4, 15–17, 19, 22, 29, 40–42, 49, 54, 55, 57, 137, 156, 157, 161, 162, 173, 178, 263
Sleeping places, 22, 128, 143
Social brain, 242, 244–246, 255, 256
Speech, 48, 114, 190, 203, 208–210, 215, 218
Stone manipulation, 230, 231

T

Terrestriality, 2–3, 23, 34–35, 49, 196, 245
Troops, 13, 20, 23, 31, 32, 52, 53, 67, 76–78, 86, 91, 111–113, 126–132, 134–137, 141–144, 150–155, 157, 158, 161–162, 164, 169, 171–175, 177–180, 183, 189, 192, 200, 201, 205, 207, 208, 218, 223–225, 231, 232, 234–236, 243, 245, 247, 248, 253, 255, 257, 261–263, 268, 269

U

Underground storage organs (USOs), 68, 70, 79, 80, 97, 229, 247, 261
Unimale groups, 130, 132, 138, 141–144, 154, 158, 163, 174, 175, 179, 250, 263

V

Vocalizations, 108, 113, 126, 187, 189–194, 196, 200, 208–211, 217, 218, 253, 256

Printed in the USA
CPSIA information can be obtained
at www.ICGtesting.com
CBHW071501290924
15089CB00003B/40

9 783031 367687